STATICS OF
DEFORMABLE SOLIDS

RAYMOND L. BISPLINGHOFF

JAMES W. MAR
Massachusetts Institute of Technology

THEODORE H. H. PIAN
Massachusetts Institute of Technology

DOVER PUBLICATIONS, INC., NEW YORK

Published in Canada by General Publishing Company, Ltd., 30 Lesmill
Road, Don Mills, Toronto, Ontario.
Published in the United Kingdom by Constable and Company, Ltd., 10
Orange Street, London WC2H 7EG.

This Dover edition, first published in 1990, is an unabridged, corrected
republication of the work first published by Addison-Wesley Publishing
Company, Inc., Reading, Mass., 1965 (in the "Addison-Wesley Series in
Mechanics and Thermodynamics"). Section 4.3 has been augmented with
an Addendum on page 92.

Manufactured in the United States of America
Dover Publications, Inc., 31 East 2nd Street, Mineola, N.Y. 11501

Library of Congress Cataloging-in-Publication Data

Bisplinghoff, Raymond L.
   Statics of deformable solids / by Raymond L. Bisplinghoff, James W. Mar,
and Theodore H. H. Pian.
         p.   cm.
   Includes bibliographical references.
   Reprint. Originally published: Reading, Mass. : Addison-Wesley, 1965.
   ISBN 0-486-66360-4
   1. Strength of materials.   2. Elastic solids.   3. Statics.   I. Mar,
James W.   II. Pian, Theodore H. H., 1919–   III. Title.
TA405.B53   1990
620.1'12—dc20   90-3057
                                                                     CIP

Preface

This book, like so many others, had its genesis in notes. These particular notes were prepared to aid in the teaching of a first course in solid mechanics to second-term sophomores in the Department of Aeronautics and Astronautics at the Massachusetts Institute of Technology. The material is intended to build on the mechanics background normally obtained from a first course in physics in an engineering curriculum.

The content of the book and the philosophy of approach which has been adopted are based upon convictions which the authors have developed over a period of several years of experiment with solid mechanics curricula. Although the book represents a solution to their particular curriculum problems, the authors are the first to admit that there is no unique solution to all the needs of mechanics curricula.

The attempt has been made throughout the book to reveal the subject in its fundamentals and at the same time to attend to details in such a way as to retain the student's interest. The teaching of elementary solid mechanics in the classical engineering college curriculum of the past tended to subdivide the subject in the student's mind to give him a facility in methods applicable to special cases, and to give him only a vague picture of the broader principles. Graduate students often found that their elementary mechanics courses had not provided a sound foundation and that no special thought or planning had gone into the integration between elementary and advanced mechanics. Introducing solid mechanics in a fundamental way at the sophomore level, just as fluid mechanics is commonly introduced, is necessary in a modern engineering curriculum in order to overcome some of these difficulties.

The instructor in mechanics is faced with the dilemma of finding a technique for teaching and illustrating the fundamental principles without getting bogged down in so much detail that the student loses sight of the main stream of thought. The art of teaching mechanics consists largely in effectively introducing fundamental principles with a minimum of detail. The fundamental principles of solid mechanics involve, for example, the concepts of force, the principles of equilibrium and compatibility, and the idea of constitutive relationships. The details embrace, for example, geometry and kinematics, mathematical notation, and moments and products of inertia.

The present book proceeds directly to a full exposition of the underlying principles. As a result, the book divides naturally into two parts. The first, consisting of Chapters 1 through 4, introduces these fundamental principles, but illustrates them only by means of simple discrete elastic systems, without becoming involved with the geometry of the continua. The second part, comprising Chapters 5 through 9, deals with the application of these principles to three-dimensional continua, both elastic and plastic. Although it may appear at first glance that the two parts of the book represent different orders of rigor and difficulty, it has been the authors' aim to drive home the fundamental principles of solid mechanics in the first and attend to details of the continua in the second.

Chapter 2 presents the laws of Newtonian mechanics fundamental to the entire field, including particle, rigid, solid, and fluid mechanics. This introductory portion of the subject is covered in its broadest form so that the student can relate it to companion courses in mechanics. In Chapter 3, the laws introduced in Chapter 2 are applied to problems concerning the static equilibrium of particles and bodies under the action of forces and moments. Chapters 2 and 3 thus present in an abbreviated manner the material generally covered in an introductory course in statics.

It has been the authors' experience that undergraduates grasp more easily the concepts of force and equilibrium than other principles of mechanics. It is believed, then, that some of the problem drill ordinarily conducted at this stage may be profitably carried out after the principles of statically indeterminate structures have been introduced. In Chapter 4, simple statically indeterminate systems are presented by supplementing the principles of static equilibrium with constitutive relations and conditions of consistent deformation. The problem drill is designed to exercise the student's ability not only to apply equilibrium principles, but also to supplement them when necessary with compatibility conditions.

In Chapters 5 and 6, strain and stress in three-dimensional solids are dealt with, and the notions of strain and stress tensors and their properties are discussed. Chapter 7 ties together Chapters 5 and 6 through Hooke's law, and introduces the theory of elasticity in its elementary form. Chapter 8 deals with the stress-strain relations for plastic solids. Finally, Chapter 9 introduces energy principles in solid continuum. This chapter embraces not only the concepts of work and energy, but also the principle of virtual work and the theorems of minimum potential energy, minimum complementary energy, least work, and Castigliano's two theorems.

Each chapter is followed by a concise summary of the important principles which have been introduced, a bibliography, and a list of problems. The bibliography is not intended to be exhaustive, but rather to include at least one reference to an original source as well as a few selected references to contemporary books for supplementary reading.   Many of the problems are taken from home assignments and examinations used at the Massachusetts Institute of Technology.

Deep appreciation is due to the many people who have aided in bringing the book to completion.   Several colleagues on the Aeronautics and Astronautics staff at M.I.T., as well as former students, read portions of the manuscript and offered valuable criticism.   These include Professors Paul Sandorff, Emmett Witmer, and John Dugundji.   Work on the figures and associated calculations was contributed by Messrs. Oscar Orringer and Charles Bruggeman. The typing, preparation, and reproduction of the manuscript were skillfully handled by Miss Ann Gorrasi.   The preparation of the class notes was supported in part by a grant made to the Massachusetts Institute of Technology by the Ford Foundation. This support is gratefully acknowledged.

Washington, D.C.   R. L. B.
Cambridge, Mass.   J. W. M.
June, 1965   T. H. H. P.

Contents

I

Foundation of Solid Mechanics

1.1 INTRODUCTION

A deformable body is distinguished from a rigid body by its susceptibility to changes in shape under the influence of forces. Strictly speaking, all of nature's bodies are deformable, but in problems of mechanics one may sometimes assume with sufficient precision that certain bodies retain their shape under load, that is, they are perfectly rigid. A deformable body may, however, be a solid or a fluid. The term "solid" is used to identify a solid deformable body in contrast to a fluid with the restriction that a solid possesses shear strength while a fluid does not. Our concern will be limited, furthermore, to the static behavior of solids, that is, to shape changes resulting from static loads only. Our interest in the behavior of solid bodies will extend to a full definition of the deformations, strains, and stresses throughout the volume and over the surface.

Why should the present-day engineer study solid mechanics? In the profession of engineering, for example in the field of aeronautics and astronautics, there is always an end product in mind—a supersonic transport, a launch vehicle, a spacecraft. All these systems, in order to fulfill their missions, must possess physical strength and rigidity. The provision of strength and rigidity can be accomplished relatively simply, but in the process the flight vehicle may become so heavy, so impractical, or so uneconomical that it will not be useful. For example, the structure of a supersonic transport airplane is subjected to widely fluctuating loads as well as temperature excursions of as much as 500° F during each flight. In addition, its users demand a lifetime of some 50,000 hours of flying. In the design of such an airplane, a structural weight of say 23 percent of the gross weight would ensure economic success, whereas a comparable figure only a few percent higher would result in economic failure. The difference between a spacecraft which achieves orbit and one which remains earthbound is sometimes measured in ounces. Thus the structural design of flight vehicles requires exotic materials, sophisticated methods of analysis, and ingenious structural design. Such demands for high structural efficiency are not peculiar to aeronautical and space systems. Because of the high temperatures usually required in energy conversion devices, solid mechanics plays a dominant role in their analysis and design.   Higher structural efficiencies are increasingly demanded for civil engineering structures as well as for ground vehicles and marine vessels.

1.2   WAYS OF THINKING

A student of solid mechanics is at first confronted with a dichotomy in his way of thinking about the subject.   There is first the physical way of thinking which relates directly to the real world in which we live and to the problems which we seek to solve.   These problems are posed to us in a physical way: to predict the deformation of a bridge under load or to compute the stresses in a rocket booster case during launch.   Then there is the mathematical way of thinking in which the theorems of mathematics are employed as tools for swift and accurate reasoning.

It is the purpose of a course in solid mechanics to interrelate in the student's mind these two ways of thinking, and to help him develop as second nature a facility for transition from the physical to the mathematical and back again. The student has already experienced this in its simplest manifestation during his study of geometry.   Plane Euclidean geometry is a legitimate branch of mathematics regardless of whether there exist in the real world the physical entities of points, lines, and planes.   The student, however, easily recognizes these as physical entities at the outset, and the axioms and theorems of geometry thereby provide him with tools for studying their behavior.   Other branches of mathematics pose a more subtle transition between physical and mathematical ways of thinking, and a methodology is required to aid in our thinking.

1.3   METHODOLOGY IN SOLID MECHANICS

In general, five successive stages are involved in the study of solid mechanics and in the solution of problems.

(1) A physical system is presented as an object of analysis or design.   This may range from a complete system such as a vehicle or a bridge to a single subsystem such as a beam, column, or fastener.

(2) An ideal model is conceived and sketched.

(3) Mathematical reasoning is applied to the ideal model which now becomes a mathematical model.   The equations which result from this application are derived and solved.

(4) The mathematical results are interpreted physically in terms pertinent to the original object of the analysis or design.

(5) The results are compared with the results of physical observations.   These observations may be made on a laboratory model or on the finished object.

Although it is the purpose of a large portion of the present book to discuss these five stages, certain remarks can be made about them at this point.

Stage 1 implies that man has a curiosity concerning the physical world about him and desires to infer information about it beyond that which he can immediately observe. This curiosity is generally motivated by a desire to construct an engineering system.

It is probably evident that the second stage, conception of the ideal model, is the key element in making the transition from physical to mathematical ways of thinking. Nature is, in general, so complex that it does not permit itself to be expressed exhaustively in a single model, equation, or thought. The scientist or engineer is invariably faced with a task of constructing simplified models of nature to represent levels of abstraction consistent with the results which are sought. If he seeks, for example, to describe the position of the atomic particles of a solid under load, the model he adopts must be vastly different from that required to describe the gross change of external shape. The most difficult task that confronts him is then one of introducing only those simplifying assumptions important to the physical process he is studying, and of neglecting all influences of lower order.

Applications of science at all levels involve conceptual models of nature which are used for mathematical or physical reasoning. The different divisions of basic science may be classified by their conceptual models. For example, to the student of solid mechanics a metal beam is a mass of homogeneous, isotropic, sometimes elastic and sometimes plastic material subjected to force and displacement boundary conditions; to the metallurgist, a collection of randomly oriented grains and crystals; to the chemist, a collection of atoms and molecules; and to the solid-state physicist, a swarm of nuclei and electrons. We can construct an approximate model of a portion of the physical world and trace the relations between its parts, but we cannot by such methods reveal the total intrinsic reality of nature. By applying the mathematical tools of reasoning to models, we can infer new information. If the model is consistent with experimental observation, it is regarded as a valid model. Pure scientists are engrossed in improving nature's models; engineers or applied scientists apply them to machines. In the student's professional life, his success in the use of mechanics will depend as much on his skill in the game of "mathematical make-believe" as on any other single factor. The art of stripping away physical complexities which are unimportant for the problem at hand, and of designing ideal models which yield to analysis is at the heart of successful engineering practice. The student will recognize that the development of these skills and viewpoints is, in fact, a requirement in all branches of the engineering sciences.

The third stage, application of mathematical reasoning, requires an application of physical laws to the ideal model. These laws, which in the present book rest upon the foundation of Newtonian mechanics, provide an experimentally verified basis for constructing mathematical equations to describe the behavior of the physical system. Mathematical reasoning is then brought into play by employing the axioms and theorems governing the mathematical equations which have been constructed.

The fourth stage, physical interpretation of mathematical results, presents no difficulty providing the physical laws introduced in the third stage are fully understood. Under these circumstances there should be a clear realization of the relationship between the parameters of the mathematical model and their counterparts in nature.

The fifth step, comparison with physical observations, is an important step in the process, although it will not be one of the objects of the present book to deal explicitly with the experimental sciences. The engineer or applied scientist must learn from experience. The tools he employs should be those which experience has shown will yield accurate, experimentally verified results. Mechanical, aeronautical, civil, and other engineers have learned from experience that certain ideal models provide accurate representations of physical systems with which they have traditionally dealt. The ideal models used to represent pulley systems, cantilever airplane wings, and bridge trusses are examples of accumulated experience based on experimental observation. The engineer working at the forefront of his technology may need to construct ideal models for systems which are entirely new and for which there is no accumulated experience. He may desire, in such cases, to construct experimental laboratory models or to make careful confirmatory measurements on the first version of the operational system in order to verify his design calculations. Thus he accumulates additional experience which gives him a unique advantage.

FIGURE 1.1

A simple example may serve to illustrate at this point the nature of the five stages. There are many structural problems associated with the design of an airplane. In principle, one ideal model or one gigantic mathematical equation covering the entire airplane (Fig. 1.1) is possible, but solving it would be impractical and even impossible. The engineer is forced to break the airplane down into its separate structural components and examine them individually. For instance, in the analysis and design of a wing structure to carry the flight loads, the five stages may be described in the following terms:

(1) The wing provides the lift necessary to sustain flight, and as such must be strong enough to resist the aerodynamic pressures created by the air flow over the wing and must be rigid enough to prevent excessive deflections and wing flutter.

(2) The wing is idealized as a one-dimensional structure rigidly attached to a wall. Such a structure, illustrated by Fig. 1.2, is called a cantilever beam. The bending-stiffness properties of the beam are condensed into a single parameter called the bending-stiffness factor, EI, and the airloads are condensed into a load per unit length, $p(x)$.

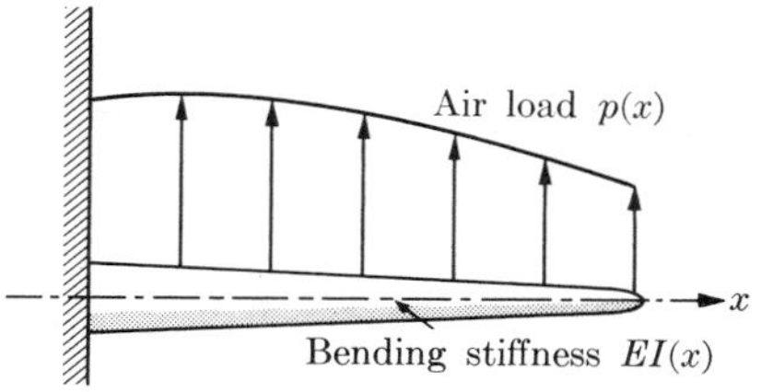

Air load $p(x)$

FIG. 1.2   Airplane wing as a cantilever beam.

Bending stiffness $EI(x)$

(3) The equation which describes the bending behavior of the idealized model is

$$\frac{d^2}{dx^2}\left(EI\frac{d^2w}{dx^2}\right) = p(x), \tag{1.1}$$

where $w(x)$ is the lateral deflection of the beam as a function of x. Since EI is also a function of x, the above equation is most easily integrated in numerical form.

(4) The lateral deflection of the wing is obtainable directly from $w(x)$, and the strains and stresses may also be computed from the result by employing additional mathematical formulas.

(5) The structural designer often verifies his computations and obtains additional information from static tests in the laboratory of models or of a full-scale wing. The ultimate verification of the analysis and design, however, is the successful flight operation of the prototype airplane.

Other examples are abundant, and as the subject is gradually developed these will become evident. It will become increasingly clear to the student that mathematics plays an important role in the study of solid mechanics. Even though the problems are generally motivated by an engineering objective, and although the ultimate verification lies in the performance of a hardware object, a reasonably high proficiency in mathematics is required for participation in the challenging engineering projects which are at the forefront of new technologies. Of course, many of the fundamental principles of solid mechanics were laid down by men blessed with tremendous intuitive gifts and mental abilities. Fortunately, it is not necessary to possess these same attributes in order to understand the principles of solid mechanics and to apply them effectively to engineering problems.

Principles of Mechanics

2.1 INTRODUCTION

We shall find in subsequent developments of the present book that four fundamental laws provide a foundation for the subject of mechanics. These laws represent in concise form an acceptable mathematical model of a segment of nature. They are listed below for the sake of convenient reference, and they are discussed in detail in the following articles of the present chapter.

(a) Law of the Parallelogram of Forces (Section 2.3)
(b) Law of Transmissibility of Forces (Section 2.4)
(c) Law of Motion (Section 2.5)
(d) Law of Action and Reaction (Section 2.6)

These four laws provide an experimentally verified foundation for Newtonian mechanics upon which, by mathematical analysis, such vast subjects as particle, rigid-body, solid, and fluid mechanics may be erected. These laws can be fully understood only by applying them to diverse problems. It is quite important that the student adopt a philosophy of seeking an understanding of their full meaning by numerous applications to problems. Several examples are given in the present and later chapters to illustrate their application, and additional problems are listed at the ends of the chapters.

2.2 THE CONCEPT OF FORCE

All normal human beings are familiar in their day-to-day activities with the notion of a force. The opening and closing of doors, the lifting of weights, and the steering of an automobile are experiences which require one to exert force. It is evident from daily activities that the notion of a force involves both magnitude and direction; that is, a force is a vector quantity. Furthermore, a force is associated with a point of application. Forces are therefore represented in mechanics by vectors acting at specific points. In mechanics as in everyday experience, forces are usually produced by the action of one body on another. Since forces are vector quantities, they will be represented by bold-face letters such as **F**, **P**, and **Q**. On the blackboard or in manuscript they are usually denoted by arrows over the letters, such as $\vec{F}$, $\vec{P}$, and $\vec{Q}$.

The vector qualities of a force permit it to be represented in space by a directed line segment. In Fig. 2.1 a force **F** acting at a point p is schematically represented in three-dimensional space by the line segment pa. As explained in the Appendix, for the summation notation and tensor notation to be used in this text, the three rectangular Cartesian coordinates are represented by y_1, y_2, and y_3, respectively.

The force **F** can be written as the vector sum of three independent components,

$$\mathbf{F} = F_1\mathbf{i}_1 + F_2\mathbf{i}_2 + F_3\mathbf{i}_3 = \sum_{m=1}^{3} F_m\mathbf{i}_m,$$

$$(2.2.1a)$$

where $\mathbf{i}_1$, $\mathbf{i}_2$, and $\mathbf{i}_3$ are the unit vectors associated with the rectangular Cartesian coordinates y_1, y_2, and y_3. According to the summation notation (see Appendix, Section A.2) this is represented simply as

$$\mathbf{F} = F_m\mathbf{i}_m. \qquad (2.2.1b)$$

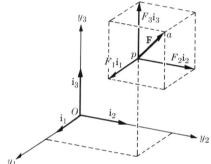

Fig. 2.1   Cartesian representation of a force vector.

The repeated index, such as the m in Eq. (2.2.1b), is to be summed from one to three. Geometrically, the vector components $F_1\mathbf{i}_1$, $F_2\mathbf{i}_2$, and $F_3\mathbf{i}_3$ are seen to form the edges of a rectangular parallelepiped which has **F** as its main diagonal. The magnitude of **F** is given by the scalar product (see Appendix, Section A.12):

$$|\mathbf{F}| = \sqrt{\mathbf{F} \cdot \mathbf{F}} = [(F_m\mathbf{i}_m) \cdot (F_n\mathbf{i}_n)]^{1/2},$$

$$|\mathbf{F}| = (F_mF_n\delta_{mn})^{1/2}, \qquad (2.2.2)$$

where δ_{mn} is called the Kronecker delta (Appendix, Section A.3), which is equal to unity if the two indices are equal, and is zero if the two indices are unequal. The vertical bars on either side of a vector, e.g., $|\mathbf{F}|$, signify the magnitude or length of a vector.

Equation (2.2.2) is written in expanded form as

$$|\mathbf{F}| = [(F_1)^2 + (F_2)^2 + (F_3)^2]^{1/2}. \qquad (2.2.3)$$

Hence the scalar quantities F_1, F_2, F_3 are seen to be the magnitudes of the vector components of **F**.

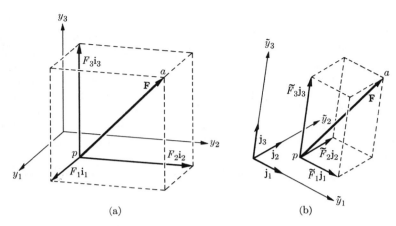

FIG. 2.2   Components of the same vector in different Cartesian coordinate systems.

The coordinate system consisting of the axes y_1, y_2, and y_3 has been arbitrarily chosen. It should be evident that the force **F** does not depend upon the choice of a coordinate system. In another coordinate system, for example, $\tilde{y}_1$, $\tilde{y}_2$, and $\tilde{y}_3$, the force is written as

$$\mathbf{F} = \tilde{F}_m \mathbf{j}_m, \tag{2.2.4}$$

where the $\mathbf{j}_m$ are the unit vectors associated with the new coordinate system, and where $\tilde{F}_1 \mathbf{j}_1$, $\tilde{F}_2 \mathbf{j}_2$, and $\tilde{F}_3 \mathbf{j}_3$ are the vector components which describe **F** relative to the $\tilde{y}_n$-axes (see Fig. 2.2).

Since Eqs. (2.2.1b) and (2.2.4) describe the same vector, we may combine them to obtain

$$\tilde{F}_m \mathbf{j}_m = F_n \mathbf{i}_n. \tag{2.2.5}$$

In order to express the $\tilde{F}_m$-components in terms of the F_n-components, we need only take the scalar product of both sides of Eq. (2.2.5) with $\mathbf{j}_r$:

$$\mathbf{j}_r \cdot \tilde{F}_m \mathbf{j}_m = \tilde{F}_m \delta_{mr} = \mathbf{j}_r \cdot \mathbf{i}_n F_n. \tag{2.2.6}$$

The scalar products on the right-hand side of Eq. (2.2.6) are the cosines of the angles between the axes of the two axis systems (see Appendix, Section A.12). These are called direction cosines, and will be denoted by $l_{\tilde{r}n}$, that is,

$$l_{\tilde{r}n} = \mathbf{j}_r \cdot \mathbf{i}_n. \tag{2.2.7}$$

Then Eq. (2.2.6) becomes

$$\tilde{F}_r = l_{\tilde{r}n} F_n. \tag{2.2.8}$$

Equation (2.2.8) is the transformation law (see Appendix, Section A.6) for a first-order tensor, and therefore the components of a force form a first-order tensor.

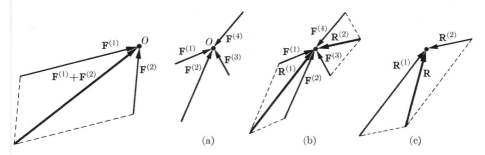

FIG. 2.3 Addition of vectors   FIG. 2.4 Addition of forces by repeated application of
by parallelogram method.   parallelogram law.

2.3 LAW OF THE PARALLELOGRAM OF FORCES—RESULTANT OF CONCURRENT FORCES

In dealing with the subject of mechanics, one is very often confronted with situations in which several forces act at once at a specified point. The first question to be raised in such a situation is, what is the resultant of all of these forces? This question is answered traditionally by the *Law of the Parallelogram of Forces: When two forces,* $\mathbf{F}^{(1)}$ *and* $\mathbf{F}^{(2)}$, *act simultaneously at a specified point, they are together equivalent to a single force,* $\mathbf{F}^{(1)} + \mathbf{F}^{(2)}$, *acting at that point,* as shown by the parallelogram construction of Fig. 2.3. It is seen that the so-called Law of the Parallelogram of Forces may be interpreted as a consequence of the vector addition of two concurrent forces (see preceding section). When more than two forces are acting, the Parallelogram Law may be applied repeatedly to obtain the resultant. For example, in Fig. 2.4, the forces $\mathbf{F}^{(1)}$, $\mathbf{F}^{(2)}$, $\mathbf{F}^{(3)}$, and $\mathbf{F}^{(4)}$ acting at O may be replaced by the force $\mathbf{R}$ at O. The latter is obtained by constructing $\mathbf{R}^{(1)}$ from a parallelogram formed with $\mathbf{F}^{(1)}$ and $\mathbf{F}^{(2)}$, $\mathbf{R}^{(2)}$ from a parallelogram with $\mathbf{F}^{(3)}$ and $\mathbf{F}^{(4)}$, and finally $\mathbf{R}$ from a parallelogram with $\mathbf{R}^{(1)}$ and $\mathbf{R}^{(2)}$. These repeated steps are represented by the vector equation

$$\mathbf{F}^{(1)} + \mathbf{F}^{(2)} + \mathbf{F}^{(3)} + \mathbf{F}^{(4)} = \mathbf{R}. \qquad (2.3.1)$$

The vector sum represented by Eq. (2.3.1) may be found graphically by the polygon construction shown in Fig. 2.5, provided that all the forces are in a plane. When the forces are not coplanar, a space construction of the vector sum is required.

If a coordinate system y_1, y_2, y_3 is chosen, then each of the four forces in Eq. (2.3.1) can be written in the form

$$\mathbf{F}^{(n)} = F_m^{(n)}\mathbf{i}_m, \qquad n = 1, 2, 3, 4, \qquad (2.3.2)$$

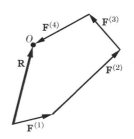

FIG. 2.5 Addition of vectors by polygon construction.

where the enclosed superscript is an identifying tag. The resultant can also be written in vector form with respect to the same axes,

$$\mathbf{R} = R_m \mathbf{i}_m,$$ (2.3.3)

and therefore Eq. (2.3.1) yields the result

$$R_m = \sum_{n=1}^{4} F_m^{(n)}.$$ (2.3.4)

2.4 LAW OF TRANSMISSIBILITY OF FORCES

Another fundamental feature of the action of a force on a rigid body is the *Law of Transmissibility of a Force: When a force is acting upon a rigid body, it may be translated along its line of action so that it acts on any point of the rigid body which coincides with the line of action.* It may be seen from the above that two forces of equal magnitude and opposite direction acting upon opposite sides of a rigid body and along a common line of action, as illustrated by Fig. 2.6, cancel each other entirely. This is evident since the two forces may be slid along their common line of action to the same point of application. When the parallelogram law is applied to this pair, their resultant is seen to vanish.

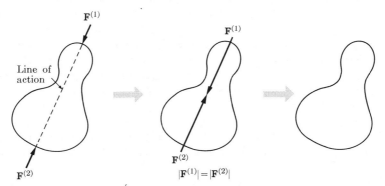

Fig. 2.6   One consequence of the principle of force transmissibility.

This law applies only to rigid bodies, i.e., to bodies in which the deformations may be considered negligible. We will see that such an assumption is justified for some analyses and not for others.

2.5 LAW OF MOTION

Let us assume that experiments are being conducted to determine the influence of forces on bodies. Perhaps the simplest of all experiments is that of observing the falling of a body in the gravitational field of the earth. Suppose that the weight of a body at a given location on the surface of the earth is measured by a very sensitive spring scale, and that the acceleration of gravity in a vacuum is also measured by sensitive timing devices at the same location. If this process is repeated with the same object at other locations on the surface of the earth, it will be found that

$$\frac{w^{(1)}}{g^{(1)}} = \frac{w^{(2)}}{g^{(2)}} = \cdots = \frac{w^{(n)}}{g^{(n)}} = m, \qquad (2.5.1)$$

where $w^{(i)}$ and $g^{(i)}$ are the weights and accelerations of gravity, respectively. The ratio of weight to acceleration of gravity, called the mass of the body, m, is invariant with respect to location on the earth's surface.

The action of the earth's gravitational field upon a body can be summarized by the formula

$$\mathbf{W} = m\mathbf{g}, \qquad (2.5.2)$$

where $\mathbf{W}$ is a force vector with magnitude equal to the body weight and direction along a plumb line, $\mathbf{g}$ is an acceleration vector with magnitude equal to the local acceleration of gravity and direction also along the local plumb line.

Experience has shown that formula (2.5.2) can be generalized to other forces and accelerations, providing one selects a suitable frame of reference relative to which the accelerations are measured. This generalization, known as *Newton's Second Law of Motion*, may be stated as follows: *A particle which is unrestrained and acted upon by a force vector* $\mathbf{F}$, *accelerates at a rate* $\mathbf{a}$ *which is proportional in magnitude to the magnitude of* $\mathbf{F}$ *and along the line of action and in the sense of direction of* $\mathbf{F}$. This proportionality may be represented by the equation

$$\mathbf{F} = m\mathbf{a}, \qquad (2.5.3)$$

where m is a scalar mass, the value of which depends upon the body and on the choice of units of force, length, and time.

The first requirement which must be placed upon the frame of reference is that it be either fixed or moving with constant velocity. Such a frame of reference is called Newtonian. Experimental evidence shows that a suitable system is the astronomical frame of reference in which the mass center of the solar system is fixed and which is without rotation relative to the fixed stars as a whole.* Another frame of reference which is nearly Newtonian is that of the earth's surface. Strictly speaking, there are differences between experimental evidence and predictions of Eq. (2.5.3) due to the earth's rotation, when the earth's surface is used as a reference frame. However, these differences are negligible in most engineering calculations.

---

* Equation (2.5.3) provides excellent although not perfect agreement with astronomical observations in the solar system. The agreement is, however, sufficiently good so that it is usually used as a basis for the study of the mechanics of the solar system.

In scientific and engineering work involving the use of forces, it is of fundamental importance to be able to describe the magnitude of a given force. This requires that some definite force be selected as a standard unit. For example, one might employ a spring scale for this purpose. The scale could be calibrated by measuring the deflection produced by the weight of a standard body at a particular location on the earth's surface. Calibration units frequently employed in scientific and engineering work are the British gravitational unit of the pound and the metric gravitational unit of the kilogram. The former is defined as the weight, measured at sea level and at a latitude of 45°, of a standard pound platinum body kept at the Standards Office of the London Board of Trade. The latter is the weight, also measured at sea level and at a latitude of 45°, of a standard kilogram platinum body kept at the International Bureau of Weights and Measures, Sèvres, France.

In mechanics, three fundamental dimensions are necessary to form a complete mechanical dimension system. With length and time as a common basis, a single additional independent dimension is required. One may choose from among mass, force, power, energy and others. We shall consider two of these possibilities.

(a) Dynamical or physical systems. In these systems, we select the fundamental dimensions of length, mass, and time. Perhaps the best known system in this category is the *CGS* (*centimeter-gram-second*) *system*. In this system the unit of mass is taken as the gram-mass, which is $1/1000$ of the mass of the standard kilogram body. The force which is required to produce an acceleration of 1 cm/sec^2 of 1-gm mass is termed a *dyne*. Thus the fundamental units of the CGS system are the centimeter, the gram-mass, and the second. A derived quantity is the dyne force having the units of g-cm/sec^2. Another system similar to the cgs-system is the mks-system in which the fundamental units are the meter, the kilogram-mass, and the second. The force is again a derived quantity; the force necessary to give a 1-kg mass an acceleration of 1 m/sec^2 is defined as the *newton*. In terms of the fundamental units of the mks-system, the newton has the units of kg-m/sec^2. The *English dynamical system* employs the fundamental units of the foot, the pound-mass, and the second. The force is derived from these quantities; the force necessary to give a 1-lb mass an acceleration of 1 ft/sec^2 is defined as the *poundal*. The poundal has the fundamental units of lb-ft/sec^2.

(b) Gravitational or technical systems. In contrast to the dynamical or physical systems of units, we now select as the fundamental units in the gravitational or technical systems the units of length, force, and time. Mass is left as a derived quantity. Such systems are primarily used by engineers, with some advantages accruing from the fact that force is a fundamental rather than a derived quantity. The *English gravitational system* is probably the most widely used of all such systems, and it employs as its fundamental units feet, pounds-force, and seconds. Thus, if a force of 1 lb produces an acceleration of 1 ft/sec^2, the body

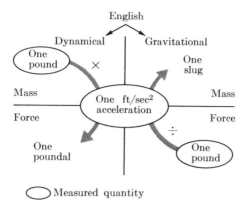

Fig. 2.7   Comparison of the English dynamical and gravitational systems of unity.

in question has a unit mass. This unit of mass is termed a *slug,* and in terms of the fundamental units of the system it has the units lb-sec^2/ft. The two English systems are compared in Fig. 2.7.

2.6   LAW OF ACTION AND REACTION

Finally, we state the important *Law of Action and Reaction: When two particles exert forces on each other, whether in direct contact or at a distance from each other, these forces are equal in magnitude and opposite in sense, and act along a line joining the two particles.* According to the Law of Action and Reaction, the mutual effect of one particle on another, whether it be by direct contact or at a distance through the agency of magnetism, gravitation, or electrostatic action, involves pairs of forces of equal magnitude and opposite direction. Thus, if the gravitational pull of the earth on an orbiting satellite is a force **F** along a line joining the satellite with the center of the earth, the gravitational pull of the satellite on the earth is a force **F** acting along the same line. If a landing airplane exerts a maximum downward force of 100,000 lb by its tire on the runway, the runway exerts an equal and opposite upward force of 100,000 lb on the tire. These statements may all be summed up succinctly by saying that action and reaction are equal and opposite.

2.7   EQUILIBRIUM OF A PARTICLE

A particle is regarded in mechanics as a body which has dimensions sufficiently small that it may be regarded as situated at a point. Thus, when a particle is situated at a point O, and acted on by several forces, it may be said that the forces are concurrent at O. We have seen (Section 2.5) that the motion of a particle under the action of a resultant external force **F** is governed by the Law of

Motion, Eq. (2.5.3). Application of this law to the prediction of particle motion constitutes the field of particle dynamics. It is not our intention to enter into a discussion of this field here, but rather to enunciate the fundamental principle of equilibrium of an unrestrained particle acted on by external forces. This principle, known as *d'Alembert's Principle*, reads: *The system of forces composed of the external forces,* **F**, *and the reversed inertial forces,* $-m\mathbf{a}$, *taken together in vector addition, add up to zero for equilibrium.* Thus, we find that if there are a number of external force vectors $\mathbf{F}^{(1)}$, $\mathbf{F}^{(2)}$, $\mathbf{F}^{(3)}$, ..., $\mathbf{F}^{(n)}$ acting on a particle at O, the particle is in equilibrium pro-viding

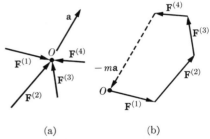

$$\sum_{i=1}^{n} \mathbf{F}^{(i)} + (-m\mathbf{a}) = 0. \quad (2.7.1)$$

This is equivalent to stating that the particle is in equilibrium when and only when the vectors $\mathbf{F}^{(i)}$ and $-m\mathbf{a}$ drawn successively, beginning to end, form a closed polygon. Such a polygon is illus-trated by Fig. 2.8(b). Note that the vector representing the inertia force $[m\mathbf{a}]$

FIG. 2.8   Equilibrium of a particle.

in the polygon construction has a sense opposite to the acceleration vector of Fig. 2.8(a). The force polygon will, of course, be a plane figure only when the forces $\mathbf{F}^{(i)}$ are coplanar. Otherwise, it will be a space figure, and methods of descriptive geometry will be required to represent it on a plane.

It is evident that in a statical system where constraints are applied to the particle so that it cannot move, thus requiring that $\mathbf{a} = 0$, the condition of equilibrium simplifies to

$$\sum_{i=1}^{n} \mathbf{F}^{(i)} = 0. \quad (2.7.2)$$

The preceding formula is a vector equation which expresses the equilibrium of the particle; by expressing each force $\mathbf{F}^{(i)}$ in component form, Eq. (2.7.2) becomes

$$\sum_{i=1}^{n} F_r^{(i)} \mathbf{i}_r = 0. \quad (2.7.3)$$

The scalar product of $\mathbf{i}_p$ with Eq. (2.7.3) leads to three scalar equations of equilibrium,

$$\sum_{i=1}^{n} F_p^{(i)} = 0, \quad p = 1, 2, 3. \quad (2.7.4)$$

When p is set equal to one, Eq. (2.7.4) states that the sum of the y_1-components of all the forces acting on the particle is zero. Similarly, the other two values of p state that the sum of y_2- and y_3-components must be zero independently. Thus we may speak of the "three" independent equations for the equilibrium of a particle.

2.8   SUMMARY OF THE PRINCIPLES OF MECHANICS

(a) A *force* has direction, magnitude and a point of application.

(b) *Law of the Parallelogram of Forces.* When two forces, $\mathbf{F}^{(1)}$ and $\mathbf{F}^{(2)}$, act simultaneously at a specified point, they are together equivalent to a single force, $\mathbf{F}^{(1)} + \mathbf{F}^{(2)}$, acting at that point.

(c) *Law of Transmissibility of Forces.* When a force is acting upon a rigid body, it may be translated along its line of action so that it acts on any point of the rigid body which coincides with the line of action.

(d) *Law of Motion.* A particle which is unrestrained and acted upon by a force vector $\mathbf{F}$ acquires an acceleration $\mathbf{a}$ which is proportional in magnitude to the magnitude of $\mathbf{F}$ and along the line of action and in the sense of direction of $\mathbf{F}$.

(e) *Law of Action and Reaction.* When two particles exert forces, one upon the other, whether in direct contact or at a distance from each other, these forces are equal in magnitude and opposite in sense and act along a line joining the two particles.

PROBLEMS

2.1   Expand each of the following sets of equations written with the indicial notation:

(a) $\sigma_m \mathbf{i}_m = k_{mn}\sigma_{mn}$,

(b) $\alpha_m \sigma_{am} + \beta_q \sigma_{aq} = \gamma_r \sigma_{ar}$,

(c) $A_{mn}\sigma_m = C_{mr}\sigma_{nr}\mathbf{i}_m$,

(d) $\delta_{mn}a_{ms}b_{nt} = C_{st}$.

FIGURE P.2.2

2.2   What is the consequence of the Law of Transmissibility of Forces in the situation illustrated by Fig. P.2.2?

2.3   Write a vector expression for the acceleration of the particle shown in Fig. P.2.3 (forces are expressed in pounds):

$$\mathbf{F}^{(1)} = \mathbf{i}_1 + \mathbf{i}_2 + \mathbf{i}_3,$$
$$\mathbf{F}^{(2)} = 3\mathbf{i}_1 + \mathbf{i}_2 - 2\mathbf{i}_3,$$
$$\mathbf{F}^{(3)} = 4\mathbf{i}_1 - 3\mathbf{i}_2 + 7\mathbf{i}_3,$$
$$\mathbf{F}^{(4)} = 2\mathbf{i}_1 + 5\mathbf{i}_2 + 2\mathbf{i}_3.$$

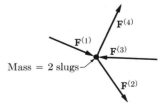

Mass = 2 slugs

FIGURE P.2.3

2.4   A 5000-lb airplane is in a steady glide with the flight path at an angle θ below the horizontal (Fig. P.2.4). The drag force in the direction of the flight path is 750 lb. Find the lift force L normal to the flight path, and angle θ.

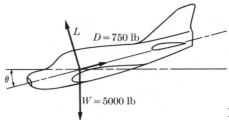

FIGURE P.2.4

2.5   Three coplanar forces $\mathbf{F}^{(1)}$, $\mathbf{F}^{(2)}$, and $\mathbf{F}^{(3)}$ are acting on a particle in equilibrium (Fig. P.2.5). If $\mathbf{F}^{(1)}$ and α are fixed quantities, determine $\mathbf{F}^{(2)}$, $\mathbf{F}^{(3)}$, and β such that $|\mathbf{F}^{(3)}|$ is a minimum.

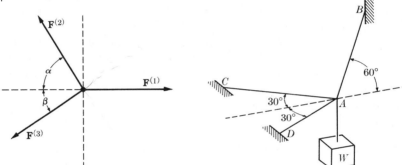

FIGURE P.2.5   FIGURE P.2.6

2.6   A weight W is supported by three strings AB, AC, and AD (Fig. P.2.6). A, D, and C are on one horizontal plane; AB is at an angle of 60° with respect to this plane. Calculate the forces in the strings. [Neglect the change in geometry, i.e., the deflection of point A due to the elongation of the strings.]

2.7   Given the two coordinate systems shown in Fig. P.2.7, and that $\tilde{y}_1$ and $\tilde{y}_2$ lie in the y_1y_2-plane.
(a) Fill in the table of nine direction cosines $l_{\tilde{m}n}$.

	y_1	y_2	y_3
$\tilde{y}_1$			
$\tilde{y}_2$			
$\tilde{y}_3$			

(b) Given the force vector $\mathbf{F} = \mathbf{i}_1 - 2\mathbf{i}_2 + \mathbf{i}_3$ expressed in the $y_1y_2y_3$-system. Express this vector in the new $\tilde{y}_1\tilde{y}_2\tilde{y}_3$-system.

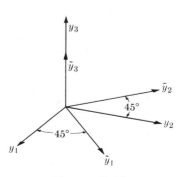

FIGURE P.2.7

2.8   Three mutually perpendicular strings, OP, OQ, and OR, are attached to the corner, O, of a cube.   The edges of the cube are represented by OA, OB, and OC. Each string carries a tensile force.   The magnitudes of the string forces are shown in Fig. P.2.8. The   direction cosines between the strings and the cube edges are:

	OP	OQ	OR
OA	$-\dfrac{1}{\sqrt{2}}$	$\dfrac{1}{\sqrt{2}}$	0
OB	$-\dfrac{1}{2}$	$-\dfrac{1}{2}$	$\dfrac{1}{\sqrt{2}}$
OC	$-\dfrac{1}{2}$	$-\dfrac{1}{2}$	$-\dfrac{1}{\sqrt{2}}$

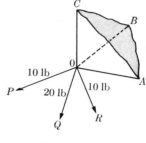

FIGURE P.2.8

(a) Find the magnitude of the resultant force which acts on the corner of the cube.
(b) Find the components of the resultant force with respect to the edges of the cube.
[All results may be left in radical form.]

2.9   A weight W is suspended by two identical springs, each with a spring constant k (lb/in). The length of each spring in its unstretched position is L. The upper ends of the springs are located a distance L apart (Fig. P.2.9). Write the equations which can be used to determine H, the final length of the spring, in terms of L, k, and W. [*Note:* In this problem the change in geometry has been included in the equilibrium equations. What is the parameter which must be large before the change in geometry can be neglected?]

FIGURE P.2.9

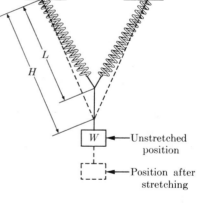

Unstretched position

Position after stretching

Statics

3.1 INTRODUCTION

In the preceding chapter we discussed four laws basic to the whole subject of mechanics. We shall take up now the application of these laws to problems of the static equilibrium of particles and bodies under the action of forces and moments. We introduce this subject by discussing some of the special properties of forces and moments and their interrelations. Finally, we examine in detail some important scientific and engineering applications of statics.

3.2 PROPERTIES OF FORCE AND MOMENTS

(a) The moment of a force about a point. We will consider a force vector $\mathbf{F}$ which is arbitrarily oriented with respect to a point p, and calculate the moment produced by $\mathbf{F}$ about p. Let us construct the plane $ABCD$ defined by the vector $\mathbf{F}$ and the point p, as shown by Fig. 3.1, and draw in the plane the perpendicular line a from p to $\mathbf{F}$. Then the magnitude of the moment of $\mathbf{F}$ about p is given by the product of $\mathbf{F}$ and the distance a.

$$|\mathbf{M}| = a|\mathbf{F}|. \tag{3.2.1}$$

It is convenient to assign a positive direction to the moment, and we do this by regarding it as a vector defined by

$$\mathbf{M} = \mathbf{r} \times \mathbf{F}, \tag{3.2.2}$$

where $\mathbf{r}$ is any vector drawn in the plane $ABCD$ from p to a point on the line of action of $\mathbf{F}$. The positive direction of the moment vector $\mathbf{M}$, according to the rule for vector products (see Appendix, Section A.13), is as shown by Fig. 3.1 and is along a line perpendicular to the plane $ABCD$. It is in the direction in which a right-hand screw will advance if the vectors $\mathbf{r}$ and $\mathbf{F}$ are translated along their lines of action so that they act at a common point, and the vector $\mathbf{r}$ is turned into the vector $\mathbf{F}$ through the smallest possible angle.

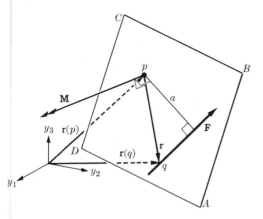

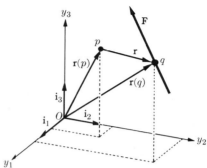

Fig. 3.1   Moment of a force about a point.   Fig. 3.2   The component form of the moment of **F** about p is required.

Since **M** is a vector, it can also be written in component form:

$$\mathbf{M} = M_m \mathbf{i}_m. \tag{3.2.3}$$

The meaning of the components $M_1\mathbf{i}_1$, $M_2\mathbf{i}_2$, and $M_3\mathbf{i}_3$ can be determined with the aid of the right-hand rule.

It is evident that the moment of **F** about p is unchanged by moving **F** along its line of action, and that the moment will vanish only when **F** passes through p which means $a = |\mathbf{r}| = 0$.

EXAMPLE: To compute in component form the moment of an arbitrary force about a point.

Solution: Referring to Fig. 3.2 we designate the force vector by **F**, and the point about which the moment is to be taken by p. Both quantities are assumed to be embedded in an orthogonal Cartesian coordinate system y_1, y_2, and y_3 with the unit vectors $\mathbf{i}_1$, $\mathbf{i}_2$, and $\mathbf{i}_3$. The point p is located by the position vector $\mathbf{r}(p)$. Suppose that the vector **r** is drawn from the point p to any point q along the line of action of **F**. Then the moment of **F** about p is given by

$$\mathbf{M} = \mathbf{r} \times \mathbf{F} = [\mathbf{r}(q) - \mathbf{r}(p)] \times \mathbf{F}, \tag{3.2.4}$$

where $\mathbf{r}(q)$ is the vector connecting the origin of coordinates with q. The component forms of these vectors are

$$\begin{aligned}
\mathbf{r} &= y_1\mathbf{i}_1 + y_2\mathbf{i}_2 + y_3\mathbf{i}_3 = y_m\mathbf{i}_m, \\
\mathbf{r}(p) &= y_1(p)\mathbf{i}_1 + y_2(p)\mathbf{i}_2 + y_3(p)\mathbf{i}_3 = y_m(p)\mathbf{i}_m, \\
\mathbf{r}(q) &= y_1(q)\mathbf{i}_1 + y_2(q)\mathbf{i}_2 + y_3(q)\mathbf{i}_3 = y_m(q)\mathbf{i}_m, \\
\mathbf{F} &= F_1\mathbf{i}_1 + F_2\mathbf{i}_2 + F_3\mathbf{i}_3 = F_m\mathbf{i}_m.
\end{aligned} \tag{3.2.5}$$

Inserting Eq. (3.2.5) into (3.2.4) yields the following component form of the moment vector:

$$\mathbf{M} = \{[y_2(q) - y_2(p)]F_3 - [y_3(q) - y_3(p)]F_2\}\mathbf{i}_1$$
$$+ \{[y_3(q) - y_3(p)]F_1 - [y_1(q) - y_1(p)]F_3\}\mathbf{i}_2$$
$$+ \{[y_1(q) - y_1(p)]F_2 - [y_2(q) - y_2(p)]F_1\}\mathbf{i}_3. \qquad (3.2.6)$$

Although it may appear at first glance that the magnitude of $\mathbf{M}$ is dependent upon the location of the terminus of $\mathbf{r}(q)$ on the line of action of $\mathbf{F}$, closer inspection will show that the result is independent of this location.

The use of the permutation tensor, e_{mnp} (see Appendix, Section A.3) very neatly yields the results shown in Eqs. (3.2.4), (3.2.5), and (3.2.6). Thus Eq. (3.2.4) can be written as

$$\mathbf{M} = [y_m(q)\mathbf{i}_m - y_m(p)\mathbf{i}_m] \times F_r\mathbf{i}_r, \qquad (3.2.7a)$$

or

$$\mathbf{M} = [y_m(q) - y_m(p)]F_r(\mathbf{i}_m \times \mathbf{i}_r). \qquad (3.2.7b)$$

The vector products of the unit vectors (see Appendix, Section A.13) can be expressed as

$$\mathbf{i}_m \times \mathbf{i}_r = e_{mrk}\mathbf{i}_k, \qquad (3.2.8)$$

and hence

$$\mathbf{M} = [y_m(q) - y_m(p)]F_r e_{mrk} \mathbf{i}_k. \qquad (3.2.9)$$

The components of $\mathbf{M}$ are seen to be

$$M_k = [y_m(q) - y_m(p)]F_r e_{mrk}. \qquad (3.2.10)$$

(b) The moment of a force about an axis. In the previous subsection we saw that the moment of a force $\mathbf{F}$ about a point is given by Eq. (3.2.2), in which any position vector $\mathbf{r}$ drawn from the point to the line of action of the force is crossed into the force vector. Suppose that a line l is drawn through the point p, and we now compute the moment of $\mathbf{F}$ about this line. Of the two possible senses of l, we take one as positive and designate it by a unit vector $\mathbf{e}$ collinear with l, as illustrated by Fig. 3.3. Since the line l is assigned a positive sense of direction, we refer to it as an axis. The scalar component of the moment of $\mathbf{F}$ about l is defined as the component M_l of $\mathbf{M}$ in the positive sense of l. Then

$$M_l = \mathbf{e} \cdot \mathbf{M} = \mathbf{e} \cdot (\mathbf{r} \times \mathbf{F}).$$
$$(3.2.11)$$

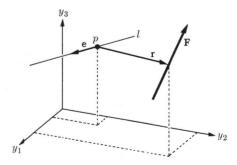

Fig. 3.3   Moment of a force about an axis.

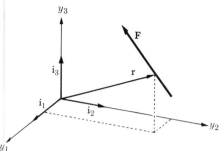

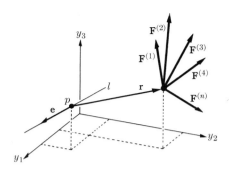

FIG. 3.4   Moment of a force about the   FIG. 3.5   Moment of several force vec-
y_3-axis.   tors about a line.

Equation (3.2.11) announces that the moment of a force $\mathbf{F}$ about an axis is equal to the component of $\mathbf{r} \times \mathbf{F}$ along that axis. Clearly the vector $\mathbf{r}$ may be any vector drawn from the axis to the line of action of $\mathbf{F}$.

EXAMPLE: To compute in component form the moment of an arbitrary force about the y_3-axis.

Solution: We take the vector $\mathbf{r}$ which connects the y_3-axis with the line of action of $\mathbf{F}$, as illustrated by Fig. 3.4. Then the solution is obtained by substituting into

$$M_3 = \mathbf{i}_3 \cdot (\mathbf{r} \times \mathbf{F}) \qquad (3.2.12)$$

the component forms

$$\mathbf{r} = y_m \mathbf{i}_m, \qquad \mathbf{F} = F_r \mathbf{i}_r. \qquad (3.2.13)$$

This yields the final result,

$$M_3 = y_m F_r e_{rm3} = y_1 F_2 - y_2 F_1. \qquad (3.2.14)$$

(c) Theorem of Varignon.   We consider next the question of computing the moment about the line l of several force vectors having a common origin A as illustrated by Fig. 3.5 for the forces $\mathbf{F}^{(1)}, \mathbf{F}^{(2)}, \mathbf{F}^{(3)}, \ldots, \mathbf{F}^{(n)}$. This moment is the sum of the moments of the individual vectors, and

$$M_l = \mathbf{e} \cdot (\mathbf{r} \times \mathbf{F}^{(1)}) + \mathbf{e} \cdot (\mathbf{r} \times \mathbf{F}^{(2)}) + \cdots + \mathbf{e} \cdot (\mathbf{r} \times \mathbf{F}^{(n)}), \qquad (3.2.15)$$

which reduces, since the vector-product law is distributive, to

$$M_l = \mathbf{e} \cdot [\mathbf{r} \times (\mathbf{F}^{(1)} + \mathbf{F}^{(2)} + \cdots + \mathbf{F}^{(n)})]. \qquad (3.2.16)$$

It is evident that if $\mathbf{R}$ is the resultant of the forces $\mathbf{F}^{(k)}$,

$$\mathbf{R} = \mathbf{F}^{(1)} + \mathbf{F}^{(2)} + \cdots + \mathbf{F}^{(n)}, \qquad (3.2.17)$$

then we may conclude that the moment of the resultant is equal to the sum of the moments of the individual forces. This result, known as the Theorem of

Varignon, is stated in detail as follows: *The sum of the moments about an axis of several force vectors with a common origin is equal to the moment about the axis of the resultant of these vectors.*

(d) Couples. A pair of forces of equal magnitude, opposite direction, and different but parallel lines of action is known as a *couple.* Figure 3.6 illustrates a couple consisting of forces $\mathbf{F}$ and $-\mathbf{F}$ lying in a plane designated by $ABCD$. It is evident that the vector sum of a couple is zero. Let us consider the moment of the couple about an arbitrary line s perpendicular to this plane. Referring to the notation of Fig. 3.6, we see that this moment is

$$\mathbf{C} = \mathbf{r}(q) \times \mathbf{F} - \mathbf{r}(p) \times \mathbf{F} = [\mathbf{r}(q) - \mathbf{r}(p)] \times \mathbf{F} = \mathbf{r} \times \mathbf{F}. \qquad (3.2.18)$$

Then

$$\mathbf{C} = a|\mathbf{F}|\mathbf{n}, \qquad (3.2.19)$$

where a is the perpendicular distance between the parallel force vectors and $\mathbf{n}$ is a unit vector which is perpendicular to the plane $ABCD$. The positive sense of $\mathbf{n}$ is given by the right-hand rule. Equation (3.2.19) indicates that the total moment of the forces forming a couple is independent of the location of s. The symbol $\mathbf{C}$ is used to distinguish a couple from the moment vector $\mathbf{M}$, which is a function of location. Note also that the couple vector does not have a definite point of application. We refer therefore to the magnitude of the moment of a couple in an absolute sense as the product of the magnitude of the forces of the couple times the perpendicular distance between these forces. The direction of the moment of the couple is determined by the same rule of direction that governs the moments of forces. For example, in Fig. 3.6 the moment vector associated with the couple points in the plus $\mathbf{n}$-direction, since a right-hand screw turned in the direction of rotation of the couple would advance in this direction.

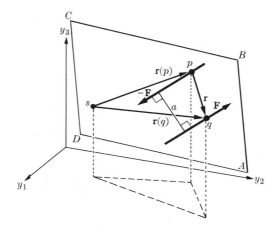

Fig. 3.6   Moment produced by a couple.

(e) Equipollent force systems. Two systems of forces are said to be *equipollent* when either system may be replaced by the other. (Pollent is Greek, meaning "powerful," and thus, equipollent means equally powerful.) In more detail we say that the two systems are equipollent if

    (1) the resultant of the forces of one system is equal to the resultant of forces of the other system, and

    (2) the sum of the moments of the forces of one system about an arbitrary base point is equal to the sum of the moments of the forces of the other system about the same point.

It is evident, for example, that system A which consists of a force vector $\mathbf{F}$ and a position vector $\mathbf{r}$ is equipollent to system B in which the force vector $\mathbf{F}$ is moved to the origin and a moment $\mathbf{r} \times \mathbf{F}$ is added.

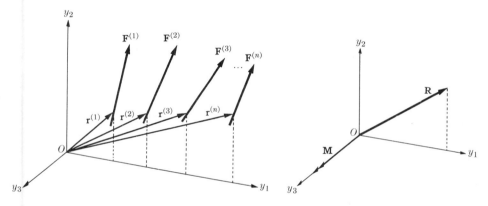

FIG. 3.7   Arbitrary system of coplanar forces.   FIG. 3.8   Reduced system of co-planar forces.

(f) Reduction of coplanar force systems. Let us consider next for simplicity a number of forces lying in the $y_1 y_2$-plane, as illustrated by Fig. 3.7. Let $\mathbf{R}$ be their sum and $\mathbf{M}$ their total moment about the y_3-axis. The system consisting of $\mathbf{R}$ and $\mathbf{M}$ shown in Fig. 3.8 is evidently plane equipollent to the given system of forces.

We may also state in a slightly different way that any system of forces in a plane may be reduced to a single force located at some arbitrary point s and a moment about s. This reduction process may be carried out analytically. Let us suppose, for example, that we wish to reduce the system of Fig. 3.7 to a single force at the origin and a moment about the origin. This result is illustrated by Fig. 3.8, where

$$\mathbf{R} = \sum_{i=1}^{n} \mathbf{F}^{(i)}, \tag{3.2.20}$$

$$\mathbf{M} = \sum_{i=1}^{n} \mathbf{r}^{(i)} \times \mathbf{F}^{(i)}, \tag{3.2.21}$$

and $\mathbf{r}^{(i)}$ are position vectors from the origin to points on the lines of action of the force vectors $\mathbf{F}^{(i)}$. In component form, Eq. (3.2.20) becomes

$$R_1 = \sum_{i=1}^{n} F_1^{(i)}, \qquad R_2 = \sum_{i=1}^{n} F_2^{(i)}, \qquad (3.2.22)$$

where the subscripts 1 and 2 indicate components along the y_1- and y_2-axes, respectively. Similarly, Eq. (3.2.21) has the scalar form of

$$|\mathbf{M}| = \sum_{i=1}^{n} (y_1^{(i)}F_2^{(i)} - y_2^{(i)}F_1^{(i)}). \qquad (3.2.23)$$

We have so far demonstrated how a system of forces may be reduced to a single force at a point and a moment about this same point. There can be further reductions. For example, let us consider a case where $\mathbf{R} \neq 0$ and $\mathbf{M} \neq 0$, and replace $\mathbf{M}$ in Fig. 3.8 by a couple with forces $\pm\mathbf{R}$. Thus we may replace the system of Fig. 3.8 by another equipollent system shown in Fig. 3.9. The perpendicular distance OA in Fig. 3.9, that is, the arm of the couple, is equal to $|\mathbf{M}|/|\mathbf{R}|$. Since the vectors $\pm\mathbf{R}$ at the origin cancel each other, it is evident that the system represented by a single force vector $\mathbf{R}$ located at the perpendicular distance OA from the origin is plane equipollent to the system of Fig. 3.8 and the original system of Fig. 3.7.

Let us consider next a case when $\mathbf{R} = 0$ and $\mathbf{M} \neq 0$. It is obvious that in this case the equipollent system consists of a couple. Finally, we summarize the results stated above by saying that an arbitrary coplanar force system may be reduced to (1) a single force at an arbitrary point plus a moment about this point, (2) to a single force at a unique point if there are no couples, or (3) to a couple if there is no resultant force.

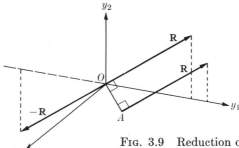

FIG. 3.9   Reduction of force and couple to single force.

(g) Reduction of force systems in space. In the preceding section we discussed the reduction of arbitrary coplanar force systems to a single force and a couple about an arbitrary point. The same conclusion may be drawn also for force systems in space. We may, for example, reduce any system of forces in space to its equipollent system consisting of a single force and moment as represented

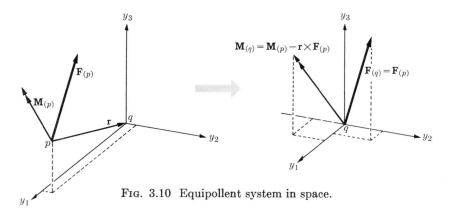

FIG. 3.10 Equipollent system in space.

by the vectors $\mathbf{F}(p)$ and $\mathbf{M}(p)$, respectively, on the left-hand side of Fig. 3.10. If we change the base point p to another base point q, and if we wish to construct another equipollent system at q, we move the force $\mathbf{F}$ without alteration to q, but affix the new moment vector

$$\mathbf{M}(q) = \mathbf{M}(p) - \mathbf{r} \times \mathbf{F}(p), \qquad (3.2.24)$$

where $\mathbf{r}$ is the displacement vector connecting base points p and q. Thus, under a change in base point, the original force $\mathbf{F}$ and moment $\mathbf{M}$ at p are transformed at q to $\mathbf{F}(q)$ and $\mathbf{M}(q)$ given by

$$\mathbf{F}(q) = \mathbf{F}(p), \qquad \mathbf{M}(q) = \mathbf{M}(p) - \mathbf{r} \times \mathbf{F}(p). \qquad (3.2.25)$$

3.3 EQUILIBRIUM OF A PARTICLE

(a) Condition of equilibrium. In mechanics a particle is regarded as a body which is concentrated in a single point. It may be of such small dimensions that it can be considered to occupy a vanishingly small area represented by a point P. The forces on the particle may, therefore, be regarded as concurrent at the point P. However, of greater practical importance, we may regard any body as a particle so long as the forces are applied such that when they are shifted along their lines of action they may be made concurrent at some point P in the body.

A particle is said to be in static equilibrium when the vector sum of the concurrent forces acting at the point P is zero (cf. Eq. 2.7.2). That is, if the forces acting on the particle are represented by $\mathbf{F}^{(1)}, \mathbf{F}^{(2)}, \ldots, \mathbf{F}^{(n)}$, the condition for equilibrium is represented by the vector equation,

$$\sum_{i=1}^{n} \mathbf{F}^{(i)} = 0. \qquad (3.3.1)$$

The vector equation (3.3.1) may also be expressed in component form. If we take components along the orthogonal y_1-y_2-y_3 triad, we have as the scalar form of Eq. (3.3.1):

$$\sum_{i=1}^{n} F_1^{(i)} = 0, \qquad \sum_{i=1}^{n} F_2^{(i)} = 0, \qquad \sum_{i=1}^{n} F_3^{(i)} = 0. \qquad (3.3.2)$$

Equations (3.3.2) require that the sums of the forces in each of the three orthogonal directions be individually equal to zero.

Equation (3.3.1) may be satisfied in a geometric sense by constructing a polygon of forces where the vectors $\mathbf{F}^{(i)}$ are drawn successively tail-to-head to form a space polygon.

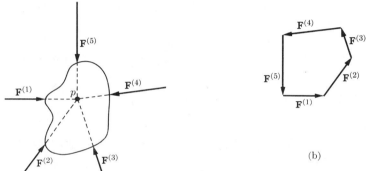

(a)

(b)

Fig. 3.11 Body acted upon by concurrent forces in static equilibrium: (a) free-body diagram; (b) polygon of forces.

(b) Free-body diagram. A free-body diagram in mechanics represents a diagram in which the body is drawn free and clear of all its supports or reacting bodies, with their influence represented by appropriate reactive forces. Figure 3.11(a) illustrates a coplanar free-body diagram with forces concurrent at a single point. Figure 3.11(b) shows the corresponding coplanar force polygon which represents the particle equilibrium.

As a simple illustration of how one arrives at a free-body diagram, we refer to Fig. 3.12, which shows a circular cylinder resting on a flat surface. The physical diagram of Fig. 3.12(a) is replaced by the free-body diagram of Fig. 3.12(b) by removing the horizontal surface and replacing it by the normal reaction R. This step puts one in position to write the equations of equilibrium of the particle, which in this case are

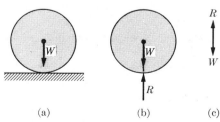

(a)   (b)   (c)

Fig. 3.12 Circular cylinder resting on table: (a) physical diagram; (b) free-body diagram; (c) force polygon.

obtained merely by summing the vertical forces (see Fig. 3.12c):

$$R - W = 0, \tag{3.3.3}$$

where W is the weight of the cylinder. Equation (3.3.3) allows one to compute the reaction of the circular cylinder on the plane as $R = W$.

EXAMPLE: To compute the reactions on a circular cylinder resting in a smooth trough, as illustrated by Fig. 3.13(a).

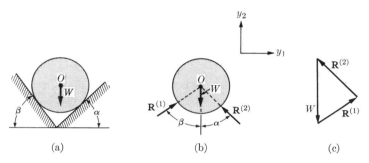

FIG. 3.13 Circular cylinder resting in a smooth trough: (a) physical diagram; (b) free-body diagram; (c) force polygon.

Solution: We may construct the free-body diagram of Fig. 3.13(b) by removing the trough and replacing it by two reactions $\mathbf{R}^{(1)}$ and $\mathbf{R}^{(2)}$ perpendicular to the surfaces of the trough. Since the surfaces of the trough are assumed smooth these reactions pass through the center O of the circular cylinder. We have a system of forces which are concurrent at O, and the cylinder may be treated as a particle. The two unknowns of the problem, $\mathbf{R}^{(1)}$ and $\mathbf{R}^{(2)}$, may be computed by applying Eqs. (3.3.2), that is, by requiring that the sum of the horizontal and the sum of the vertical forces each be zero. We find that

$$\sum_{i=1}^{3} F_1^{(i)} = R^{(1)} \sin \beta - R^{(2)} \sin \alpha = 0,$$

$$\sum_{i=1}^{3} F_2^{(i)} = R^{(1)} \cos \beta + R^{(2)} \cos \alpha - W = 0, \tag{3.3.4}$$

from which we obtain the reactions

$$|\mathbf{R}^{(1)}| = \frac{W \sin \alpha}{\sin (\alpha + \beta)}, \qquad |\mathbf{R}^{(2)}| = \frac{W \sin \beta}{\sin (\alpha + \beta)}. \tag{3.3.5}$$

(c) Supports. There are some symbols employed in solid mechanics to represent special kinds of supports which are used to restrain structures. The symbol ⬚ or ⬚ represents a support which can only supply a force of either sense along a line perpendicular to the surface. The symbol ⬚ represents a support

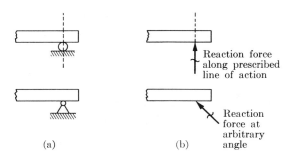

FIG. 3.14 Schematic representation of supports: (a) physical diagrams; (b) free-body diagrams.

which fixes the point of application but not the line of action of the force. Generally these supports are used only in planar structures and such supports are actually used. (The next time you sail under a simple bridge, look at the supports for each span.) Their properties are shown schematically in Fig. 3.14. In this figure the symbol ⚡ is used to indicate a reaction force.

3.4   EQUILIBRIUM OF A SYSTEM OF PARTICLES

(a) Conditions of equilibrium.   Let us consider in detail the conditions for static equilibrium of a system of particles which may be interconnected in any manner whatsoever. In such a problem we must be very careful and clear at the outset as to which particles constitute the total system. For example, we might regard the solar system as a system of several particles consisting of the sun and the planets; or we might regard the earth and moon as a two-particle system. A pile of n books lying on a table, where we regard each book and the table as a separate particle, may comprise a system of $n + 1$ particles. It is quite clear that the choice of individual particles and definition of their aggregate system from a given arrangement in nature is an important preliminary step in the discussion of a system of particles.

In examining a given particle in a system of particles, we find two kinds of forces, internal and external. A force on a particle is internal when it is exerted by another particle of that system. The Law of Action and Reaction comes into play in dealing with internal forces. The law asserts that internal forces occur in equal and opposite pairs, each pair representing the mutual interaction of a pair of particles, one upon the other. These internal forces may be due, for example, to surface contact of the particles, mechanical connections between the particles, or electromagnetic, electrostatic, and gravitational forces. Figure 3.15 shows an example of a three-particle system under the influence of external forces. $\mathbf{P}^{(i)}$ is the vector sum of all external forces on the ith particle, $\mathbf{R}^{(ij)}$ is the internal force on the ith particle due to jth particle. For example, mass (1) is acted on by a resultant external force $\mathbf{P}^{(1)}$ and two internal forces $\mathbf{R}^{(12)}$ and

$\mathbf{R}^{(13)}$. In general, for a system of n particles, the equation of equilibrium for the ith particle is

$$\mathbf{P}^{(i)} + \sum_{i \neq j}^{n} \mathbf{R}^{(ij)} = 0, \quad i, j = 1, 2, \ldots, n,$$

$$(3.4.1)$$

or the force polygon formed by these force vectors must be closed. Note that

$$\mathbf{R}^{(ij)} = -\mathbf{R}^{(ji)}, \quad i, j = 1, 2, \ldots, n,$$

$$(3.4.2)$$

because action equals reaction, and quantities such as $\mathbf{R}^{(11)}$, $\mathbf{R}^{(22)}$, etc., do not exist. Equation (3.4.2), which states that the in-

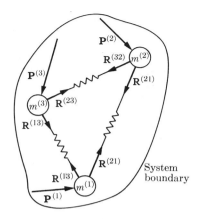

FIG. 3.15 A system of particles.

ternal forces between any two particles are equal in magnitude and opposite in direction, applies irrespective of how the particles are interconnected.

By summing the forces on all the particles of the system, that is, by summing Eq. (3.4.1) on i, we obtain

$$\sum_{i=1}^{n} \mathbf{P}^{(i)} + \sum_{i=1}^{n} \sum_{j=1}^{n} \mathbf{R}^{(ij)} = 0.$$

$$(3.4.3)$$

This merely represents the addition of all n of the equations represented by Eq. (3.4.1). If the double summation term is written out explicitly, it will be observed that the terms can be grouped in pairs such as $\mathbf{R}^{(12)} + \mathbf{R}^{(21)}$, $\mathbf{R}^{(23)} + \mathbf{R}^{(32)}$, etc., and by virtue of Eq. (3.4.2) these terms will vanish; hence the equilibrium equation (3.4.3) reduces to

$$\sum_{i=1}^{n} \mathbf{P}^{(i)} = 0.$$

$$(3.4.4)$$

Equation (3.4.4) states one of the requirements of equilibrium; that the sum of the external forces on the system be equal to zero. Another requirement for equilibrium is derived from Eq. (3.4.1). Let $\mathbf{r}^{(i)}$ be the position vector of the ith particle with respect to some axis system located at point O. Then we can form the vector product of $\mathbf{r}^{(i)}$ with Eq. (3.4.1),

$$\mathbf{r}^{(i)} \times \mathbf{P}^{(i)} + \sum_{j=1}^{n} \mathbf{r}^{(i)} \times \mathbf{R}^{(ij)} = 0,$$

$$(3.4.5)$$

which becomes, after summing on i (that is, adding up all n equations),

$$\sum_{i=1}^{n} \mathbf{r}^{(i)} \times \mathbf{P}^{(i)} + \sum_{i=1}^{n} \sum_{j=1}^{n} \mathbf{r}^{(i)} \times \mathbf{R}^{(ij)} = 0.$$

$$(3.4.6)$$

We may regroup the second term in the above equation as

$$\frac{1}{2} \sum_{i=1}^{n} \sum_{j=1}^{n} (\mathbf{r}^{(i)} - \mathbf{r}^{(j)}) \times \mathbf{R}^{(ij)},$$

by applying Eq. (3.4.2). Since the two vectors, $(\mathbf{r}^{(i)} - \mathbf{r}^{(j)})$ and $\mathbf{R}^{(ij)}$, are parallel, this term is identically zero. We obtain

$$\sum_{i=1}^{n} \mathbf{r}^{(i)} \times \mathbf{P}^{(i)} = 0. \tag{3.4.7}$$

Equation (3.4.7) states that the total moment about O of the external force system is zero.

Thus the two requirements for a system of particles to be in equilibrium are that (1) the sum of the external forces on the system be equal to zero [vector equation (3.4.4)] and (2) the sum of the moments of the external forces about some arbitrary point be equal to zero [vector equation (3.4.7)]. If we resolve the vectors in their component form with respect to an orthogonal triad y_1-y_2-y_3 with origin at O, we have the following six scalar equations of equilibrium:

$$\sum_{i=1}^{n} P_1^{(i)} = 0, \tag{3.4.8}$$

$$\sum_{i=1}^{n} P_2^{(i)} = 0, \tag{3.4.9}$$

$$\sum_{i=1}^{n} P_3^{(i)} = 0, \tag{3.4.10}$$

$$\sum_{i=1}^{n} (y_2^{(i)} P_3^{(i)} - y_3^{(i)} P_2^{(i)}) = 0, \tag{3.4.11}$$

$$\sum_{i=1}^{n} (y_3^{(i)} P_1^{(i)} - y_1^{(i)} P_3^{(i)}) = 0, \tag{3.4.12}$$

$$\sum_{i=1}^{n} (y_1^{(i)} P_2^{(i)} - y_2^{(i)} P_1^{(i)}) = 0. \tag{3.4.13}$$

The first three equations state that the resultants of the y_1-, y_2-, and y_3-components, respectively, of all the external forces must be equal to zero. The last three equations state that the moments of all the forces about the y_1-, y_2-, and y_3-axes, respectively, must be zero.

The entire theory of statics rests on the vector equations (3.4.4) and (3.4.7), or alternatively on their scalar representation, Eqs. (3.4.8) through (3.4.13). We have stated previously that the particles of the body may be connected in any manner whatsoever. Thus these equations may be applied to a rigid body, an elastic or plastic deformable body, or a volume of fluid. We shall illustrate several examples of the equilibrium of bodies of various kinds.

EXAMPLE 1: As a current example of a system of particles let us consider the reentering Mercury capsule shown in Fig. 3.16. All forces and accelerations are parallel to the y_1-axis, and it is assumed that the parachute and capsule centers of gravity lie on a common line parallel to the y_1-axis. For such a system the three equilibrium equations for moment [Eqs. (3.4.11) through (3.4.13)] do not appear in the problem, and only one equilibrium equation for the external forces is required. The external forces acting on the system are

$$\mathbf{D}^{(i)} \quad \text{(drag forces on parachute and capsule)},$$
$$\mathbf{W}^{(i)} \quad \text{(weights of parachute and capsule)},$$

where $i = p, c$. In addition, d'Alembert forces $-m^{(i)}\mathbf{a}$ may also be considered as external forces acting on the capsule and parachute because of the deceleration of the system.

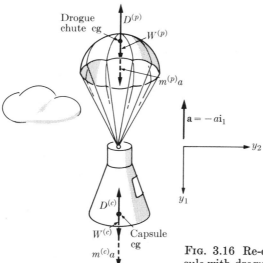

FIG. 3.16 Re-entering Mercury capsule with drogue parachute deployed.

All the forces acting may be transferred to the center of gravity of the entire system, and the following equilibrium equation may be written in the y_1-direction:

$$\mathbf{W}^{(p)} + \mathbf{W}^{(c)} - (m^{(p)} + m^{(c)})\mathbf{a} - \mathbf{D}^{(p)} - \mathbf{D}^{(c)} = 0. \qquad (3.4.14)$$

Equation (3.4.14) completes the analysis of the parachute and capsule as a one-particle system. Figure 3.17 illustrates the manner in which we may represent the capsule and parachute as a two-particle system. In addition to

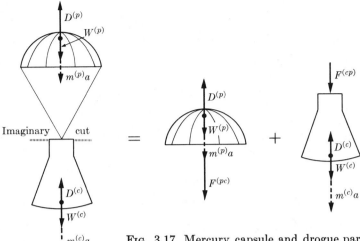

Fig. 3.17 Mercury capsule and drogue parachute as a two-particle system.

the external and d'Alembert forces, we must now delineate the internal forces,

$\mathbf{F}^{(pc)}$ on the parachute from the capsule,

$\mathbf{F}^{(cp)}$ on the capsule from the parachute.

The equations of equilibrium for each of the particles now read

$$\mathbf{F}^{(pc)} + \mathbf{W}^{(p)} - m^{(p)}\mathbf{a} - \mathbf{D}^{(p)} = 0 \quad \text{(parachute)},$$
$$\mathbf{F}^{(cp)} + \mathbf{W}^{(c)} - m^{(c)}\mathbf{a} - \mathbf{D}^{(c)} = 0 \quad \text{(capsule)}. \qquad (3.4.15)$$

If we make an imaginary cut to obtain the two-particle representation, as shown at the left of Figure 3.17, we see that the internal force, $\mathbf{F}^{(pc)}$, has been added to the parachute to account for the effects of the external forces on the parachute through the capsule, that is,

$$\mathbf{F}^{(pc)} = \mathbf{W}^{(c)} - m^{(c)}\mathbf{a} - \mathbf{D}^{(c)}. \qquad (3.4.16)$$

Similarly, the internal force, $\mathbf{F}^{(cp)}$, has been added to the capsule to account for the effects of the external forces on the capsule through the parachute,

$$\mathbf{F}^{(cp)} = \mathbf{W}^{(p)} - m^{(p)}\mathbf{a} - \mathbf{D}^{(p)}. \qquad (3.4.17)$$

The equilibrium equation (3.4.14) for the original one-particle representation may be rewritten

$$\mathbf{W}^{(p)} - m^{(p)}\mathbf{a} - \mathbf{D}^{(p)} = -[\mathbf{W}^{(c)} - m^{(c)}\mathbf{a} - \mathbf{D}^{(c)}], \qquad (3.4.18)$$

which, when combined with Eqs. (3.4.16) and (3.4.17), yields the expected

$$\mathbf{F}^{(cp)} = -\mathbf{F}^{(pc)}. \qquad (3.4.19)$$

Thus we have developed an alternative view of internal forces. We begin with an entire system in equilibrium with a set of applied external forces. When a particular particle of the system is isolated for detailed study, internal forces arise at the particle-system interfaces from the transmission of the action of external forces on the rest of the system.

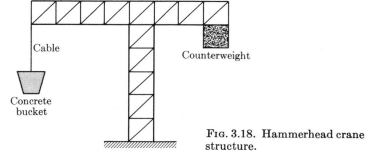

Cable

Counterweight

Concrete
bucket

FIG. 3.18. Hammerhead crane
structure.

EXAMPLE 2: As another example which may be considered as a collection of particles, let us consider the so-called hammerhead crane which is used in the erection of buildings (see Fig. 3.18). The crane consists of a vertical tower (the handle of a hammer) and a long horizontal boom (the head). The concrete bucket (or other useful load) is suspended on a cable at one end of the boom, and a large mass of concrete which serves as a counterweight is fixed at the other end. There are many possible subdivisions which can be made of the structural system depicted in Fig. 3.18. One such system is shown in Fig. 3.19. An explanation follows.

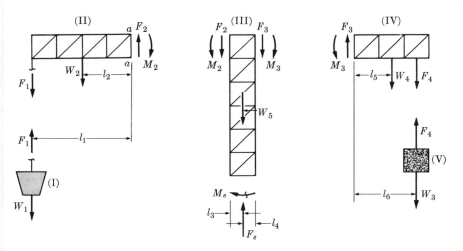

FIG. 3.19   Free-body diagram of subdivided hammerhead crane structure.

In subcollection I, W_1 represents the weight of the bucket plus contents and a portion of the cable, F_1 is the force in the cable at this point, and for equilibrium

$$F_1 = W_1. \tag{3.4.20}$$

For subcollection II, F_2 and M_2 are a force and moment equipollent to the internal forces at section a–a of the boom, and F_1 is the cable force which by the Law of Action and Reaction (see Section 2.6) is the equal and opposite of the F_1-force shown in subcollection I.

For the vertical equilibrium of subsystem II

$$F_2 = F_1 + W_2, \tag{3.4.21}$$

and for rotational or moment equilibrium,

$$M_2 = F_1 l_1 + W_2 l_2, \tag{3.4.22}$$

where W_2 is the weight of the boom and cable included in II, l_2 locates its center of gravity, and l_1 is the length of the boom. The horizontal equilibrium of part II is assured because there are no externally applied forces with horizontal components.

For subcollections IV and V the corresponding equations are

$$F_3 = W_4 + F_4, \tag{3.4.23}$$

$$M_3 = F_4 l_6 + W_4 l_5, \tag{3.4.24}$$

$$F_4 = W_3. \tag{3.4.25}$$

The central vertical tower has acting on it the internal forces transmitted from the two portions of the horizontal boom, the dead weight W_5 of the tower, and the support force and moment F_s and M_s. Equilibrium of subcollection III requires that

$$F_s = F_2 + F_3 + W_5, \tag{3.4.26}$$

$$M_s = M_2 + F_2 l_3 - M_3 - F_3 l_4. \tag{3.4.27}$$

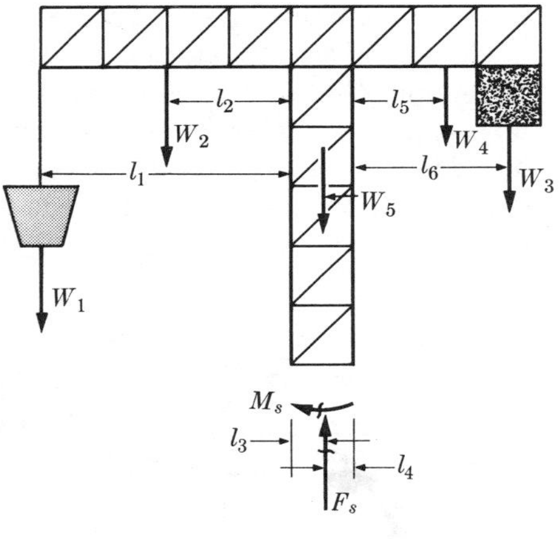

FIG. 3.20 Free-body diagram of complete hammerhead crane structure.

We can obtain the support reactions more directly by reassembling the five subcollections into the total structure (see Fig. 3.20). Then the internal forces are no longer visible. The equations of equilibrium obtained by examining Fig. 3.20 are

$$F_s = W_1 + W_2 + W_3 + W_4 + W_5, \tag{3.4.28}$$

$$M_s = W_1(l_1 + l_3) + W_2(l_2 + l_3) - W_3(l_4 + l_6) - W_4(l_4 + l_5). \tag{3.4.29}$$

It can be easily verified that the simultaneous solution of Eqs. (3.4.20) through (3.4.27) will yield Eqs. (3.4.28) and (3.4.29).

EXAMPLE 3: Let us next compute the forces required to maintain an airplane in equilibrium in steady level flight. Figure 3.21 shows a simplified free-body diagram of an airplane of weight W in steady level flight. Let it be assumed that at the flight speed of the airplane the drag D is known. It is required to compute the lift on the wing $L^{(w)}$, the lift on the tail $L^{(t)}$, and the engine thrust T for steady level flight.

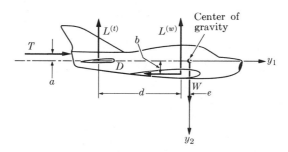

FIG. 3.21   Free-body diagram of an airplane in steady level flight.

Solution: This is a problem of plane statics, and we assume airplane body axes y_1 and y_2, as illustrated by Fig. 3.21. Summing forces along y_1 and y_2 we obtain respectively

$$T - D = 0, \tag{3.4.30}$$

$$-L^{(w)} - L^{(t)} + W = 0. \tag{3.4.31}$$

Summing moments about the center of gravity of the airplane yields

$$Ta + Db + L^{(t)}(d+e) + L^{(w)}e = 0. \tag{3.4.32}$$

From Eq. (3.4.30) we obtain the thrust,

$$T = D. \tag{3.4.33}$$

Solving Eqs. (3.4.31) and (3.4.32) simultaneously, we compute the following for the tail lift $L^{(t)}$, and the wing lift $L^{(w)}$:

$$L^{(t)} = -\frac{D(a+b) + We}{d}, \tag{3.4.34}$$

$$L^{(w)} = W + \frac{D(a+b) + We}{d}. \tag{3.4.35}$$

It is evident in this example that a downward tail load is required for equilibrium.

(b) Special force systems. Equations (3.4.8) through (3.4.13) show that in general the external forces on a system of particles or a rigid body are governed by six scalar equations of equilibrium. In many special force systems, some of these conditions are satisfied identically so that fewer equations of equilibrium remain to be satisfied. Some of these special cases are listed below:

(1) *Coplanar force systems, where all force vectors are in one plane.* If this is the $y_1 y_2$-plane, then

$$y_3^{(i)} = 0, \tag{3.4.36}$$

and

$$P_3^{(i)} = 0. \tag{3.4.37}$$

Thus Eqs. (3.4.10), (3.4.11), and (3.4.12) are satisfied identically, and the equations which remain to be satisfied are

$$\sum_{i=1}^{n} P_1^{(i)} = 0, \tag{3.4.38}$$

$$\sum_{i=1}^{n} P_2^{(i)} = 0, \tag{3.4.39}$$

$$\sum_{i=1}^{n} (y_1^{(i)} P_2^{(i)} - y_2^{(i)} P_1^{(i)}) = 0. \tag{3.4.40}$$

(2) *Parallel force systems, where all the forces are parallel to the y_2-axis.* Then

$$P_1^{(i)} = 0, \tag{3.4.41}$$

$$P_3^{(i)} = 0. \tag{3.4.42}$$

Thus Eqs. (3.4.8), (3.4.10), and (3.4.12) are satisfied identically, and the equations which remain to be satisfied are

$$\sum_{i=1}^{n} P_2^{(i)} = 0, \tag{3.4.43}$$

$$\sum_{i=1}^{n} y_3^{(i)} P_2^{(i)} = 0, \tag{3.4.44}$$

$$\sum_{i=1}^{n} y_1^{(i)} P_2^{(i)} = 0. \tag{3.4.45}$$

Note the reduction of the moment equations (3.4.11) and (3.4.13) to only one group of terms.

(3) *Concurrent force systems, where all forces pass through one single point.* If this point is chosen as the origin, then the sums of moments of all forces about

the three coordinate axes are identically zero. The remaining condition to be satisfied is that the sums of forces along each of the three perpendicular directions be zero. When the concurrent force system is also coplanar, the number of equations of equilibrium is reduced to two.

(4) *Systems where the lines of action of all forces pass through the same axis.* It is obvious that the sum of moments of all forces about this axis is identically zero, and only five equations of equilibrium remain to be satisfied.

3.5   EXAMPLES OF THE USE OF THE FREE-BODY DIAGRAM

Before we proceed to the examples, it is desirable to reiterate the importance of the free-body diagram and lay down some general principles for its construction. The first step in any problem in statics is to isolate that physical object or portion of a physical object which is to be studied. A diagram is then drawn showing this object free and clear of its supports or connecting members, these latter having been replaced by appropriate reactive forces. Such a diagram is a *free-body diagram* (read Section 3.3b). All the forces supplied by the supports or connecting members to the isolated free body are then considered as external forces.

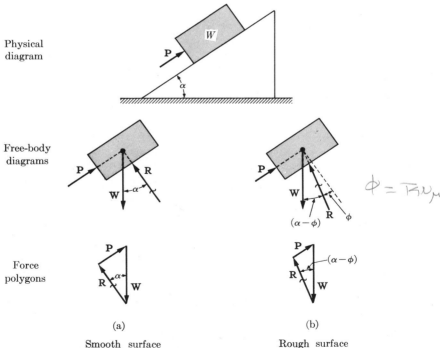

FIG. 3.22   Free-body diagrams and force polygons of block on inclined surface: (a) smooth surface; (b) rough surface.

When a body is resting on a surface, removal of the surface requires that it be replaced by a reaction **R**. When the surface is smooth, **R** is normal to the surface. If the surface is rough, **R** makes an angle with the surface normal. For example, if the body is on the verge of slipping on a surface having a coefficient of sliding friction, μ, **R** will make an angle, $\phi = \tan \mu$, with the normal and in a direction opposite to the motion. Figure 3.22 illustrates, for example, free-body diagrams of a block on a smooth and on a rough surface. The force **P** in the diagram is externally applied, and is of just sufficient magnitude to keep the block from sliding down. The reaction **R** is normal to the surface in free-body diagram (a), whereas it makes an angle with the surface normal in (b). The force polygons corresponding to each free-body diagram are shown also in Fig. 3.22. In these polygons the directions of all the force vectors are known; however, only the magnitude of **W** is known, and the magnitudes of **P** and **R** remain to be found.

In the force polygon of free-body diagram (a) the vectors **R** and **W** have an included angle α, whereas in (b) the angle is $\alpha - \phi$, where ϕ is the friction angle. It is evident that the magnitude of **P** is reduced as ϕ approaches α, and goes to zero for $\phi = \alpha$. In the latter case, the surface is sufficiently rough to prevent the block from sliding without application of an external force **P**.

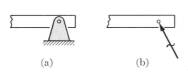
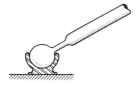

(a)   (b)

FIG. 3.23 Frictionless pin joint: (a) physical diagram; (b) free-body diagram.

FIG. 3.24 Cutaway view of ball-and-socket joint.

In some cases a body is supported by frictionless pin joints. For example, in Fig. 3.23 the attachment of the end of a beam to its support through a pin is shown. Such a joint obviously is not capable of resisting any moment about the axis of the pin. Thus the resultant of the reactive forces must pass through the center of the pin. The corresponding free-body diagram is shown at the right of the figure. Such a support has also been described in a previous section (see Fig. 3.13).

A pin-joint support, of course, can transmit a moment about any axis which is normal to the axis of the pin. A joint which will not transmit any moment at all is a ball-and-socket joint, as shown in Fig. 3.24.

A body is often supported by or attached to cables. By definition, a cable can transmit only tension (although one must admit that the cables of the Golden Gate Bridge or the George Washington Bridge can resist bending and compression, the efficiency of a suspension bridge derives from designing the cables to carry only tension). In a free-body diagram these cables are shown

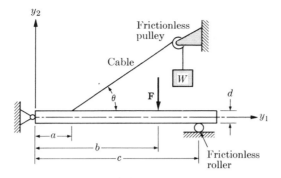

FIG. 3.25 Physical diagram of beam-cable system.

cut and replaced by reactive forces along their lines of action. Figure 3.25 illustrates a planar uniform member supported by a pin joint at one end and a frictionless roller at the other. The member is acted upon by a vertical force **F** and is attached to a weight **W** through a cable-pulley system as shown. If the pulley is frictionless, the tension in the cable becomes a known force equal to the weight **W**. The unknown forces are the reactions $\mathbf{R}^{(1)}$ and $\mathbf{R}^{(2)}$ at both ends of the member.

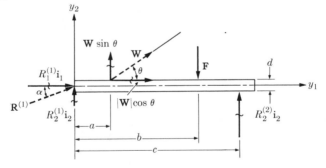

FIG. 3.26   Free-body diagram of beam-cable system.

Since only coplanar forces are involved, the vector expressions for $\mathbf{R}^{(1)}$ and $\mathbf{R}^{(2)}$ are

$$\mathbf{R}^{(1)} = R_1^{(1)}\mathbf{i}_1 + R_2^{(1)}\mathbf{i}_2, \tag{3.5.1}$$

$$\mathbf{R}^{(2)} = R_2^{(2)}\mathbf{i}_2. \tag{3.5.2}$$

These two reactions involve the three unknown quantities $R_1^{(1)}$, $R_2^{(1)}$, and $R_2^{(2)}$. The three equations which will yield the solution are the three equations of coplanar force systems given by (3.4.38), (3.4.39), and (3.4.40). The free-body diagram is shown in Fig. 3.26.

We first resolve the tension in the cable into its two components, $|\mathbf{W}| \cos \theta$ and $|\mathbf{W}| \sin \theta$. From the first two equilibrium conditions we have

$$\sum_{i=1}^{n} P_1^{(i)} = R_1^{(1)} + |\mathbf{W}| \cos \theta = 0 \tag{3.5.3}$$

and

$$\sum_{i=1}^{n} P_2^{(i)} = R_2^{(1)} + |\mathbf{W}| \sin \theta - |\mathbf{F}| + R_2^{(2)} = 0. \tag{3.5.4}$$

From (3.5.3) we obtain

$$R_1^{(1)} = - |\mathbf{W}| \cos \theta. \tag{3.5.5}$$

If we choose the reference axis to coincide with the axis of the pin joint at the left-hand end of the member, we obtain

$$\sum_{i=1}^{n} (y_1^{(i)} F_2^{(i)} - y_2^{(i)} F_1^{(i)}) = - |\mathbf{W}| \frac{d}{2} \cos \theta + |\mathbf{W}| a \sin \theta - |\mathbf{F}| b + R_2^{(2)} c = 0. \tag{3.5.6}$$

Solving for $R_2^{(2)}$, we have

$$R_2^{(2)} = \frac{1}{c} \left[|\mathbf{W}| \frac{d}{2} \cos \theta - |\mathbf{W}| a \sin \theta + |\mathbf{F}| b \right]. \tag{3.5.7}$$

From (3.5.4),

$$R_2^{(1)} = |\mathbf{F}| - |\mathbf{W}| \sin \theta - R_2^{(2)}, \tag{3.5.8}$$

or

$$R_2^{(1)} = |\mathbf{F}| \left(1 - \frac{b}{c} \right) - |\mathbf{W}| \sin \theta \left(1 - \frac{a}{c} \right) - \frac{|\mathbf{W}| d}{2c} \cos \theta. \tag{3.5.9}$$

It should be noted that the third equilibrium equation applies to the moment about any arbitrary axis perpendicular to the $y_1 y_2$-plane. For example, the sum of moments about an axis at the right-hand end of the member should also be zero,

$$\sum_{i=1}^{n} M_3^{(i)} = - |\mathbf{W}| \frac{d}{2} \cos \theta - |\mathbf{W}| (c - a) \sin \theta + |\mathbf{F}| (c - b) - R_2^{(1)} c = 0, \tag{3.5.10}$$

which is an easy method of checking the accuracy of the previous calculations.

It can easily be verified that the solutions for the forces $R_2^{(2)}$ and $R_2^{(1)}$ can be obtained by using Eqs. (3.5.4) and (3.5.10), or by using the two moment equations (3.5.6) and (3.5.10). Thus although one may have infinitely many ways to write the moment equilibrium conditions, one can find only three independent equations of equilibrium for this coplanar force system. The most efficient procedure depends upon the specific problem, the "art" of solving such problems will come with experience.

We shall consider another typical support which consists of a weightless bar with hinges at both ends, as shown in Fig. 3.27(a). The rigid bar AB is isolated, and a free-body diagram for this bar is drawn as shown in Fig. 3.27(b). The reactions at the end points are $\mathbf{F}^{(1)}$ and $\mathbf{F}^{(2)}$. We shall show first that when a rigid body is in equilibrium under only two external forces, these must be collinear, equal in magnitude, and opposite in direction.

From the equilibrium condition $\sum \mathbf{F}^{(i)} = 0$ we obtain

$$\mathbf{F}^{(1)} + \mathbf{F}^{(2)} = 0 \quad \text{or} \quad \mathbf{F}^{(1)} = -\mathbf{F}^{(2)}. \tag{3.5.11}$$

Since the sum of the moments of all forces about A must be zero,

$$\mathbf{r}^{(12)} \times \mathbf{F}^{(2)} = 0. \tag{3.5.12}$$

This means that $\mathbf{r}^{(12)}$ and $\mathbf{F}^{(2)}$ are collinear. Similarly, since the sum of the moments of all forces about B is zero, $\mathbf{r}^{(12)} \times \mathbf{F}^{(1)} = 0$, or $\mathbf{r}^{(12)}$ and $\mathbf{F}^{(1)}$ are also collinear. We have thus proved that the force at A must be in line with the axis of the bar AB. This demonstrates a very important rule concerning pin-ended bars: *for a weightless bar which has two pin ends, and is loaded only at its ends, the reactive force must be in line with the bar.*

Figure 3.28 shows a vertical bar at the right-hand end of the uniform member. When this bar is removed, it must be replaced by a reaction along AB, which is in this case perpendicular to the member. Thus the effect of a vertical bar with hinged ends is identical to that of the frictionless roller shown in Fig. 3.25.

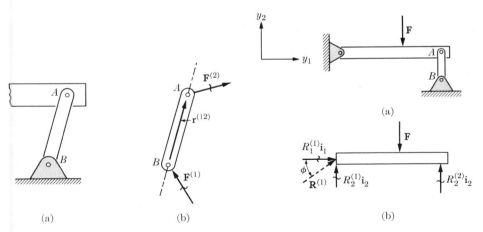

(a)   (b)   (b)

Fig. 3.27 Equilibrium of a pin-ended bar: (a) physical diagram; (b) free-body diagram.

Fig. 3.28 Equilibrium of a pin-ended beam: (a) physical diagram; (b) free-body diagram.

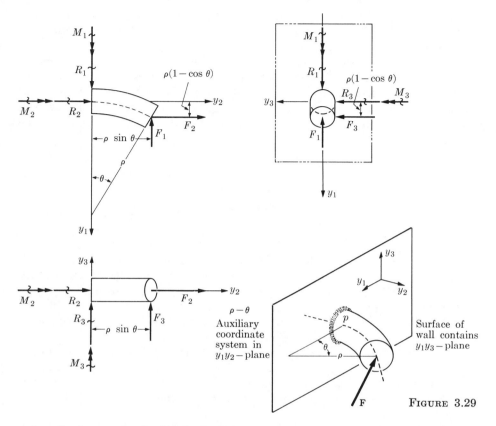

FIGURE 3.29

As a final example of a free-body diagram, let us consider a curved rod clamped to the wall and loaded by a force **F** at the other end, as shown in Fig. 3.29. We want to find the force system that must be applied at the support when the wall is removed.

The junction of the wall and the bar is a plane surface, but it is more convenient to represent the action of the wall on the bar by an equipollent system acting at the center of the bar. We saw in Section 3.2(e) that a force system in space can be reduced to an equipollent system consisting of a single force vector and a single moment vector. Thus the equipollent system representing the action of the wall on the bar at point p is a force vector and a moment vector. These have been resolved into their components R_1, R_2, R_3, and M_1, M_2, M_3. Note that the unit vectors have been omitted from the labels attached to the various force and moment components. This generally leads to better clarity, since the location of the arrowheads on the lines will pictorially give the sense of the vector.

A straightforward application of the six scalar equations of equilibrium, Eqs. (3.4.8) through (3.4.13), will yield the solutions for the reactions.

First we resolve the applied load $\mathbf{F}$ into three components, F_1, F_2, F_3,

$$\sum P_1^{(i)} = 0 \quad \text{or} \quad F_1 + R_1 = 0,$$
$$\therefore R_1 = -F_1; \tag{3.5.13}$$

$$\sum P_2^{(i)} = 0 \quad \text{or} \quad F_2 + R_2 = 0,$$
$$\therefore R_2 = -F_2; \tag{3.5.14}$$

$$\sum P_3^{(i)} = 0 \quad \text{or} \quad F_3 + R_3 = 0,$$
$$\therefore R_3 = -F_3; \tag{3.5.15}$$

$$\sum M_1^{(i)} = 0 \quad \text{or} \quad F_3\rho \sin \theta + M_1 = 0,$$
$$\therefore M_1 = -F_3\rho \sin \theta; \tag{3.5.16}$$

$$\sum M_2^{(i)} = 0 \quad \text{or} \quad M_2 - F_3\rho(1 - \cos \theta) = 0,$$
$$\therefore M_2 = +F_3\rho(1 - \cos \theta); \tag{3.5.17}$$

$$\sum M_3^{(i)} = 0 \quad \text{or} \quad F_1\rho \sin \theta + F_2\rho(1 - \cos \theta) + M_3 = 0,$$
$$\therefore M_3 = -F_1\rho \sin \theta - F_2\rho(1 - \cos \theta). \tag{3.5.18}$$

Thus the reaction at the clamped end can be replaced by three moments and three forces, these being the components of a moment vector and a force vector, respectively.

3.6 SYSTEMS OF PARALLEL FORCES—CENTER OF GRAVITY

When all the forces $\mathbf{F}^{(i)}$ in a system are parallel (Fig. 3.30a), they are also parallel to their resultant,

$$\mathbf{F} = \sum_{i=1}^{n} \mathbf{F}^{(i)},$$

and the sum of the moments about any point O due to these forces is always a moment vector perpendicular to $\mathbf{F}$. The equipollent force system (Fig. 3.30b)

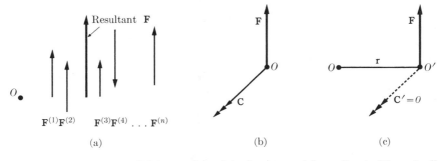

(a)   (b)   (c)

FIG. 3.30 System of parallel forces: (a) original system with resultant; (b) equipollent system with force at O; (c) equipollent system with force at O'.

is thus a force **F** at O and a couple **C**. If we choose another point O', the equipollent force system becomes a force **F** along the same direction and a couple **C'**, given by

$$\mathbf{C'} = \mathbf{C} - \mathbf{r} \times \mathbf{F} \qquad (3.6.1)$$

(see Fig. 3.30c). Here $\mathbf{r} \times \mathbf{F}$ is also perpendicular to **F**, and we can always arrange the magnitude and direction of **r** such that $\mathbf{C} = \mathbf{r} \times \mathbf{F}$ and hence **C'** becomes zero. Thus the system of parallel forces can be replaced by a single equipollent resultant force with its line of action passing through O'.

Consider a system of parallel forces $\mathbf{F}^{(i)}$ $(i = 1, 2, \ldots, n)$ with their direction oriented along the y_3-axis (Fig. 3.31). We intend to determine the location **r** of the resultant force **F**. We can write each force as

$$\mathbf{F}^{(i)} = F_3^{(i)}\mathbf{i}_3. \qquad (3.6.2)$$

Thus the resultant is

$$\mathbf{F} = \sum_{i=1}^{n} \mathbf{F}^{(i)} = \left(\sum_{i=1}^{n} F_3^{(i)} \right)\mathbf{i}_3.$$

$$(3.6.3)$$

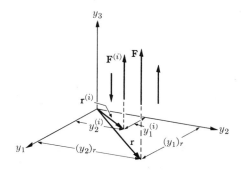

Fig. 3.31   System of parallel forces in space.

Let us choose the origin O as the reference point and let all position vectors be on the y_1y_2-plane. Thus we can locate each force by

$$\mathbf{r}^{(i)} = y_1^{(i)}\mathbf{i}_1 + y_2^{(i)}\mathbf{i}_2. \qquad (3.6.4)$$

The location of the resultant force is given by

$$\mathbf{r} = (y_1)_r\mathbf{i}_1 + (y_2)_r\mathbf{i}_2. \qquad (3.6.5)$$

We have

$$\mathbf{M} = \sum_{i=1}^{n} (\mathbf{r}^{(i)} \times \mathbf{F}^{(i)}) = \mathbf{r} \times \mathbf{F};$$

therefore

$$\sum_{i=1}^{n} (-y_1^{(i)}F_3^{(i)}\mathbf{i}_2 + y_2^{(i)}F_3^{(i)}\mathbf{i}_1) = -(y_1)_r\left(\sum_{i=1}^{n} F_3^{(i)} \right)\mathbf{i}_2 + (y_2)_r\left(\sum_{i=1}^{n} F_3^{(i)} \right)\mathbf{i}_1.$$

This is equivalent to

$$(y_1)_r = \frac{\sum_{i=1}^{n} y_1^{(i)}F_3^{(i)}}{\sum_{i=1}^{n} F_3^{(i)}}, \qquad (3.6.6)$$

$$(y_2)_r = \frac{\sum_{i=1}^{n} y_2^{(i)}F_3^{(i)}}{\sum_{i=1}^{n} F_3^{(i)}}. \qquad (3.6.7)$$

An application of the above discussion is found when the parallel forces are taken as the gravitational forces acting on the various particles of a body. The center of gravity (cg) of a body is defined as the point of action of a single resultant force equipollent to the system of gravitational forces. Here

$$F_3^{(i)} = -m^{(i)}g,$$

and the cg position is given by

$$(y_1)_c = \frac{\sum y_1^{(i)} m^{(i)}}{\sum m^{(i)}}, \tag{3.6.8}$$

$$(y_2)_c = \frac{\sum y_2^{(i)} m^{(i)}}{\sum m^{(i)}}. \tag{3.6.9}$$

We may rotate the body as well as the axes such that the other axis (say y_1) becomes vertical; by following the same treatment, we obtain

$$(y_3)_c = \frac{\sum y_3^{(i)} m^{(i)}}{\sum m^{(i)}}. \tag{3.6.10}$$

Equations (3.6.8), (3.6.9), and (3.6.10) can be written as

$$(\mathbf{r})_c = \frac{\sum \mathbf{r}^{(i)} m^{(i)}}{\sum m^{(i)}}, \tag{3.6.11}$$

and the point which locates $(\mathbf{r})_c$ may also be called the *center of mass*. For a continuous system,

$$(\mathbf{r})_c = \frac{\int_{\text{volume}} \mathbf{r}\, dm}{\int_{\text{volume}} dm}. \tag{3.6.12}$$

If the density distribution is $\rho = \rho(y_k)$, $k = 1, 2, 3$, we have $dm = \rho\, dV$ and

$$(\mathbf{r})_c = \frac{\iiint \rho(y_1, y_2, y_3)\mathbf{r}\, dy_1\, dy_2\, dy_3}{\iiint \rho(y_1, y_2, y_3)\, dy_1\, dy_2\, dy_3}. \tag{3.6.13}$$

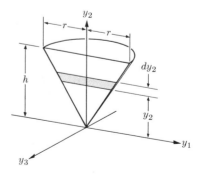

FIG. 3.32   Right circular cone.

EXAMPLE: To compute the center of mass of a homogeneous right circular cone of height h and base radius r.

Solution: Let the axis of symmetry of the cone be placed along the y_2-axis with the origin at the apex as illustrated by Fig. 3.32. Since the center of mass is on the y_2-axis, that is, $(y_1)_c = (y_3)_c = 0$, we need to find only the distance $(y_2)_c$ from the vertex to the center of mass. Applying Eq. (3.6.13), we have

$$(y_2)_c = \frac{\iiint_v \rho y_2 \, dV}{\iiint_v \rho \, dV} \, . \tag{3.6.14}$$

Taking a slice parallel to the base of thickness dy_2, we have for the volume element

$$dV = \pi y_1^2 \, dy_2. \tag{3.6.15}$$

Putting

$$y_1 = \frac{r}{h} y_2$$

and introducing Eq. (3.6.15) in (3.6.14) we find that

$$(y_2)_c = \frac{\rho(r^2/h^2)\pi \int_0^h (y_2)^3 \, dy_2}{\rho(r^2/h^2)\pi \int_0^h (y_2)^2 \, dy_2} = \frac{3}{4} h. \tag{3.6.16}$$

3.7 PLANE AND SPACE TRUSSES

A truss is a structure which consists of a group of straight members connected to each other only at their ends by frictionless joints (Fig. 3.27). In a plane truss where all the members and the loads are coplanar, these joints (Fig. 3.33) are of pin connections; in a space truss, they are ball-and-socket joints (Fig. 3.34). For an ideal truss, all external loads are applied only at the joints. Under such conditions each member of the truss is subjected only to two forces at its two end points. These two forces must necessarily be equal and opposite, and must act along the direction of the member (see discussion relative to Fig. 3.27). Thus each member is subjected to either an axial tension or compression, constant along its length, but to no other force. The so-called truss analysis is to determine the force in each member of the truss under externally applied loads.

In the general practical truss the members are usually bolted, riveted, or welded at the ends. However, it has been found that an analysis of such a

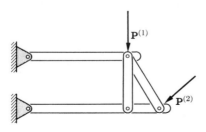

FIG. 3.33 Plane truss.

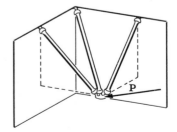

FIG. 3.34 Space truss.

structure based on the assumption of an idealized truss yields approximately the correct values for the forces in the members. This is true mainly because the lengths of the truss members are much greater (say 10 or more times) than the dimensions of the cross sections.

Before proceeding with the details of truss analysis, let us summarize the characteristics of an ideal truss as follows:

(1) The truss members are connected at their ends (called the joints) by frictionless pins.
(2) The loads and reactions are applied only at the joints.
(3) The axis of each bar is straight, and coincides with the line connecting the joints.
(4) The force system equipollent to the internal forces in a truss member consists only of a component coinciding with the centroidal axis.

In this chapter we shall consider only statically determinate trusses, for which all the forces can be determined by the equations of equilibrium. This means that when we isolate the complete structure or a portion of the structure, and draw the corresponding free-body diagram, we expect the number of unknown quantities to be equal to the available number of equations of equilibrium. Thus, in general, for a space truss we have six equations, and for a plane truss or a concurrent force system we have only three equations. For a coplanar and concurrent force system we have only two equations of equilibrium to be satisfied.

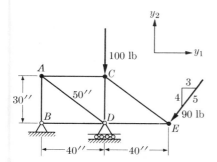

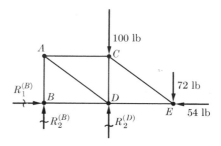

Fig. 3.35 Statically determinate plane truss.

Fig. 3.36 Free-body diagram of entire truss.

We shall demonstrate the method of truss analysis by a typical plane truss problem illustrated in Fig. 3.35. Note that the members of the ideal truss are represented by lines which coincide with the axes of the members. The truss is supported at B and D, and two loads are applied at C and E. It is desired to evaluate the axial forces carried by each member of the truss. The first step in the analysis is to draw the free-body diagram (Fig. 3.36) of the complete system, replacing the supports by the three unknown reactions, $R_1^{(B)}$, $R_2^{(B)}$, and

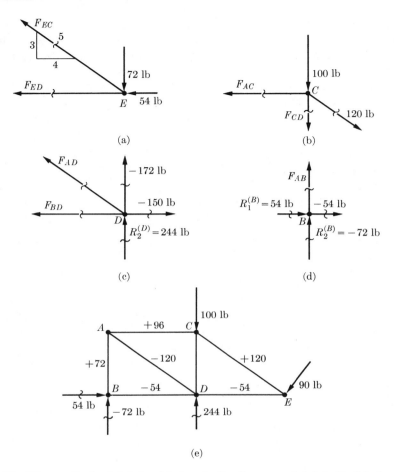

(a)

(b)

(c)

(d)

(e)

FIG. 3.37   The method of joints: (a) free-body diagram of joint E; (b) free-body diagram of joint C; (c) free-body diagram of joint D; (d) free-body diagram of joint B; (e) truss diagram with bar forces.

$R_2^{(D)}$. It is seen that these unknown reactions can be determined by the three equations of equilibrium for a coplanar force system as follows:

$$\sum_{i=1}^{n} F_1^{(i)} = R_1^{(B)} - 54 = 0, \tag{3.7.1}$$

$$\sum_{i=1}^{n} F_2^{(i)} = R_2^{(B)} + R_2^{(D)} - 100 - 72 = 0, \tag{3.7.2}$$

$$\sum_{i=1}^{n} M_3^{(i)} = -(100)(40) + 40R_2^{(D)} - (72)(80) = 0. \tag{3.7.3}$$

From these equations we obtain

$$R_1^{(B)} = 54 \text{ lb}, \qquad R_2^{(D)} = 244 \text{ lb}, \qquad \text{and} \qquad R_2^{(B)} = -72 \text{ lb}.$$

We can proceed to determine the axial load in each member by isolating individual joints. For example, when we consider joint E (Fig. 3.37a), we replace the members EC and ED by unknown axial forces, F_{EC} and F_{ED}, which are assumed to be positive in tension, as shown in the diagram. If the numerical answer is positive, then the *bar force* (as the force in the truss member is called) is tension. Conversely, if the numerical answer is negative, then the bar force is compression. For this coplanar and concurrent force system the two equations of equilibrium are

$$\sum_{i=1}^{n} F_1^{(i)} = 0 \qquad \text{or} \qquad -F_{ED} - \frac{4}{5} F_{EC} - 54 = 0, \tag{3.7.4}$$

$$\sum_{i=1}^{n} F_2^{(i)} = 0 \qquad \text{or} \qquad \frac{3}{5} F_{EC} - 72 = 0. \tag{3.7.5}$$

From these equations we obtain $F_{EC} = 120$ lb, and $F_{ED} = -150$ lb. This means that member EC is under a tension force of 120 lb, while member ED is under a compressive force of 150 lb.

The determination of axial forces in various members by considering equilibrium at a joint is called the *method of joints*. We can now proceed to isolate joint C (Fig. 3.37b) and replace member CE by a known tension of 120 lb, and members CD and AC by unknown external forces F_{CD} and F_{AC}. Again we apply the two equations of equilibrium,

$$\sum_{i=1}^{n} F_1^{(i)} = -F_{AC} + (\tfrac{4}{5})(120) = 0, \tag{3.7.6}$$

$$\sum_{i=1}^{n} F_2^{(i)} = -F_{CD} - 100 - (120)(\tfrac{3}{5}) = 0. \tag{3.7.7}$$

From these we obtain $F_{AC} = 96$ lb, and $F_{CD} = -172$ lb. When we proceed to joint D (Fig. 3.37c), we see that we have already evaluated the reaction $R_2^{(D)}$, and we again have only two unknown axial forces to be determined. The two equilibrium equations are

$$\sum_{i=1}^{n} F_1^{(i)} = -F_{BD} - \frac{4}{5} F_{AD} - 150 = 0, \tag{3.7.8}$$

$$\sum_{i=1}^{n} F_2^{(i)} = \frac{3}{5} F_{AD} - 172 + 244 = 0, \tag{3.7.9}$$

from which we obtain

$$F_{AD} = -120 \text{ lb},$$

$$F_{BD} = -150 + 96 = -54 \text{ lb}.$$

The only remaining unknown quantity F_{AB} can be determined by considering either joint B or joint A. If we consider B (Fig. 3.37d), we find from

$$\sum_{i=1}^{n} F_2^{(i)} = F_{AB} - 72 = 0 \qquad (3.7.10)$$

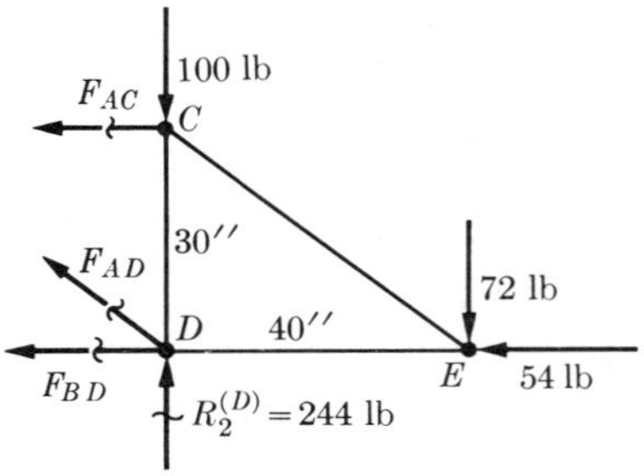

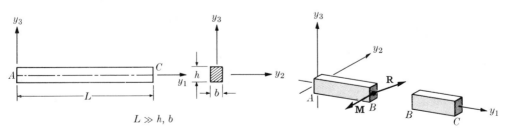

FIG. 3.38 The method of sections.

that $F_{AB} = 72$ lb. Figure 3.37(e) shows the completely solved truss.

It is possible to determine the forces in some members of the truss without analyzing the complete truss. For example, if we are interested only in the axial force F_{AC}, we can cut AC, AD, and BD and isolate the structure at the right-hand side of the cut. The free body is shown in Fig. 3.38. Such a procedure, of course, is tantamount to adding up only a selected few of the equations given by Eq. (3.4.1). Here three of the four unknown quantities F_{AC}, F_{AD}, F_{BD}, and $R_2^{(D)}$ act at joint D. Hence, by setting the sum of moments about D equal to zero, we can calculate the unknown force F_{AC} immediately. We have, then, $(F_{AC})(30) - (72)(40) = 0$, $F_{AC} = 96$ lb. This last method is called the *method of sections*. For a more complete exposition of truss analysis the reader is referred to Norris and Wilbur (Ref. 3.1).

3.8 INTERNAL FORCES AND MOMENTS IN SLENDER BEAMS

A very common structural component is called a beam. Just as in the case of an ideal truss, a structural member is called an ideal beam because certain restrictions are first placed on its geometry. Then, because of the restrictions on the geometry, further restrictions are placed on the types of forces and moments which can be developed. In Fig. 3.39, the geometrical restrictions on a beam are shown. A beam is a three-dimensional structure with one dimension (called the *span* or *length*) much larger (say 10 times) than the other two dimensions

FIG. 3.39 Beam geometry.

FIG. 3.40 Beam as a two-particle system—equipollent forces on portion AB.

(i.e. dimensions of the *cross section*). It is assumed that the externally applied loads are coplanar with the y_1y_3-plane, where y_1 is in the span direction of the beam.

In order to apply the results of Section 3.4 to a beam, we need only to visualize the beam as a collection of particles. The particles in this case are the atoms, but the atoms are so numerous that the usual engineering structure can be treated as a continuum. The internal forces in a beam are the forces between atoms, but the notion of equipollent force systems makes it possible to ignore the details of the actual distribution of the internal forces. We can form a subcollection of particles by figuratively cutting the beam into two parts by a plane perpendicular to the y_1-axis of the beam (see Fig. 3.40). The action of the portion BC on the portion AB can be represented by an equipollent system consisting of a force vector **R** and a moment vector **M**. Since the beam is assumed to be loaded only by coplanar forces, the vector forms of **R** and **M** are

$$\mathbf{R} = F\mathbf{i}_1 - S\mathbf{i}_3, \tag{3.8.1}$$

$$\mathbf{M} = -M\mathbf{i}_2. \tag{3.8.2}$$

The somewhat irrational combination of signs attached to the preceding two equations is an outgrowth of the manner in which beam analysis has developed. There is such a tremendous quantity of numerical data which conform to the above sign conventions that it is not feasible to adopt a more rational approach. The component F is called the *axial force*, S is called the *shear force*, and M is called the *bending moment*. These are shown in Fig. 3.41, where in conformity with the usual practice, the moment component M is shown as a curved vector.

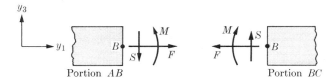

Portion AB Portion BC

FIG. 3.41   Sign conventions for axial force, shear force, and bending moment.

Note carefully that the positive directions for F, S, and M are defined on the positive face of the cut at B, the positive face being the face which has an outward normal vector in the positive $\mathbf{i}_1$-direction. On the other face of the cut, which has an outward normal in the negative $\mathbf{i}_1$-direction, the forces F and S and the moment M must all point in the opposite direction by virtue of Eq. (3.4.2).

The sign conventions associated with beam theory are apt to be confusing. Another scheme for remembering what constitutes positive and negative is shown in Fig. 3.42.

The assumption that the forces acting on a beam are coplanar means that the cross-sectional shape is unimportant to the analysis of internal forces, F, S, and M. Thus a beam can be schematically represented by a thick line (see Fig.

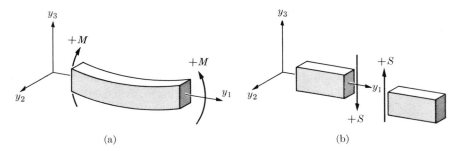

FIG. 3.42 Sign conventions, shear force and bending moment: (a) positive bending moment tends to apply compression to the upper fibers of the beam, tension, to the lower; (b) positive shear tends to lower the right half of the beam relative to the left half.

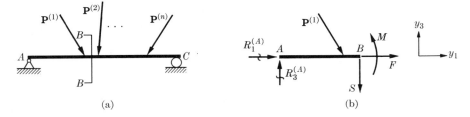

FIG. 3.43 Schematic representation of a beam: (a) physical diagram; (b) free-body diagram.

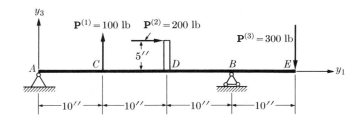

FIG. 3.44 Physical diagram of a beam under a system of coplanar forces.

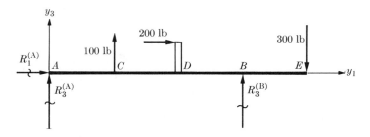

FIG. 3.45 Free-body diagram of the complete beam.

3.43). However, the cross-sectional shape of a beam is important to the determination of the distribution of the forces, i.e., of internal stresses (see Chapter 6, "Analysis of Stress").

EXAMPLE: Consider a beam (Fig. 3.44) supported at A by a pin joint and at B by a roller. Two of the applied forces are in the vertical direction, while the third one, $P^{(2)}$, is in the horizontal direction. The force $P^{(2)}$ acts at the end of a vertical post which is 5 in. in height. We make the assumption that the deformations of the beam under the action of the loads will not significantly change the specified dimensional relations between the loads and the structure. This will later be shown to be true for most problems of structural analysis. It is desired to determine the distribution of internal forces and moment along the beam.

Referring to Fig. 3.45 we observe that the three reactions $R_1^{(A)}$, $R_3^{(A)}$, and $R_3^{(B)}$ can be determined by using three equations of statics:

$$\sum_{i=1}^{n} F_1^{(i)} = R_1^{(A)} + 200 = 0, \tag{3.8.3}$$

$$\sum_{i=1}^{n} F_3^{(i)} = R_3^{(A)} + 100 + R_3^{(B)} - 300 = 0, \tag{3.8.4}$$

$$\sum_{i=1}^{n} M_2^{(i)} = (100)(10) - (200)(5) + (R_3^{(B)})(30) - (300)(40)$$
$$= 0 \quad \text{(about } A\text{)}. \tag{3.8.5}$$

From these equations we obtain

$$R_1^{(A)} = -200 \text{ lb}, \qquad R_3^{(A)} = -200 \text{ lb}, \qquad R_3^{(B)} = 400 \text{ lb}. \tag{3.8.6}$$

In order to determine the internal forces and moments at a given station we need to make a cut perpendicular to the axis of the beam at this station, and isolate either part of the beam. Let us consider first, a station located between A and C (i.e., for $0 < y_1 < 10$), and isolate the part of the beam to the left of this station, as shown in Fig. 3.46(a). The three unknowns, internal forces $F(y_1)$ and $S(y_1)$ and moment $M(y_1)$, can be determined by the three equations for static equilibrium:

$$\sum_{i=1}^{n} F_1^{(i)} = -200 + F(y_1) = 0, \tag{3.8.7}$$

$$\sum_{i=1}^{n} F_3^{(i)} = -200 - S(y_1) = 0, \tag{3.8.8}$$

$$\sum_{i=1}^{n} M_2^{(i)} = 200y_1 + M(y_1) = 0 \quad \text{(about } P\text{)}. \tag{3.8.9}$$

The last equation refers to the sum of the moments about the point P which is located on the axis of beam at the station where the cut has been made. In this case a counterclockwise moment is considered as positive. It should be realized that the reference point for the equation of moment equilibrium may be any point along the beam. However, a judicious choice of the reference point can simplify the algebra and arithmetic. The internal forces and moment obtained by these three equations are

$$F(y_1) = 200 \text{ lb}, \quad S(y_1) = -200 \text{ lb}, \quad M(y_1) = -200 \, y_1 \text{ in.-lb.} \quad (3.8.10)$$

These results can also be obtained by isolating the part of the beam to the right of the station, as shown in Fig. 3.46(b). The internal forces and moment

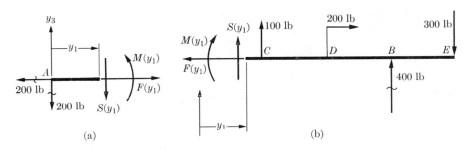

FIG. 3.46   Free-body diagrams for determination of internal forces and moment between B and C: (a) left-hand part of a cut at a station between A and C; (b) right-hand part of a cut at a station between A and C.

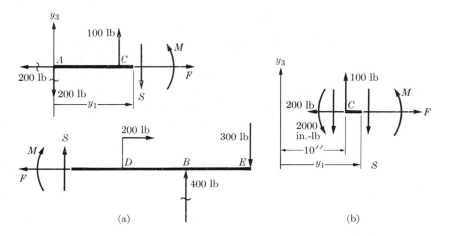

FIG. 3.47   Free-body diagrams for determination of internal forces and moment between C and D: (a) left- and right-hand parts of a cut at a station between C and D; (b) a small segment to the left of a cut between C and D.

which are in the opposite direction to those in the previous figure can also be calculated by the three equations for static equilibrium:

$$\sum_{i=1}^{n} F_1^{(i)} = 200 - F(y_1) = 0, \tag{3.8.11}$$

$$\sum_{i=1}^{n} F_3^{(i)} = S(y_1) + 500 - 300 = 0, \tag{3.8.12}$$

$$\sum_{i=1}^{n} M_2^{(i)} = -M(y_1) + 100(10 - y_1) - (200)(5) + 400(30 - y_1)$$
$$-300(40 - y_1) = 0 \quad \text{(about } P). \tag{3.8.13}$$

These equations yield

$$F(y_1) = 200 \text{ lb}, \qquad S(y_1) = -200 \text{ lb},$$
$$M(y_1) = -200 \, y_1 \text{ in-lb}. \tag{3.8.14}$$

At a station located at an infinitesimally small distance to the left of C the internal forces and moments are

$$F(10-) = 200 \text{ lb}, \qquad S(10-) = -200 \text{ lb},$$
$$M(10-) = -2000 \text{ in-lb}. \tag{3.8.15}$$

The symbol "$10-$" is used to indicate that y_1 is slightly smaller than 10, and similarly "$10+$" that y_1 is slightly larger than 10.

To determine the internal forces and moment at a station between C and D we again make a cut at this station and isolate either part of the beam shown in Fig. 3.47(a). It can also be seen that since we have already determined the internal forces and moment at the left of station C, we may make a cut there and replace the left-hand part of that cut by the known forces and moment. The free-body diagram of this shorter segment of the beam is shown in Fig. 3.47(b). Let us consider only the last diagram and write the following conditions of static equilibrium:

$$\sum_{i=1}^{n} F_1^{(i)} = -200 + F = 0, \tag{3.8.16}$$

$$\sum_{i=1}^{n} F_3^{(i)} = -200 + 100 - S = 0, \tag{3.8.17}$$

$$\sum_{i=1}^{n} M_2^{(i)} = 2000 + M - S(y_1 - 10) = 0 \quad \text{(about } C). \tag{3.8.18}$$

We obtain from these equations, for $10 < y_1 < 20$,

$F(y_1) = 200$ lb,

$S(y_1) = -100$ lb,   (3.8.19)

$M(y_1) = -100\, y_1 - 1000,$

and

$F(20-) = 200$ lb,

$S(20-) = -100$ lb,   (3.8.20)

$M(20-) = -3000$ in-lb.

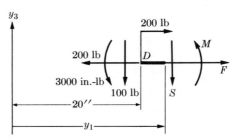

FIG. 3.48 A small segment to the left of a cut between D and B.

To determine the internal forces and moment at a station between D and B we consider a short segment to the left of this station, as shown in Fig. 3.48. The equilibrium conditions are

$$\sum_{i=1}^{n} F_1^{(i)} = -200 + 200 + F = 0,$$ (3.8.21)

$$\sum_{i=1}^{n} F_3^{(i)} = -100 - S = 0,$$ (3.8.22)

$$\sum_{i=1}^{n} M_2^{(i)} = 3000 - (200)(5) + M - S(y_1 - 20)$$
$$= 0 \quad (\text{about } D).$$ (3.8.23)

These yield, for $20 < y_1 < 30$,

$F(y_1) = 0,$ $S(y_1) = -100$ lb, $M(y_1) = -100\, y_1$ in-lb,   (3.8.24)

$F(30-) = 0,$ $S(30-) = 100$ lb, $M(30-) = -3000$ in-lb.   (3.8.25)

Finally, when we calculate the sectional forces and moment at a station to the left of B, we isolate the right-hand part of the cut as shown in Fig. 3.49. The equations of static equilibrium are

$$\sum_{i=1}^{n} F_1^{(i)} = -F + 0 = 0,$$ (3.8.26)

$$\sum_{i=1}^{n} F_3^{(i)} = S - 300 = 0,$$ (3.8.27)

$$\sum_{i=1}^{n} M_2^{(i)} = -M - S(40 - y_1) = 0 \quad (\text{about } E).$$ (3.8.28)

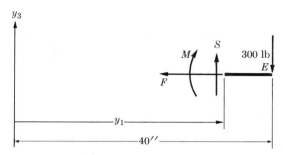

FIG. 3.49 A segment to the right of a cut between B and E.

We obtain, for $30 < y_1 < 40$,

$$F(y_1) = 0, \qquad S(y_1) = 300, \qquad M(y_1) = 300\,y_1 - 12,000, \quad (3.8.29)$$

and

$$F(40-) = 0, \qquad S(40-) = 300, \qquad M(40-) = 0. \qquad (3.8.30)$$

The distributions of axial force, shear force, and bending moment along the beam are shown in Fig. 3.50. These diagrams are called the axial-force, shear, and bending-moment diagrams.

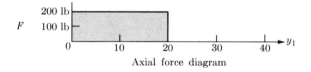

Axial force diagram

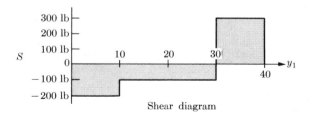

Shear diagram

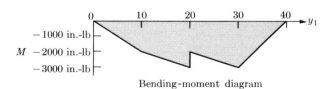

Bending-moment diagram

FIGURE 3.50

(handwritten notes at top)
$$E = S + \frac{p y_1\, dy_1}{2} + S + \frac{dS}{dy_1}\, dy_1$$
$$-m + \frac{p_3\, dy_2}{2} + m + \frac{dm}{dy_3}\, dy_3$$

3.9 RELATIONS BETWEEN LOAD, SHEAR, AND BENDING MOMENT

The beam, as has been pointed out, is treated as a continuum and hence the applied loads, the internal forces F and S, and the bending moment M, must obey certain differential relationships. In general, we consider a slender beam (Fig. 3.51) under a continuously distributed load $p(y_1)$ per unit length of the beam. Note that loads with lines of action parallel to the axis of the beam are omitted. Beams are seldom called upon to transmit loads of this kind. If there are loads of this nature, then the application of the force equilibrium equation in the y_1-direction will yield the internal axial force.

Let us isolate a very small segment of length dy_1 and draw a free-body diagram as shown in Fig. 3.52. When the distributed load $p(y_1)$ is continuous and because the element dy_1 is an infinitesimal, the resultant of the distributed load within this segment may be represented approximately by $p(y_1)\, dy_1$ and may be considered to act at the center of this segment. The force system acting on this segment is shown in Fig. 3.53. The two equations of static equilibrium are

$$\sum_{i=1}^{n} F_3^{(i)} = S + p(y_1)\, dy_1 - \left(S + \frac{dS}{dy_1}\, dy_1\right) = 0, \qquad (3.9.1)$$

and

(handwritten) $p(y_1)\, dy_1 = dS / dy_1$

$$\sum_{i=1}^{n} M_2^{(i)} = -M - S\frac{dy_1}{2} - (S + dS)\frac{dy_1}{2} + \left(M + \frac{dM}{dy_1}\, dy_1\right) = 0$$

$$\text{(about center of element).} \qquad (3.9.2)$$

Equation (3.9.1) may be simplified to

$$\frac{dS}{dy_1} = p(y_1), \qquad (3.9.3)$$

and Eq. (3.9.2) may be reduced to

$$\frac{dM}{dy_1} = S + \frac{dS}{2}. \qquad (3.9.4)$$

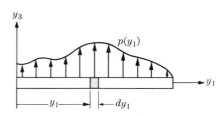

FIG. 3.51 Beam under distributed load.

Consider the case where the distribution of shear is continuous. Then, in the limit as dy_1 approaches zero, dS becomes negligibly small in comparison to S, and we obtain

$$\frac{dM}{dy_1} = S(y_1). \qquad (3.9.5)$$

Equations (3.9.3) and (3.9.5) are the two basic differential equations relating the load distribution $p(y_1)$, shear distribution $S(y_1)$, and bending-moment distribution $M(y_1)$. They state respectively that *the slope of the shear diagram is equal to the magnitude of the distributed load p,* and *the slope of the bending-*

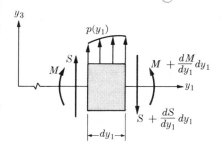

FIG. 3.52 Free-body diagram of a small segment of the beam.

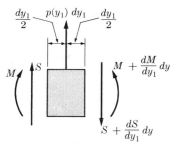

FIG. 3.53 Free-body diagram of the segment with the distributed load replaced by its resultant.

moment diagram is equal to the magnitude of the shear force S. Thus, if the load distribution is given, the shear and bending-moment distribution can be obtained by integration.

Integrating Eq. (3.9.3), we obtain

$$S(y_1^{(B)}) = \int_{y_1^{(A)}}^{y_1^{(B)}} p(y_1)\, dy_1 + S(y_1^{(A)}), \qquad (3.9.6)$$

where $S(y_1^{(A)})$ and $S(y_1^{(B)})$ are the values of the shear at stations A and B, respectively. Similarly, from Eq. (3.9.5) we have

$$M(y_1^{(B)}) = \int_{y_1^{(A)}}^{y_1^{(B)}} S(y_1)\, dy_1 + M(y_1^{(A)}), \qquad (3.9.7)$$

where $M(y_1^{(A)})$ and $M(y_1^{(B)})$ are the values of the bending moment at stations A and B, respectively.

EXAMPLE: Consider a cantilever beam with the tapered load distribution shown in Fig. 3.54. It is desired to determine the shear and bending-moment diagrams.

The free-body diagram of the beam is shown in Fig. 3.55. The shear force S_0 and the bending moment M_0 at the clamped end of the beam can be obtained

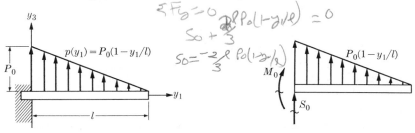

FIG. 3.54 Physical diagram of cantilever beam with tapered load.

FIG. 3.55 Free-body diagram of the complete beam.

first by the two equations of static equilibrium,

$$\sum_{i=1}^{n} F_3^{(i)} = S_0 + \int_0^l p(y_1)\,dy_1 = 0, \tag{3.9.8}$$

$$\sum_{i=1}^{n} M_2^{(i)} = -M_0 + \int_0^l p(y_1)y_1\,dy_1 \quad \text{(about clamped end).} \tag{3.9.9}$$

From these we obtain

$$S_0 = -\int_0^l \frac{p_0}{l}(l - y_1)\,dy_1 = -\frac{p_0 l}{2},$$
$$M_0 = \int_0^l \frac{p_0}{l}(l - y_1)y_1\,dy_1 = +\frac{p_0 l^2}{6}. \tag{3.9.10}$$

The shear and bending-moment distributions can thus be calculated by Eqs. (3.9.6) and (3.9.7):

$$S(y_1) = \int_0^{y_1} \frac{p_0}{l}(l - y_1)\,dy_1 - \frac{p_0 l}{2}, \qquad S(y_1) = -\frac{p_0}{2l}(l - y_1)^2,$$
$$M(y_1) = -\int_0^{y_1} \frac{p_0}{2l}(l - y_1)^2\,dy_1 + \frac{p_0 l^2}{6}, \qquad M(y_1) = \frac{p_0}{6l}(l - y_1)^3. \tag{3.9.11}$$

The shear and bending-moment diagrams are shown in Fig. 3.56.

A remark should now be made concerning concentrated loads. A *concentrated load* may be considered as a load distributed within a segment of infinitesimally small length. We can write the magnitude of a concentrated load as

$$P = p(y_1)\Delta y_1 \tag{3.9.12}$$

when Δy_1 approaches zero or

$$p(y_1) = \lim_{\Delta y_1 \to 0} \left(\frac{P}{\Delta y_1}\right). \tag{3.9.13}$$

It is seen that when P is finite, $p(y_1)$ must be infinitely large. From Eq. (3.9.3) we can conclude that at a station where a concentrated load is acting, the slope of the shear diagram is infinite, i.e., the shear diagram is discontinuous. It is obvious that if a concentrated load P

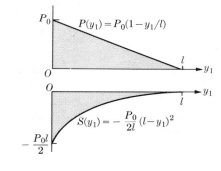

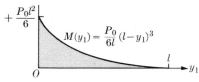

FIG. 3.56   Cantilever beam with tapered load.

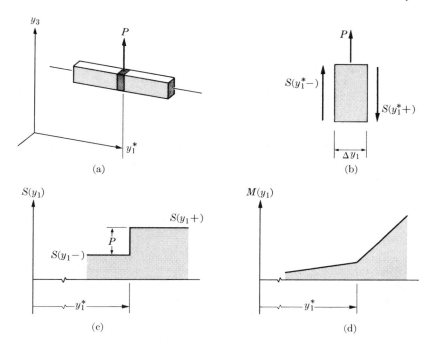

FIG. 3.57   The effects of a concentrated load: (a) physical diagram of beam subjected to concentrated load; (b) free-body diagram of a small segment at a concentrated load; (c) infinite slope and finite jump of shear diagram at concentrated load; (d) discontinuity in slope of bending-moment diagram at concentrated load.

acts at y_1^*, the shear force at the right-hand side of this station is equal to the shear force at the left-hand side plus P. This is shown in Fig. 3.57. The bending moment at a station is given by the differential relation $dM/dy_1 = S$. Since the shear forces are different on either side of a concentrated load, the slope of the bending-moment diagram will be discontinuous at the loading station, as shown in Fig. 3.57(d).

3.10 GENERAL BEAM THEORY

The results of planar beam theory can be extended to the more general case in which loads are also applied parallel to the y_1y_2-plane. Then the equipollent force and moment vectors will each have three components (see Fig. 3.58):

$$\mathbf{F} = F\mathbf{i}_1 - S_2\mathbf{i}_2 - S_3\mathbf{i}_3, \tag{3.10.1}$$

$$\mathbf{M} = T\mathbf{i}_1 - M_2\mathbf{i}_2 + M_3\mathbf{i}_3, \tag{3.10.2}$$

where, by the application of the right-hand rule, it is seen that T represents a torsional moment.

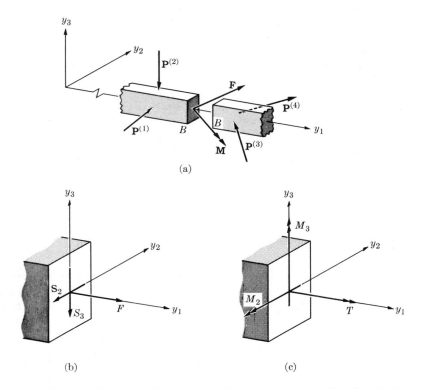

FIG. 3.58 (a) Internal force and moment in a beam under spatial loads; (b) shear and axial force conventions; (c) bending and torsion moment conventions.

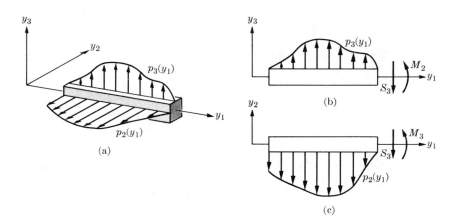

FIG. 3.59 Beam under lateral forces in y_1y_2- and y_1y_3-planes: (a) perspective view of beam under lateral loads; (b) side elevation; (c) plan view.

The differential relations between load and shear, and between shear and bending moment may be extended to include lateral forces in both the y_1y_2- and y_1y_3-planes, as shown in Fig. 3.59.

These relations are

$$\frac{dS_3}{dy_1} = p_3(y_1), \tag{3.10.3}$$

$$\frac{dM_2}{dy_1} = S_3(y_1), \tag{3.10.4}$$

$$\frac{dS_2}{dy_1} = p_2(y_1), \tag{3.10.5}$$

$$\frac{dM_3}{dy_1} = S_2(y_1). \tag{3.10.6}$$

3.11 TORSION OF A ROD

Differential relations can also be obtained between distributed torque and internal torsional moment. Consider a slender rod under a distributed torque of $t(y_1)$ per unit length, shown in Fig. 3.60(a). Let us isolate a small segment of length dy_1, and draw the free-body diagram shown in Fig. 3.60(b). The equilibrium of all moments about the y_1-axis yields

$$\sum_{i=1}^{n} M_1^{(i)} = -T + t\,dy_1 + T + \frac{dT}{dy_1}\,dy_1 = 0. \tag{3.11.1}$$

We obtain from this

$$\frac{dT}{dy_1} = -t(y_1). \tag{3.11.2}$$

Equation (3.11.2) states that *the slope of the torsional-moment diagram is equal to the negative value of the torque distribution at that station.*

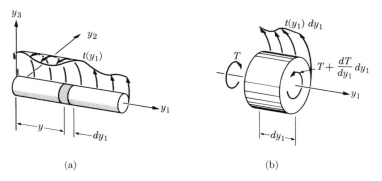

FIG. 3.60 Slender cylindrical rod under distributed torque load: (a) physical diagram; (b) free-body diagram of a small segment.

3.12 SUMMARY

Moment of a force about a point p

$$\mathbf{M} = \mathbf{r} \times \mathbf{F}, \tag{3.12.1}$$

where $\mathbf{r}$ is position vector from p to any point along line of action of $\mathbf{F}$.

$$\mathbf{M} = [y_m(q) - y_m(p)]F_r\, e_{mrk}\, \mathbf{i}_k, \tag{3.12.2}$$

where $y_m(q)$ represents coordinates of a point on line of action of $\mathbf{F}$, $y_m(p)$ coordinates of point p, and F_r components of $\mathbf{F}$.

Moment of a force about an axis collinear with the unit vector $\mathbf{e}$

$$M_l = \mathbf{e} \cdot (\mathbf{r} \times \mathbf{F}), \tag{3.12.3}$$

where $\mathbf{r}$ is the vector from line of axis of $\mathbf{e}$ to line of action of $\mathbf{F}$.

Equipollent force systems
   (1) The resultant of the forces of one system is equal to the resultant of forces of the other system.
   (2) The sum of the moments of the forces of one system about an arbitrary base point is equal to the sum of the moments of the forces of the other system about the same point.

Equilibrium of a system of particles
   Vector sum of external forces acting on system is zero:

$$\sum_{i=1}^{n} \mathbf{P}^{(i)} = 0. \tag{3.12.4}$$

Total moment about arbitrary origin O of the external forces acting on system is zero:

$$\sum_{i=1}^{n} \mathbf{r}^{(i)} \times \mathbf{P}^{(i)} = 0. \tag{3.12.5}$$

The scalar form of Eq. (3.12.4) is

$$\sum_{i=1}^{n} P_1^{(i)} = 0, \tag{3.12.6}$$

$$\sum_{i=1}^{n} P_2^{(i)} = 0, \tag{3.12.7}$$

$$\sum_{i=1}^{n} P_3^{(i)} = 0, \tag{3.12.8}$$

and similarly, the scalar form of Eq. (3.12.5) is

$$\sum_{i=1}^{n} (y_2^{(i)}P_3^{(i)} - y_3^{(i)}P_2^{(i)}) = 0, \tag{3.12.9}$$

$$\sum_{i=1}^{n} (y_3^{(i)}P_1^{(i)} - y_1^{(i)}P_3^{(i)}) = 0, \tag{3.12.10}$$

$$\sum_{i=1}^{n} (y_1^{(i)}P_2^{(i)} - y_2^{(i)}P_1^{(i)}) = 0, \tag{3.12.11}$$

where

$$\mathbf{P}^{(i)} = P_m^{(i)}\mathbf{i}_m, \tag{3.12.12}$$

$$\mathbf{r}^{(i)} = y_m^{(i)}\mathbf{i}_m. \tag{3.12.13}$$

Equilibrium of a force system coplanar with the y_1y_2-plane

$$\sum_{i=1}^{n} P_1^{(i)} = 0, \tag{3.12.14}$$

$$\sum_{i=1}^{n} P_2^{(i)} = 0, \tag{3.12.15}$$

$$\sum_{i=1}^{n} (y_1^{(i)}P_2^{(i)} - y_2^{(i)}P_1^{(i)}) = 0, \tag{3.12.16}$$

where

$$\mathbf{P}^{(i)} = P_1^{(i)}\mathbf{i}_1 + P_2^{(i)}\mathbf{i}_2, \tag{3.12.17}$$

$$\mathbf{r}^{(i)} = y_1^{(i)}\mathbf{i}_1 + y_2^{(i)}\mathbf{i}_2. \tag{3.12.18}$$

Equilibrium of a force system parallel to y_2-axis

$$\sum_{i=1}^{n} P_2^{(i)} = 0, \tag{3.12.19}$$

$$\sum_{i=1}^{n} y_3^{(i)}P_2^{(i)} = 0, \tag{3.12.20}$$

$$\sum_{i=1}^{n} y_1^{(i)}P_2^{(i)} = 0. \tag{3.12.21}$$

Characteristics of ideal truss

(1) Members are connected at their ends by frictionless joints.
(2) Loads and reactions are applied only at the joints.

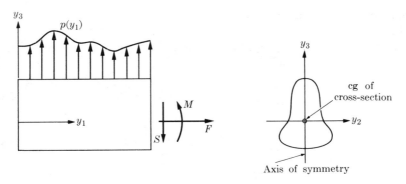

FIG. 3.61   Beam under lateral forces in y_1y_3-plane.

(3) Centroidal axis of each bar is straight and passes through the joints at the two ends.
(4) Internal forces in each truss member may be represented by a force along the centroidal axis.

Force analysis of ideal truss

(1) Method of joints
(2) Method of sections

Characteristics of ideal beam

(1) Span y_1 is much larger than dimensions of cross section y_2, y_3.
(2) Externally applied loads are coplanar in the y_1y_3-plane (see Figs. 3.61 and 3.39).
(3) Internal forces are replaced by equipollent system consisting of axial force F parallel to y_1, transverse shear force S parallel to y_3, and bending moment M about y_2-axis.

Relations between load, shear, and bending moment

$$\frac{dS}{dy_1} = p, \tag{3.12.22}$$

$$\frac{dM}{dy_1} = S. \tag{3.12.23}$$

PROBLEMS

3.1   (a) Referring to Fig. 3.2 (in the text), use vector analysis to prove that the moment of **F** about point p is independent of the location of point q on the line of action of **F**.   (b) Referring to Fig. 3.3, prove that the moment of **F** about the axis l is independent of the choice of **r** from l to **F**.

3.2   Reduce each of the following force and moment systems (a) through (e) to the simplest equipollent system.

	Force	Acts at
(a)	$\mathbf{F}^{(1)} = -3(\mathbf{i}_1 + \mathbf{i}_2)$	$(0, 3)$
	$\mathbf{F}^{(2)} = +3\mathbf{i}_2$	$(2, 0)$
	$\mathbf{F}^{(3)} = 4(\mathbf{i}_1 + \mathbf{i}_2)$	$(3, 0)$

	Force	Acts at
(b)	$\mathbf{F}^{(1)} = 3\mathbf{i}_2$	$(3, 3)$
	$\mathbf{F}^{(2)} = 2\mathbf{i}_1 + 3\mathbf{i}_2$	$(3, 3)$
	$\mathbf{F}^{(3)} = 3\mathbf{i}_1 + 2\mathbf{i}_2$	$(3, 3)$
	$\mathbf{F}^{(4)} = 4\mathbf{i}_1$	$(3, 3)$
	$\mathbf{F}^{(5)} = -3\mathbf{i}_1 + 4\mathbf{i}_2$	$(0, 4)$
	$\mathbf{F}^{(6)} = 2\mathbf{i}_1$	$(0, 2)$

	Force	Acts at
(c)	$\mathbf{F}^{(1)} = \mathbf{i}_1 + 2\mathbf{i}_2$	$(0, 1, 2)$
	$\mathbf{F}^{(2)} = 3\mathbf{i}_1 - \mathbf{i}_2$	$(1, 2, 0)$
	$\mathbf{F}^{(3)} = 4\mathbf{i}_1 + 3\mathbf{i}_2$	$(3, 2, 2)$

	Force	Acts at
(d)	$\mathbf{F}^{(1)} = 2\mathbf{i}_1 + 3\mathbf{i}_2$	$(-2, 3)$
	$\mathbf{F}^{(2)} = -5\mathbf{i}_1 + 2\mathbf{i}_2$	$(3, 1)$
	$\mathbf{F}^{(3)} = -3\mathbf{i}_1 - \mathbf{i}_2$	$(6, -1)$

	Force or moment	Acts at
(e)	$\mathbf{F}^{(1)} = \mathbf{i}_1 + \mathbf{i}_2 + \mathbf{i}_3$	$(0, 0, 0)$
	$\mathbf{F}^{(2)} = \mathbf{i}_1 - 2\mathbf{i}_2 + 3\mathbf{i}_3$	$(1, 1, 1)$
	$\mathbf{C}^{(1)} = \mathbf{i}_1 - \mathbf{i}_2$	$(4, 4, 4)$

3.3   Given the force vector $\mathbf{F}$ of magnitude 5 lb, lying in the y_2y_3-plane, and passing through the point q at $(0, 1, 1)$ as shown in Fig. P.3.3. (a) Determine the moment $\mathbf{M}$ of this force about point p at $(1, 1, 0)$. (b) Determine the magnitude of the moment about the axis Op. (c) Find the couple $\mathbf{C}$ that would be produced by this force $\mathbf{F}$, together with an equal and opposite force passing through the point a at $(0, 1, 2)$.

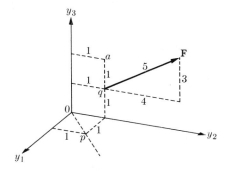

FIGURE P.3.3

3.4   An airplane, as shown in Fig. P.3.4, weighs 2500 lb, develops a maximum thrust of 143 lb, and sustains a tail load of 10.2 lb while cruising in level unaccelerated flight at 150 mph.   Determine the lift, drag, and moment about the aerodynamic center a.c. (so called because it is the reference point for a set of forces equipollent to the aerodynamic pressures acting on the wing).

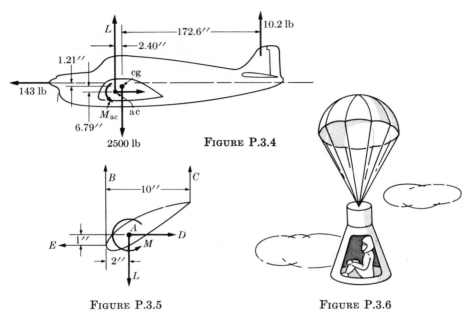

FIGURE P.3.4

FIGURE P.3.5

FIGURE P.3.6

3.5   A wind-tunnel model of an airplane wing is suspended as shown by Fig. P.3.5. Find the loads in members B, C, and E if the forces at A are $L = 50$ lb, $D = 5$ lb, and the moment about A is $M = 25$ in-lb.

3.6   The astronaut is sitting calmly in the capsule, waiting for the parachute to collapse when the capsule hits the water, as shown in Fig. P.3.6.   Assuming that the capsule is descending at constant velocity, draw and define the parachute, capsule, and astronaut as a three-particle system.   Isolate and show all the internal and external forces acting on each particle.   In another drawing, show the additional forces acting on each particle at the instant the capsule hits the water.

3.7   The rocket shown in Fig. P.3.7(a) is rising vertically with acceleration

$$\mathbf{a} = \frac{T - (D + W)}{m} \mathbf{i}_2,$$

where T is the thrust of the motor, D the total aerodynamic drag, m the total mass of the rocket, and $W = mg$ its weight.   An idealized model of the rocket is shown in Fig. P.3.7(b), in which the mass m is "lumped" into three particles, m_1, m_2, and m_3, such that

$$m_1 + m_2 + m_3 = m.$$

The lumped masses are connected by rigid, massless rods.   Treat the idealized model

Actual rocket   Idealized rocket

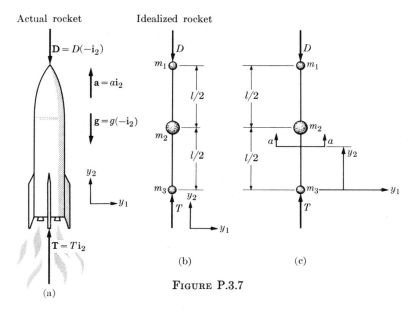

FIGURE P.3.7

as a two-particle system by making the imaginary cut represented by section a–a in Fig. P.3.7(c). Plot the internal forces $\mathbf{F}^{(LU)}$ on the lower particle from the upper, and $\mathbf{F}^{(UL)}$ on the upper particle from the lower, as the position of section a–a sweeps up the rocket ($0 \le y_2 \le l$).

3.8 Determine the center-of-gravity locations y_1, y_2, and y_3 (see Fig. P.3.8) of (a) three masses:

Mass	Weight, lb	Location
$m^{(1)}$	1	$(1, 1, 0)$
$m^{(2)}$	1	$(2, 2, 0)$
$m^{(3)}$	2	$(-1, 3, 0)$

(b) a triangular plate, where h = uniform thickness; and (c) a half-cylinder, where R = radius.

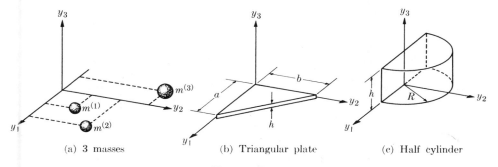

(a) 3 masses   (b) Triangular plate   (c) Half cylinder

FIGURE P.3.8

3.9   Using the *method of joints*, determine the internal forces in all the bars of the truss shown in Fig. P.3.9.

3.10   Find the reactions at the points of support, and the bar forces in all the members of the truss (Fig. P.3.10).

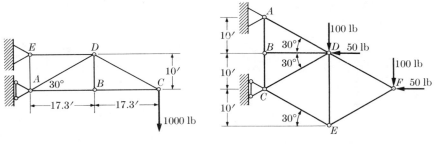

FIGURE P.3.9   FIGURE P.3.10

3.11   Using the *method of sections*, determine the internal forces in bars AB, BD, DC (these bars probably have the greatest loads). See Fig. P.3.11.

3.12   Find the reactions on the truss, and determine the axial forces in bars CK, JD, CD by the *method of sections*. Only the lettered intersections are joints (Fig. P.3.12).

3.13   In the space truss shown in Fig. P.3.13 the members AB, BC, CD, and AD form a square base in the y_1y_2-plane while the member AE coincides with the y_3-axis. $AB = BC = CD = AD = AE = 5$ ft. The truss is supported by a ball-and-socket joint at A, and by members BB', DD', and EE' at B, D, and E, respectively. BB' is parallel with the y_3-axis, DD' is parallel with the y_2-axis, and EE' is parallel with the

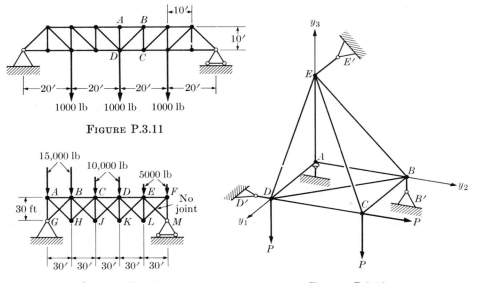

FIGURE P.3.11

FIGURE P.3.12   FIGURE P.3.13

y_1-axis. Under the external load shown in the figure, calculate forces in EE' and AB. [*Hint*: For each solution try to cut and isolate the truss in such a manner that the equation of moment equilibrium about a certain line will contain only one unknown force, i.e., the one which you are seeking.]

3.14 A beam CE (see Fig. P.3.14) which is supported by pin-ended members BD, BE, and AB is acted upon by a concentrated load W and a couple M (about an axis parallel to the y_2-axis). C is a pin joint. Determine the axial forces in AB, BD, and BE, and the reaction at C.

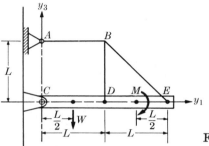

FIGURE P.3.14

3.15 Calculate the reactions, and plot the shear and bending-moment diagrams for each of the beams shown in Fig. P.3.15. Verify your solution for (d) by superposing solutions (a) and (b).

3.16 Determine the load, shear, and bending-moment diagrams for a beam of the configuration shown in Fig. P.3.16.

3.17 Determine the shear, bending-moment, and axial load distributions along the **beam** shown in Fig. P.3.17.

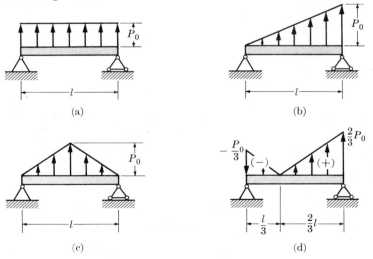

FIGURE P.3.15

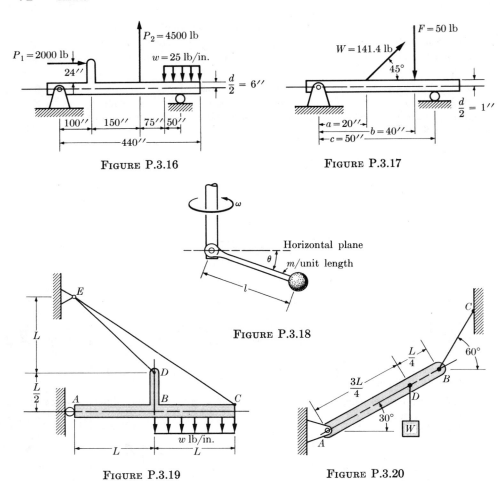

FIGURE P.3.16

FIGURE P.3.17

FIGURE P.3.18

FIGURE P.3.19

FIGURE P.3.20

3.18 A beam (see Fig. P.3.18) with a heavy mass at the tip is attached to the end of a vertical shaft by a pin joint. The shaft with the attached beam is rotating at a constant angular velocity ω. The mass distribution of the beam is m/unit length. The tip mass M is equal to one-third of the beam mass $(M = ml/3)$. Determine the angle of inclination θ, and the spanwise location at which the bending moment is a maximum.

3.19 (a) Draw a free-body diagram of the beam and kingpost, ABC and BD, respectively (Fig. P.3.19). (b) Calculate the tension forces in wires DE and CE. (c) Calculate and plot the shear and bending-moment diagrams for the beam ABC.

3.20 A beam is supported by a pin at one end and by a wire at the other end as shown in Fig. P.3.20. The weight of the beam per unit length is w(lb/in). A weight W is hanging at point D. Determine the tension in the wire BC, and the reaction at A. What is the bending moment at section D?

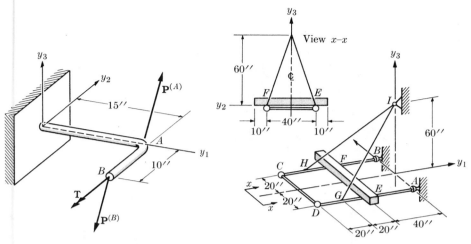

FIGURE P.3.21   FIGURE P.3.22

3.21   A three-dimensional, cantilevered beam with an L-shaped planform (Fig. P.3.21) is acted on by forces $\mathbf{P}^{(A)} = 10\mathbf{i}_2 + 5\mathbf{i}_3$, $\mathbf{P}^{(B)} = 10\mathbf{i}_1 - 10\mathbf{i}_3$, and a torque $T = -20\mathbf{i}_2$ parallel to the y_2-axis. The forces are in units of pounds, and the torque is in inch-pounds. Find the reactions at the wall, and draw shear and bending-moment diagrams for both sections of the beam.

3.22   A horizontal frame shown in Fig. P.3.22 consists of three rods AD, CD, and BC, which are connected at D and C by ball-and-socket joints. The frame is supported at A and B by ball-and-socket joints, and at G and H by cables which are attached to the wall at point I. A beam 60 in. long is placed on top of this frame. The weight of the beam is 2 lb/in., and the weight of the rigid rods AD and BC is 1 lb/in. Rod DC may be considered weightless. Determine the tension in the wire HI (or GI), and the maximum bending moment in the bar BC (or AD).

3.23   A curved beam whose axis is a circular arc of radius r is acted on by an upward (y_3-direction) distributed load $p_3(\theta)$ (lb/unit arc length) and a radial (y_2-direction) distributed load $p_2(\theta)$ (lb/unit arc length). See Fig. P.3.23. For a given section the three axes are defined by x_1 = circumferential direction, x_2 = radial direction, x_3 = vertical direction. The corresponding internal forces F, S_2, S_3 and moments T, M_2, M_3 are defined as shown in Fig. P.3.23(a). Consider the equilibrium of an element of arc length $r \times d\theta$ (Fig. P.3.23b), and show that the following differential relations are valid:

$$(1) \quad \frac{dS_3}{d\theta} = p_3 r, \qquad\qquad (4) \quad -\frac{dM_2}{d\theta} + S_3 r - T = 0,$$

$$(2) \quad \frac{dS_2}{d\theta} = p_2 r - F, \qquad\qquad (5) \quad \frac{dM_3}{d\theta} = S_2 r,$$

$$(3) \quad \frac{dF}{d\theta} = S_2, \qquad\qquad (6) \quad \frac{dT}{d\theta} - M_2 = 0.$$

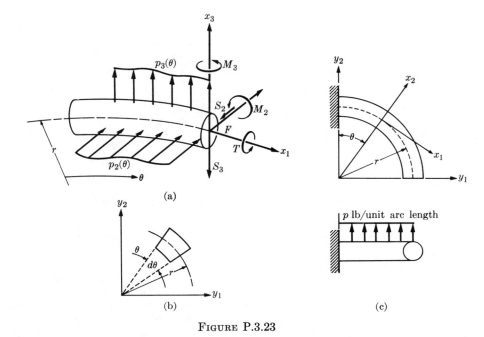

FIGURE P.3.23

A beam in the form of a quadrant of a circle (Fig. P.3.23.c) is clamped at one end and is acted on by uniform loads of p lb/unit arc length as shown in the sketch. Determine the bending moment $M_2(\theta)$ and torsional moment $T(\theta)$. Check to see whether your results satisfy Eq. (6) above.

REFERENCE

(1) C. H. NORRIS and J. B. WILBUR, *Elementary Structural Analysis*, 2nd Ed., Chaps. 3 and 4, McGraw-Hill, New York, 1960.

4

Simple Statically
Indeterminate Systems

4.1   INTRODUCTION

Chapter 3 was devoted to the principles of statics and to the analysis of statically determinate systems. We saw that many systems can be analyzed completely by merely carrying out the two steps:

(1) constructing a free-body diagram, and

(2) applying the principles of static equilibrium.

There are, however, many problems which involve a number of unknown reactions and/or internal forces greater than the number of independent equations that can be found from the principles of static equilibrium. For example, Fig. 4.1 shows a cantilever beam with three unknown reactions, $R_1^{(A)}$, $R_3^{(A)}$, and $M_2^{(A)}$, at the clamped end; since there are three equations of static equilibrium for this coplanar force system, the problem is statically

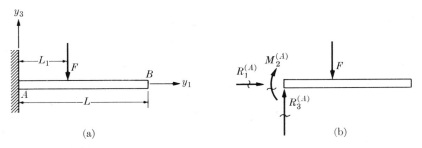

FIG. 4.1   Statically determinate beam: (a) physical diagram; (b) free-body diagram.

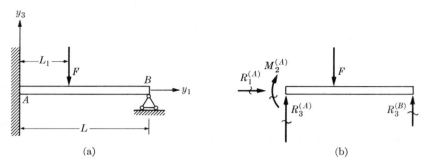

Fig. 4.2   Statically indeterminate beam: (a) physical diagram (b) free-body diagram.

determinate. Now if the beam is additionally supported by a hinge on a roller at the other end B (see Fig. 4.2), there is one additional reaction $R_3^{(B)}$. But there are still only three equations of static equilibrium that can be used.

For another example, let us consider a plane truss (Fig. 4.3) for which the members AD and BC are not connected. It is apparent that there are only three unknown reactions, $R_1^{(B)}$, $R_3^{(B)}$, and $R_3^{(D)}$, and there are three equations of static equilibrium for the complete structure. But when we begin to analyze the internal force in each member, we discover that there are three unknown forces at each joint, but only two equations of static equilibrium for each joint.

Problems which cannot be solved by using the equations of static equilibrium alone are termed *statically indeterminate*, and their analysis requires a knowledge of the deformation characteristics of the system.

Navier (Ref. 4.1) was the first to outline the method for solving statically indeterminate problems. He observed that such problems are indeterminate only if the bodies and the supports are considered as being completely rigid, but that if elasticity is considered, enough additional equations may be derived to evaluate all the unknown quantities. The present chapter is devoted to an

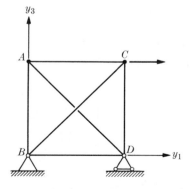

Fig. 4.3   Statically indeterminate truss.

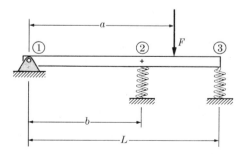

Fig. 4.4   Physical diagram of rigid hinged beam supported on two springs.

elementary exposition of the principles of analysis of statically indeterminate systems. We shall see that two additional kinds of equations are required: (1) *constitutive relations* which express the deformation response of the structure and its supports to forces and to changes in environment such as temperature variations, and (2) *consistent deformation relations* which ensure the contiguity of the system. The principles are introduced and illustrated by examples involving simple systems. More complicated problems will be deferred to a later stage in this text, after more of the fundamentals of solid mechanics have been presented.

4.2 PRINCIPLES OF ANALYSIS OF STATICALLY INDETERMINATE SYSTEMS

We shall develop these principles by means of the simple planar system illustrated by Fig. 4.4. This illustration shows a completely rigid beam of length L, hinged at one end, supported by two springs of equal spring constants k, and loaded by a single force F. The object of the problem is to compute the internal forces in the springs and the reaction supplied by the hinge.

(a) Free-body diagrams. In this problem, as in any other problem in solid mechanics, the first steps are those of drawing free-body diagrams and selecting notation for the unknown quantities. These steps are illustrated by Fig. 4.5, where $R_3^{(1)}$ is designated as the reaction provided by the hinge, and $R_3^{(2)}$ and $R_3^{(3)}$ are the internal forces in the springs.

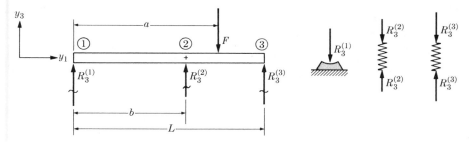

FIG. 4.5 Free-body diagram corresponding to Fig. 4.4.

(b) Equations of static equilibrium. The principles of static equilibrium require that Eqs. (3.4.8), (3.4.10), and (3.4.12) be satisfied for a planar system. We have then the following two independent equations of equilibrium:

$$\sum_{i=1}^{3} P_3^{(i)} = R_3^{(1)} + R_3^{(2)} + R_3^{(3)} - F = 0 \qquad (4.2.1)$$

and

$$\sum_{i=1}^{3} M_2^{(i)} = R_3^{(3)}L + R_3^{(2)}b - Fa = 0 \quad \text{(about the hinge).} \qquad (4.2.2)$$

Since there are three unknown quantities, namely $R_3^{(1)}$, $R_3^{(2)}$, and $R_3^{(3)}$, the two equations of statics are insufficient to provide a complete solution. The problem is said to be statically indeterminate with one *redundant* quantity or statically indeterminate to the first degree.

(c) Constitutive equations. In order to supplement the equations of static equilibrium we must turn to the deformations and seek equations which relate forces and displacements of the system. Referring to Fig. 4.5 we see that since the beam is considered rigid and the springs elastic, the only pertinent force-displacement relations are

$$R_3^{(2)} = k\delta^{(2)}, \tag{4.2.3}$$

$$R_3^{(3)} = k\delta^{(3)}, \tag{4.2.4}$$

where $\delta^{(2)}$ and $\delta^{(3)}$ are the displacements of the springs (2) and (3), respectively. Equations (4.2.3) and (4.2.4) are of a mathematical form which is applicable not only to springs* but to all structural components of any degree of complexity. For our present purposes it is assumed that k is available; actually, the difficult task lies in the determination of k, a task which is a portion of the subject of solid mechanics. Figure 4.6 illustrates graphically the linear relations between force and the displacement of the system. The balance sheet of equations and unknowns now reads: four equations, (4.2.1) through (4.2.4) and five unknowns, $R_3^{(1)}$, $R_3^{(2)}$, $R_3^{(3)}$, $\delta^{(2)}$, and $\delta^{(3)}$.

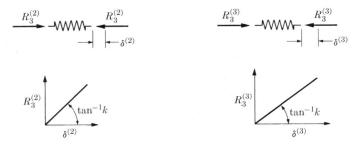

Fig. 4.6   Force-displacement properties of the system.

(d) Equations of consistent deformation. The final step in the process is to require that the deformations at points (2) and (3) be consistent with each other and with the geometric constraints which have been placed upon the system. We shall require the displacements to be small, so that the angle of rotation, θ, of the rigid bar is of a magnitude such that $\sin \theta \cong \theta$ (in radians). In Fig. 4.7

---

* Springs are usually constructed of bars with circular cross section wound into helicoidal shapes. Such a structural form presents an unusual and interesting problem for analysis.

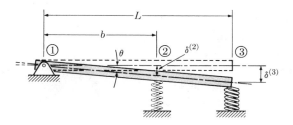

FIG. 4.7   Diagram showing consistency of deformations with geometric constraints.

it is evident that if the spring deflections are small, then

$$\frac{\delta^{(2)}}{b} = \frac{\delta^{(3)}}{L}.$$ (4.2.5)

Equation (4.2.5) is called the *equation of consistent deformation*.

We have seen that by (a) constructing a free-body diagram, and deriving (b) equilibrium equations, (c) force-displacement equations, and (d) equations of consistent deformation, we are led to a system of independent equations equal in number to the unknowns of the system. In summary, the unknowns are $R_3^{(1)}$, $R_3^{(2)}$, $R_3^{(3)}$, $\delta^{(2)}$, and $\delta^{(3)}$, and the five available equations are:

$$\left.\begin{array}{l} R_3^{(1)} + R_3^{(2)} + R_3^{(3)} - F = 0 \\[2mm] R_3^{(3)}L + R_3^{(2)}b - Fa = 0 \end{array}\right\} \quad \text{equilibrium equations,}$$

$$\left.\begin{array}{l} R_3^{(2)} = k\delta^{(2)} \\[2mm] R_3^{(3)} = k\delta^{(3)} \end{array}\right\} \quad \text{constitutive equations,}$$

$$\frac{\delta^{(2)}}{b} = \frac{\delta^{(3)}}{L}, \quad \text{equation of consistent deformation.}$$

The equations are solved by following a pattern in which the constitutive equations and the equation of consistent deformation are combined to yield a single equation relating $R_3^{(2)}$ and $R_3^{(3)}$ as follows:

$$\frac{R_3^{(2)}}{R_3^{(3)}} = \frac{b}{L}.$$ (4.2.6)

Combining Eq. (4.2.6) with the equilibrium equations (4.2.1) and (4.2.2) we find for the three unknown reactions:

$$R_3^{(1)} = F\left[1 - \frac{a(b + L)}{L^2(1 + b^2/L^2)}\right],$$

$$R_3^{(2)} = \frac{Fab}{L^2(1 + b^2/L^2)}, \qquad R_3^{(3)} = \frac{Fa}{L(1 + b^2/L^2)}.$$ (4.2.7)

4.3 EXAMPLE: AIRPLANE LANDING GEAR

We shall further illustrate the principles of Section 4.2 by applying them
to the landing gear shown by Fig. 4.8.
The rigid member ABC is assumed to
be fixed against translation at C, attached to a shock strut spring at B, and
restrained against rotation by a torsion
spring at C. The spring constant of the
linear spring at B is designated as
k_1 lb/in., and the spring constant of
the torsion spring at C is designated as
k_2 in.-lb/radian.

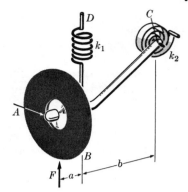

Fig. 4.8 Physical diagram of landing
gear.

(a) Free-body diagrams. Free-body diagrams of the isolated component parts
of the tripod landing gear, together with the notation for unknown reactions
and internal forces and moments, are illustrated by Fig. 4.9. The force in
the linear spring is designated by $R_3^{(1)}$, the moment supplied by the torsion
spring by $M_2^{(1)}$, and the reaction at the hinged support at C by $R_3^{(2)}$. The
reaction at C is vectored in the y_3-direction, since no y_1- or y_2-components of
force are acting on the rigid beam ABC.

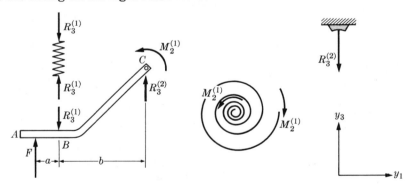

FIG. 4.9 Free-body diagram of landing gear.

(b) Equations of static equilibrium. Treating the free-body diagram of the
beam in Fig. 4.9 as a planar system, we find the following equations of static
equilibrium:

$$\sum_{i=1}^{3} P_3^{(i)} = F + R_3^{(2)} - R_3^{(1)} = 0 \tag{4.3.1}$$

and

$$\sum_{i=1}^{3} M_2^{(i)} = F(a + b) - R_3^{(1)}b - M_2^{(1)} = 0 \quad \text{(about hinge).} \tag{4.3.2}$$

The balance sheet of equations and unknowns reads: two equations, (4.3.1) and (4.3.2), and three unknowns, $R_3^{(1)}$, $R_3^{(2)}$, and $M_2^{(1)}$.

(c) Constitutive relations. The force-displacement equations of the system, when we assume complete rigidity of the beam member ABC, read:

$$R_3^{(1)} = k_1 \delta, \tag{4.3.3}$$

$$M_2^{(1)} = k_2 \theta \tag{4.3.4}*$$

In the preceding two examples the structural members have been assumed to be rigid. The same three types of equation, i.e., equilibrium, constitutive, and consistency, would be used if the flexibility of the members were taken into account; additional constitutive equations to account for the flexibility of the members would be combined with those of the springs. In real life there are no rigid members, and thus the neglect of flexibility implies that some structural component is much more rigid (or much more flexible) than some other. Thus in the examples used, the assumption of rigid members can be justified only if the flexibility of the springs (this means $1/k$) is much larger than that of the members, or conversely, the stiffness of the members is much greater than that of the springs. In solid mechanics, the terms "much larger" or "much greater" usually mean at least by one order of magnitude.

4.4 EXAMPLES OF PLANE TRUSSES

Statically Indeterminate Truss. Figure 4.10 shows a pin-jointed truss consisting of three bars meeting at the vertex D, and loaded by a force F. The bar BD has an area A_1 and modulus of elasticity E_1, and the bars AD and CD each have area A_2 and modulus of elasticity E_2. The length of the bar BD is L and the angle ABD and BDC is α.

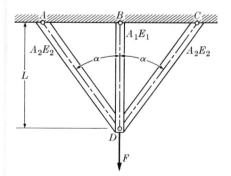

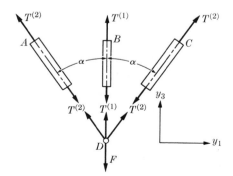

FIG. 4.10   Physical diagram of statically indeterminate planar truss.

FIG. 4.11   Free-body diagrams of planar truss.

*See Addendum, p. 92.

(a) Free-body diagrams. The appropriate free-body diagrams of the planar truss are shown in Fig. 4.11. Here we have isolated the pin at D, and each of the individual bars. Each bar is assumed to be in tension. The tension force in bar BD is designated as $T^{(1)}$, and in bars AD and CD it is designated as $T^{(2)}$. It is evident from the symmetry of the truss and its loading system that the internal forces will also be symmetrically disposed, and that bars AD and CD will therefore have equal internal forces.

(b) Equations of static equilibrium. One equation of static equilibrium may be found from the equilibrium of the pin at D in Fig. 4.11.

$$\sum_{i=1}^{3} P_3^{(i)} = T^{(1)} + 2T^{(2)} \cos \alpha - F = 0. \tag{4.4.1}$$

Symmetry has in effect used the equilibrium equation in the y_1-direction to show that $T_{AD} = T_{CD}$:

$$\sum_{i=1}^{3} P_1^{(i)} = -T_{AD} \sin \alpha + T_{CD} \sin \alpha = 0;$$

$$\therefore T_{AD} = T_{CD} = T^{(2)},$$

as stated above. The balance sheet of equations and unknowns reads: one equation, (4.4.1), and two unknowns, $T^{(1)}$ and $T^{(2)}$.

(c) Constitutive relations. In order to construct the force-displacement relations for a truss system such as that of Fig. 4.11, it is necessary to derive the relation between force and displacement for a uniform elastic bar with constant axial force. As has been stated in Section 4.2, the force-displacement relationship is of the form

$$T = k\delta. \tag{4.4.2}$$

It will be shown in subsequent chapters that k for a bar under uniform tension has the form

$$k = \frac{AE}{L}, \tag{4.4.3}$$

where A is the cross-sectional area and E is the modulus of elasticity of the material. We can then construct the following force-displacement relations for our planar truss example:

$$T^{(1)} = \frac{A_1 E_1}{L} \delta^{(1)}, \tag{4.4.4}$$

$$T^{(2)} = \frac{A_2 E_2 \cos \alpha}{L} \delta^{(2)}, \tag{4.4.5}$$

where $\delta^{(1)}$ and $\delta^{(2)}$ are the deflections of bars BD and AD (or CD), respectively. At this point in the problem there are three equations, (4.4.1), (4.4.4), and (4.4.5), and four unknowns, $T^{(1)}$, $T^{(2)}$, $\delta^{(1)}$, and $\delta^{(2)}$.

(d) Equations of consistent deformation. A single equation of consistent deformation may be derived by requiring that the stretching of the bars be compatible with their common point of attachment at D. From Fig. 4.12 it is easily seen that the equation of consistent deformation is simply

$$\delta^{(1)} \cos \alpha = \delta^{(2)}. \qquad (4.4.6)$$

Equations (4.4.1), (4.4.4), (4.4.5), and (4.4.6) are sufficient for the computations of the four unknowns $T^{(1)}$, $T^{(2)}$, $\delta^{(1)}$, and $\delta^{(2)}$. Combining Eqs. (4.4.4), (4.4.5), and (4.4.6), we find the following relation between $T^{(1)}$ and $T^{(2)}$:

$$\frac{T^{(1)}}{T^{(2)}} = \frac{A_1 E_1}{A_2 E_2 \cos^2 \alpha}. \qquad (4.4.7)$$

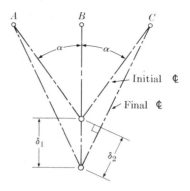

FIG. 4.12 Diagram showing consistency of stretching of bars of planar truss.

Equation (4.4.7) when combined with Eq. (4.4.1) yields

$$T^{(1)} = \frac{F}{1 + 2A_2 E_2 \cos^3 \alpha / A_1 E_1}, \qquad (4.4.8)$$

$$T^{(2)} = \frac{F \sec \alpha}{2 + A_1 E_1 / A_2 E_2 \cos^3 \alpha}. \qquad (4.4.9)$$

Statically Determinate Truss. Let us examine the behavior of a two-bar truss (Fig. 4.13). This is a statically determinate structure in which $T_{AD} = T_{CD}$, as shown previously for the similar three-bar truss. We apply the equilibrium relation for the y_3-direction:

$$\sum_{i=1}^{2} P_3^{(i)} = 2T^{(2)} \cos \alpha - F = 0$$

or

$$T^{(2)} = F/2 \cos \alpha.$$

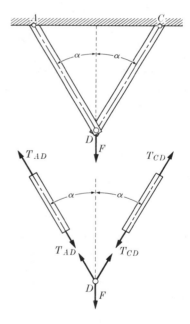

FIG. 4.13 Statically determinate plane truss.

It is clear that in the limit where $A_1 = 0$, Eqs. (4.4.8) and (4.4.9) reduce to

$$T^{(1)} = 0 \quad \text{and} \quad T^{(2)} = F/2 \cos \alpha,$$

which is the same answer that results if A_1 is set equal to zero in Eq. (4.4.9), i.e., if bar BD is removed from the indeterminate truss of Fig. 4.11.

4.5 EXAMPLE OF THERMAL STRESSES IN BOLT-AND-BUSHING ASSEMBLY

In Fig. 4.14 an assembly is shown which consists of a steel bolt inserted in a
bronze bushing. The nut is pulled up snugly against the bushing while the
whole assembly is at an ambient temperature of 50°F. It is desired to compute
the stresses in the bolt and bushing when the temperature of the assembly is
raised to 170°F.

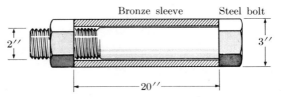

FIG. 4.14   Physical diagram of bolt-and-sleeve assembly.

(a) Free-body diagrams. Figure 4.15 illustrates the free-body diagrams which
follow from Fig. 4.14. The bolt is assumed to be in tension with load $P^{(s)}$, and
the bushing in compression with load $P^{(b)}$. The bolt head is in equilibrium under
the forces $P^{(b)}$ and $P^{(s)}$. In this discussion superscripts b refer to the bronze
bushing, and s to the steel bolt.

(b) Equations of static equilibrium. One equation of static equilibrium may
be formed from the equilibrium of the bolt head (cf. Fig. 4.15) as follows:

$$\sum_{i=1}^{2} P_1^{(i)} = P^{(s)} - P^{(b)} = 0. \tag{4.5.1}$$

The balance sheet of unknowns and equations reads: one equation (4.5.1), and
two unknowns, $P^{(b)}$ and $P^{(s)}$. The problem is therefore statically indeterminate,
and we must seek additional equations from sources other than equilibrium.

(c) Constitutive relations. In determining the deformations of components of
the assembly we must take into account the fact that the materials of which the
bolt and bushing are made expand with an increase in temperature. The
elongation of the bolt due to a change in temperature ΔT by an amount $\delta^{(s)}$ is
given by

$$\delta_T^{(s)} = \alpha_s L \, \Delta T, \tag{4.5.2}$$

where

α_s = coefficient of thermal expansion of the steel bolt material per degree
change in temperature,

L = length of the bolt,

ΔT = change in temperature.

The coefficient of thermal expansion of a material represents an experimentally

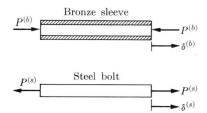

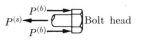

FIG. 4.15 Free-body diagrams of bolt and sleeve assembly.

determined constant which measures the expansion of the material per unit length of material per unit change of temperature.

The expansion of the bushing due to temperature change is expressed by

$$\delta_T^{(b)} = \alpha_b L\, \Delta T, \qquad (4.5.3)$$

where

α_b = coefficient of thermal expansion of the bronze bushing material per degree change in temperature.

It is evident that there will also be changes in length of the bolt and bushing due to the loads $P^{(s)}$ and $P^{(b)}$, respectively. For the bolt, the stretching due to P_s is given by

$$\delta_P^{(s)} = \frac{P^{(s)}}{k_s}, \qquad (4.5.4)$$

where the form of k_s will be the same as for the truss members of Section 4.4,

$$k_s = \frac{A_s E_s}{L}, \qquad (4.5.5)$$

and

A_s = cross-sectional area of bolt,
E_s = modulus of elasticity of the steel bolt.

The reduction in length of the bushing due to the compressive load $P^{(b)}$ is given by

$$\delta_P^{(b)} = -\frac{P^{(b)}L}{A_b E_b} = -\frac{P^{(b)}}{k_b}, \qquad (4.5.6)$$

where

$$k_b = \frac{A_b E_b}{L}, \qquad (4.5.7)$$

and

A_b = cross-sectional area of the bushing,
E_b = modulus of elasticity of the bronze bushing.

(d) Equation of consistent deformation. In order for the head and nut of the bolt to remain in contact with the ends of the bushing, the following condition on the deformation must be met:

$$\delta_T^{(b)} + \delta_P^{(b)} = \delta_P^{(s)} + \delta_T^{(s)}. \tag{4.5.8}$$

(e) Numerical solution. The six equations, (4.5.1) through (4.5.6), are sufficient to compute the six unknowns $P^{(b)}$, $P^{(s)}$, $\delta_T^{(b)}$, $\delta_P^{(b)}$, $\delta_T^{(s)}$, and $\delta_P^{(s)}$. Let us suppose that the properties of the bolt and bushing material are given by

$$\begin{aligned}
E_b &= (10)(10^6) \text{ lb/in}^2, \\
\alpha_b &= (10.1)(10^{-6})/°\text{F}, \\
E_s &= (30)(10^6) \text{ lb/in}^2, \\
\alpha_s &= (6.1)(10^{-6})/°\text{F}.
\end{aligned} \tag{4.5.9}$$

Substituting Eqs. (4.5.2) through (4.5.5) in Eq. (4.5.8), and combining with Eq. (4.5.1), we get

$$P = P^{(b)} = P^{(s)} = 13,300 \text{ lb}. \tag{4.5.10}$$

Since these answers were obtained with plus signs, we conclude that the assumed directions of $P^{(b)}$ and $P^{(s)}$ in Fig. 4.15 are correct.

4.6 EXAMPLE OF ASSEMBLY STRESSES IN BOLT-AND-NUT ASSEMBLY

Let us suppose that the bolt of the example in Section 4.5 has 20 threads to the inch, and that we wish to compute the loads in the bolt and bushing when the nut is given one-half turn after it has been brought snugly against the bushing. In such a problem the equilibrium equation is the same as (4.5.1), and the force-displacement equations are the same as (4.5.4) and (4.5.5). Since one-half turn advances the nut 0.025 in., the compatibility equation is

$$\delta_P^{(b)} = \delta_P^{(s)} - 0.025. \tag{4.6.1}$$

Combining Eqs. (4.5.1), (4.5.4), (4.5.5), and (4.6.1), and introducing the numerical data of (4.5.9), we obtain

$$P^{(b)} = P^{(s)} = 34,700 \text{ lb}. \tag{4.6.2}$$

4.7 EXAMPLE OF STATICALLY INDETERMINATE BEAM

Let us now consider the statically indeterminate beam which was discussed previously in Section 4.1, and is illustrated again in Fig. 4.16.

(a) Free-body diagram. Figure 4.16(b) illustrates the free-body diagram corresponding to Fig. 4.16(a). There are four unknown reactions, $R_1^{(A)}$, $R_3^{(A)}$, $M_2^{(A)}$, and $R_3^{(B)}$.

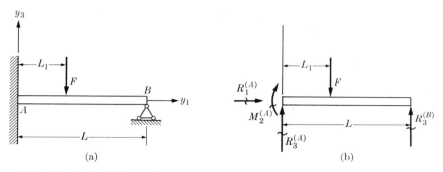

FIG. 4.16   Statically indeterminate beam: (a) physical diagram; (b) free-body diagram.

(b) Equations of static equilibrium. The three equations of equilibrium for this coplanar force system are

$$\sum P_1^{(i)} = R_1^{(A)} = 0, \tag{4.7.1}$$

$$\sum P_3^{(i)} = R_3^{(A)} - F + R_3^{(B)} = 0, \tag{4.7.2}$$

$$\sum M_2^{(i)} = M_2^{(A)} + L_1 F - L R_3^{(B)} = 0 \quad \text{(about } A\text{)}. \tag{4.7.3}$$

It is seen that $R_1^{(A)}$ is zero. The balance sheet of unknowns and equations reads: two equations, (4.7.2) and (4.7.3), and three unknowns, $R_3^{(A)}$, $M_2^{(A)}$, and $R_3^{(B)}$. The problem is therefore statically indeterminate and we must seek an additional equation.

(c) Constitutive relations. Let us assume that we can determine the tip deflection δ_B of a cantilever uniform beam, due to a vertical load P acting at any distance x from the fixed end of the beam (Fig. 4.17). The expression for the tip deflection is

$$\delta_B = \frac{Px^2}{2EI} (L - \tfrac{1}{3}x), \tag{4.7.4}$$

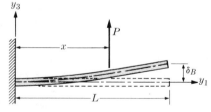

FIG. 4.17   Cantilever beam under a concentrated load.

where E is the modulus of elasticity, and I is the section moment of inertia about the y_2-axis. We can, of course, always determine the tip deflections through a series of experiments. In Chapters 7 and 9 we shall discuss analytical methods for evaluating this expression.

(d) Equations of consistent deformation. From Fig. 4.16(b) we can see that the statically indeterminate beam may be considered as a cantilever beam acted upon by a load $P = -F$ at a distance $x = L_1$ from A, and an unknown load $P = R_3^{(B)}$ acting at a distance $x = L$ from A. The deflection at B due to these two loads must necessarily be zero because B is a fixed support. The equation of consistent deformation is thus

$$\delta_B^{(F)} + \delta_B^{(R_3^B)} = 0. \tag{4.7.5}$$

Substituting

$$\delta_B^{(F)} = -\frac{FL_1^2}{2EI}\left(L - \tfrac{1}{3}L_1\right) \quad \text{and} \quad \delta_B^{(R_3^B)} = \frac{R_3^{(B)}L^2}{2EI}\left(L - \tfrac{1}{3}L\right),$$

$$\tag{4.7.6}$$

we can solve for the unknown reaction $R_3^{(B)}$. The result is

$$R_3^{(B)} = \frac{FL_1^2}{L^2}\left(\frac{3}{2} - \frac{1}{2}\frac{L_1}{L}\right). \tag{4.7.7}$$

Knowing $R_3^{(B)}$, we can evaluate the other three reactions by means of Eqs. (4.7.2) and (4.7.3).

4.8 SUMMARY

We have now completed the development of the tools necessary for the analysis of structures "in the large." We have seen that if the number of unknowns is equal to the number of equilibrium equations available, a problem may be solved with the equations of statics alone. When the number of unknowns exceeds the number of available equilibrium equations, the structure is called statically indeterminate. Such structures can be solved by simultaneous considerations of the following three groups of equations:

(1) equations of static equilibrium,
(2) constitutive equations or force displacement relations,
(3) conditions of consistent deformation.

Whether the problem is determinate or indeterminate, we observe that the unknowns are always forces and displacements, i.e., integrated-effect quantities. In the remainder of this volume we shall retrace our steps through these fundamental equations, examining the structure in greater detail. Instead of forces and displacements, we shall be interested in forces per unit area and displacements per unit length, and our general problem will be to solve for these quantities as functions of position in the structure.

PROBLEMS

4.1 Compute the reactions on the beam shown in Fig. P.4.1. Also compute the deflections at the supports. The beam may be considered rigid and weightless. The deformation of each spring is small in comparison to the length of the beam.

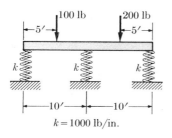

FIGURE P.4.1

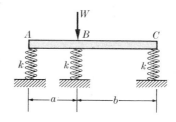

FIGURE P.4.2

4.2 A rigid beam weighing w lb/unit length is supported by three springs as shown in Fig. P.4.2. Given that the beam is loaded by an additional concentrated load W acting at point B, compute the bending moment at section B.

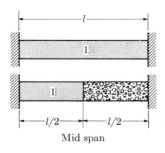

FIGURE P.4.3

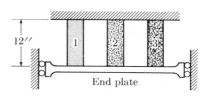

FIGURE P.4.4

4.3 Compare the effects of thermal expansion on the one-material and two-material bars shown in Fig. P.4.3. Compute the loads which arise in each material, and the deflections of the midspan stations of each bar for a rise in temperature of 100F°.

Bar	Area	Material	$\alpha/°F$ ($\times 10^{-6}$)	E, psi ($\times 10^6$)
1	A	Titanium	6	17
2	A	Concrete	6	3.4

4.4 Three bars of different materials are built into a rigid wall at one end and connected at the other by a common end plate. The end plate is rigid and is constrained by rigid tracks against rotation (Fig. P.4.4). Calculate the load in each bar, and the deflection of the end plate for a decrease in termperature of 200° F.

Bar	Area	Material	$\alpha/°F$ ($\times 10^{-6}$)	E, psi ($\times 10^7$)
1	1 in^2	Gun metal	10.2	1
2	2 in^2	Tungsten	2.4	5.2
3	3 in^2	Copper	7.8	1.8

4.5   The bar-and-plate system shown in the Fig. P.4.5 is built into rigid walls at either end, and the tie plate is constrained to horizontal motion. The system is in equilibrium under zero stress at 0°F. Compute the bar loads and the deflection, δ, at a temperature of -100°F. Each bar is 10 in. long.

Bar	Area	Material	α/°F ($\times 10^{-6}$)	E, psi ($\times 10^7$)
1	2 in^2	Molybdenum	2.7	4.3
2	1 in^2	Aluminum, 2024–T3	13	1.05
3	1 in^2	Steel, SAE 4130	6.5	2.9

4.6   A rigid beam is suspended by three elastic bars as shown by the figure. All bars are 1 in^2 in cross section and 20 in. long. Bar BE is made of steel, and bars AF and CD are made of bronze. Assuming that the bars are stress-free prior to loading, compute the forces in each bar for the loading shown by Fig. P.4.6. Suppose that the temperature of the entire assembly is raised by 100°F. Compute the forces in the bars

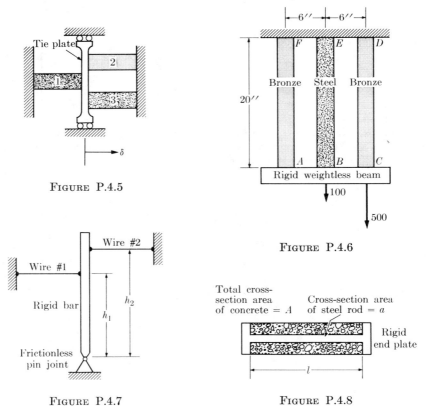

FIGURE P.4.5

FIGURE P.4.6

FIGURE P.4.7

FIGURE P.4.8

under the combined external loading and temperature environment.   Sketch the bending-moment and shear diagrams of the beam.

Moduli of elasticity: Coefficients of thermal expansion:

$E_{steel} = 30 \times 10^6$ lb/in^2, $\alpha_{steel} = 6.5 \times 10^{-6}/°F$,

$E_{bronze} = 17 \times 10^6$ lb/in^2. $\alpha_{bronze} = 10.1 \times 10^{-6}/°F$.

4.7   The wires shown in the Fig. P.4.7 are cooled by an amount $\Delta T = -100°F$. Calculate the axial tension in each wire in terms of the properties and dimensions of the wires, and the angle of rotation of the rigid bar. The dimensions and properties of the wires are

Wire No. 1: A_1, L_1, E_1, α_1,   Wire No. 2: A_2, L_2, E_2, α_2,

where A = cross section area, L = length of member, E = modulus of elasticity, and α = coefficient of thermal expansion.

4.8   Prestressed concrete is widely used as a building material today. A prestressed concrete member is formed by applying tension to a set of steel rods while concrete is poured around them. When the concrete sets, the external loads are removed from the steel rods, and the concrete is subjected to compression. The prestressed concrete member derives its great strength from the facts that (1) concrete can withstand large compressive loads but only small tensile loads, and (2) the precompression prevents an externally applied large tensile load from actually putting the concrete in tension. Figure P.4.8 illustrates an idealized prestressed concrete member, in which the steel rod is "fixed" to the concrete only by the rigid end plates.

(a) If the length of the steel rod was equal to L during the time the tensile load P was applied, calculate the length l of the prestressed member after P is removed. (b) Speculate on what might happen to a prestressed concrete floor beam in a building which has been burning long enough for the beam to become heated throughout. Given:

$E_{steel} = 30 \times 10^6$ psi, $\alpha_{steel} = 6.5 \times 10^{-6}/°F$,

$E_{concrete} = 3.4 \times 10^6$ psi, $\alpha_{concrete} = 6 \times 10^{-6}/°F$.

4.9   A cantilever beam is supported at its end by a wire of area A in tension. The beam with a moment of inertia I is loaded by a uniformly distributed load of w lb/in. Both the beam and the wire are made of the same material, with modulus of elasticity E (Fig. P.4.9). Compute the load in the wire and the reactions of the wall on the beam.

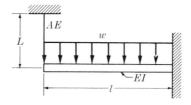

FIGURE P.4.9

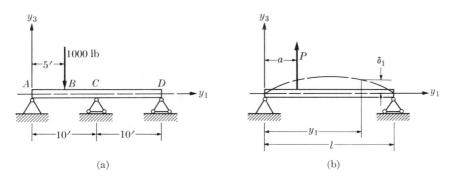

FIGURE P.4.10

4.10   Determine the reactions for the beam on three supports in Fig. P.4.10(a).
Note: For the statically determinate uniform beam loaded as shown in Fig. P.4.10(b) the deflection δ_1 at any point y_1 is:

$$\delta_1 = \begin{cases} \dfrac{Py_1(l - a)(2la - a^2 - y_1^2)}{6EIl} & \text{for} \quad y_1 \leq a, \\[4mm] \dfrac{Pa(l - y_1)(2ly_1 - a^2 - y_1^2)}{6EIl} & \text{for} \quad y_1 \geq a. \end{cases}$$

4.11   The rigid beam AB weighing w lb/unit length is pinned at its upper end A and attached at its lower end B to two cables BC and BD (Fig. P.4.11). Compute the loads in the cables and the reaction at A. The two wires have the same elastic modulus and cross-section area.

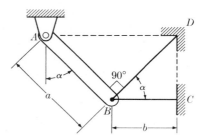

FIGURE P.4.11

REFERENCE

(1) C. L. M. H. NAVIER, *Résumé des leçons sur l'application de la méchanique*, Paris, France, 1833.

ADDENDUM TO PAGE 81

(After Eq. (4.3.4), read:)
where δ is the displacement of the linear spring BD and θ is the angular displacement of the beam about the pin at C.

(d) Equations of consistent deformation. For small displacements δ and θ are related by

$$b\theta = \delta \qquad\qquad (4.3.5)$$

Equations (4.3.1) through (4.3.5) constitute a set of five independent equations which may be used to find the five unknown quantities of the problem, namely, $R_3^{(1)}$, $R_3^{(2)}$, $M_2^{(1)}$, δ and θ.

5

Analysis of Strain

5.1 INTRODUCTION

The analytical treatment of forces has been presented in Chapters 2 and 3. In Chapters 3 and 4, the effect of the gross deformations of a body on the distribution of forces in a statically indeterminate structure has been discussed. We must now delve more deeply into the detailed nature of the deformations in deformable bodies. The information we desire is contained in the strain tensor. This chapter will be concerned with the analysis of the state of strain in a deformed body.

It is an experimental fact that bodies subjected to forces will be deformed. For a structure constructed of rubber or wood, the deformations under load are easily discernible and can be measured by relatively crude instruments. For one constructed of an aluminum or a steel alloy, the strains can be detected only with precision instruments.

The state of strain at a point P in the deformed body supplies the following information:

(a) the differences in length of all linear elements, dS, originating at a point P in the deformed structure and the corresponding elements, ds, originating at the point p in the undeformed structure (in undergoing deformation, a material particle point at point p in the undeformed body moves in a continuous fashion to its final position at point P in the deformed body), and

(b) the differences between the angles, $d\widehat{S_1 dS_2}$, formed by all pairs of linear elements originating at point P and the angles, $d\widehat{s_1 ds_2}$ formed by the corresponding linear elements ds_1 and ds_2 originating at point p in the undeformed structure.

The quantities described in (a) are called the *extensional strains*, while those of (b) are called the *shear strains*. We will shortly discover that knowledge of any combination of six strains (e.g., three extensional and three shear, or four extensional and two shear) is sufficient to completely specify the strain at a point P.

The geometrical meanings of (a) and
(b) are depicted in Fig. 5.1, wherein
point p is any point in the undeformed
body. It is assumed that point p moves
to point P in a continuous fashion (from
causes which need not be of concern
here), and this is schematically shown
by placing P within a boundary which
is labeled "deformed body." Similarly,
points q and r which are infinitesimally
close to p displace to the points labeled
Q and R. The vector joining p to P will

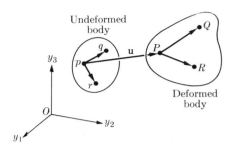

FIG. 5.1   Relation between points in
the undeformed and deformed body.

be denoted by **u**, and it can be specified as a function of the coordinates of p.
The state of strain at point P (or p, since p is uniquely related to P by **u**) is
specified if the following two items of geometry are known:

(a) $|PQ| - |pq|$ for all points q in the neighborhood of p $(dS = |PQ|$ and
$ds = |pq|),$

(b) $\angle QPR - \angle qpr$ for all pairs of line segments formed by points q and r
in the vicinity of p $(dS_1 = |PQ|, dS_2 = |PR|, ds_1 = |pq|, ds_2 = |pr|,$
$\widehat{dS_1 dS_2} = \angle QPR,$ and $\widehat{ds_1 ds_2} = \angle qpr).$

It will be shown that this information is contained in the fundamental metric
tensors of the deformed and undeformed structure.

In what follows, lower-case letters will identify points or quantities in the
undeformed body, and the same upper-case letters will identify similar quantities
in the deformed body (see Fig. 5.1).

5.2   THE FUNDAMENTAL METRIC TENSORS

Basically, the task in the analysis of strain is to chart the course of every
material particle from its original to its final position. The material particle
can be identified on the basis either of its initial coordinates or final coordinates,
the former being called the *Lagrangian representation*, the latter, the *Eulerian
representation*. In general, it is more convenient to use the Lagrangian coordinates because the undeformed structure will possess some regularities or
symmetries which make it susceptible to description by a common coordinate
frame such as rectangular Cartesian, cylindrical, or spherical. The deformed
structure usually does not possess any regularities or symmetries.

The developments are to be undertaken with respect to a rectangular Cartesian
coordinate frame y_1, y_2, and y_3. In Fig. 5.2, point p with coordinates y_1, y_2, y_3
is a material point in the undeformed structure, and point P with coordinates
Y_1, Y_2, Y_3 is the location of the same particle of material after deformation.
Both y_n and Y_n are measured with respect to the same origin O and same system
of axes. A neighboring point with coordinates $y_1 + dy_1, y_2 + dy_2, y_3 + dy_3$

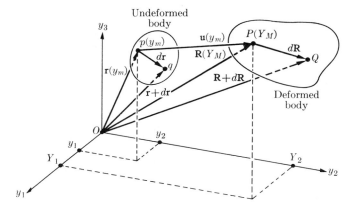

FIG. 5.2   Position vectors to undeformed and deformed body.

is identified as q, and its final position is Q with coordinates $Y_1 + dY_1$, $Y_2 + dY_2$, $Y_3 + dY_3$.

The position vector to p is denoted by $\mathbf{r}$ and its component form is

$$\mathbf{r} = y_m \mathbf{i}_m, \tag{5.2.1}$$

where the $\mathbf{i}_m$'s are the unit vectors of the y_n-frame. Similarly, $\mathbf{R}$ is the position vector to point P in the deformed body with components given by

$$\mathbf{R} = Y_m \mathbf{i}_m. \tag{5.2.2}$$

It is assumed that the particle of material moves continuously from point p to P, and that the final coordinates Y_m must be continuous and differentiable functions of the initial coordinates y_m. This is specified mathematically by writing

$$Y_m = Y_m(y_1, y_2, y_3). \tag{5.2.3}$$

If the loads or other causes of distortion do not rupture the structure, and this will be assumed, then Eq. (5.2.3) can be solved to yield the inverse relation,*

$$y_m = y_m(Y_1, Y_2, Y_3). \tag{5.2.4}$$

The differential element of length in the undeformed body is the distance between points p and q, its square being given by

$$(ds)^2 = d\mathbf{r} \cdot d\mathbf{r}. \tag{5.2.5}$$

---

* A necessary and sufficient condition that Eq. (5.2.3) be solvable for y_m, i.e., for Eq. (5.2.4) to exist, is that the Jacobian shall not vanish. Further information on tensors is to be found in the mathematical appendix.

The differential of the position vector is

$$dr = \frac{\partial r}{\partial y_m} \, dy_m. \tag{5.2.6}$$

The vectors $\partial r / \partial y_m$ are called the *base vectors* of the coordinate frame y_n. Since the y_n's form a rectangular Cartesian frame, we have

$$\frac{\partial r}{\partial y_m} = i_m \tag{5.2.7}$$

and

$$(ds)^2 = dy_m \, dy_m. \tag{5.2.8}$$

Equation (5.2.8) owes its simplicity to the use of rectangular Cartesian co-ordinates. It is a special case of the more general quadratic form

$$(ds)^2 = g_{mn} \, dx_m \, dx_n, \tag{5.2.9}$$

where the collection of the g_{mn}'s is called the *fundamental metric tensor*, and the x_m's are *general curvilinear coordinates* (which need not be orthogonal). The fundamental metric tensor has the important property that

$$g_{mn} = g_{nm}, \tag{5.2.10}$$

which means that g_{mn} is a symmetric tensor.

The general form, Eq. (5.2.9), is suitable regardless of the coordinate frame employed. Simplifications will result for special cases. Thus, if the coordinate frame is orthogonal,

$$g_{mn} = 0, \qquad \text{for} \quad m \neq n, \tag{5.2.11}$$

and if, in addition to being orthogonal, the frame is also Cartesian, then

$$g_{mn} = \delta_{mn}, \tag{5.2.12}$$

where δ_{mn} is the Kronecker delta symbol. Thus for rectangular Cartesian co-ordinates, Eq. (5.2.9) is written as

$$(ds)^2 = \delta_{mn} \, dy_m \, dy_n, \tag{5.2.13}$$

in which additional emphasis is placed on the specialized nature of the frame by using y_m instead of x_m. The form of Eq. (5.2.13) will seem to be a somewhat abstruse way to express a rather simple result (Eq. 5.2.8), but the use of Eq. (5.2.13) will put the student in the proper frame of mind for the eventual transition to general curvilinear coordinates.

Since the final coordinates are continuous functions of the initial coordinates (cf. Eq. 5.2.3), the differential of **R** can be written as

$$dR = \frac{\partial R}{\partial y_m} \, dy_m, \tag{5.2.14}$$

where the vectors $\partial\mathbf{R}/\partial y_m$ are called the *base vectors* of the deformed body. In tensor analysis these base vectors are given the symbol $\mathbf{G}_m$.

Now, the square of the line element in the deformed body is

$$(dS)^2 = \left(\frac{\partial\mathbf{R}}{\partial y_m}\,dy_m\right) \cdot \left(\frac{\partial\mathbf{R}}{\partial y_n}\,dy_n\right), \tag{5.2.15}$$

which can be put into the form

$$(dS)^2 = G_{mn}\,dy_m\,dy_n, \tag{5.2.16}$$

where the functions denoted by G_{mn} are given by the scalar products,

$$G_{mn} = \frac{\partial\mathbf{R}}{\partial y_m} \cdot \frac{\partial\mathbf{R}}{\partial y_n}. \tag{5.2.17}$$

The collection of the G_{mn}'s is the fundamental metric tensor of the deformed body with respect to the Lagrangian coordinates. It should be carefully observed that this fundamental metric tensor of the deformed body is also symmetric, that is,

$$G_{mn} = G_{nm}. \tag{5.2.18}$$

5.3   THE STRAIN TENSOR

The squares of the line elements in the undeformed and deformed body have been defined in Eqs. (5.2.13) and (5.2.16) in terms of the initial (Lagrangian) coordinates. It will be found convenient and expedient to introduce the concept of a *strain tensor*, which will be denoted by γ_{mn}. The strain tensor is introduced by writing the difference of the squares of the undeformed and deformed line elements in the convenient mathematical form

$$2\gamma_{mn}\,dy_m\,dy_n = (dS)^2 - (ds)^2. \tag{5.3.1}$$

The above definition, which at first glance may appear to be clumsy, will later prove convenient for mathematical purposes. Equation (5.3.1) means that the strain tensor is defined as one-half the difference between the deformed and undeformed metric tensors (cf. Eqs. 5.2.13 and 5.2.16),*

$$\gamma_{mn} = \tfrac{1}{2}(G_{mn} - \delta_{mn}). \tag{5.3.2}$$

---

* We will discover in our study of the strain tensor that the definition given by Eqs. (5.3.1) and (5.3.2) is well suited for most of the engineering problems concerned with small strains. For the analysis of large strains, it has been found that the Eulerian approach is more convenient. For example, the Eulerian approach has been used more extensively in the treatment of fluid mechanics.

The strain tensor, like the fundamental metric tensors, is symmetric, so that

$$\gamma_{mn} = \gamma_{nm}, \tag{5.3.3}$$

because both G_{mn} and δ_{mn} are symmetric. Thus, although there are nine elements in the strain tensor, there are only six independent ones.

5.4 THE GEOMETRICAL MEANING OF THE STRAIN TENSOR

The vector, $d\mathbf{r}$, which joins two neighboring points in the undeformed body can be enclosed by a rectangular parallelepiped with edges represented by

$$d\mathbf{r} = dy_m \frac{\partial \mathbf{r}}{\partial y_m} = dy_m \mathbf{i}_m. \tag{5.4.1}$$

This is shown in Fig. 5.3 with a perspective which is different from that used in Fig. 5.2. The lengths of the vectors $dy_1\mathbf{i}_1$, $dy_2\mathbf{i}_2$, and $dy_3\mathbf{i}_3$ are shown in Fig. 5.3 as the edges pa, pb, and pc, respectively. Similarly, the vector $d\mathbf{R}$, which joins the points P and Q in the deformed body, is enclosed in a parallelepiped which is no longer rectangular.*

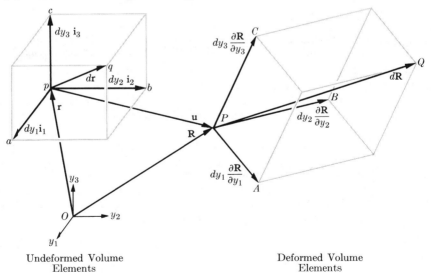

Undeformed Volume
Elements

Deformed Volume
Elements

FIG. 5.3 Vector representations of edges of undeformed and deformed volume elements.

---

* The element of volume surrounding $d\mathbf{R}$ is a parallelepiped only in the limit as $|d\mathbf{R}|$ approaches zero.

Since the Lagrangian approach has been adopted, it is informative to consider the rectangular parallelepiped $pabc$ as the reference element of volume. The process of deformation distorts this element of volume such that the corners $pabc$ become $PABC$, respectively. It follows that this figure must be considered the element of volume in the deformed body.

The geometrical meaning of the components of the strain tensor can best be appreciated by finding the changes in length of the edges of the element of volume, and the changes in angles formed by the edges. Thus the change in length of edge pa as it is deformed into PA is given by $dS_1 - ds_1$, where dS_1 denotes the length of PA and ds_1 is dy_1, the length of pa. It is more convenient to deal with the relative elongation defined as

$$E_1 = \frac{dS_1 - ds_1}{ds_1} = \frac{dS_1}{ds_1} - 1 \tag{5.4.2}$$

than with the actual change in length. Note carefully that the reference length is the original length which is natural for the Lagrangian formulation.* In words, Eq. (5.4.2) states

$$E_1 = \frac{\text{change in length of } pa}{\text{original length of } pa} = \frac{|PA| - |pa|}{|pa|}, \tag{5.4.3}$$

where $|PA|$ and $|pa|$ represent the lengths of the edges PA and pa, respectively, of Fig. 5.3. By making use of the scalar product and Eq. (5.2.17) we have

$$|PA|^2 = \frac{\partial \mathbf{R}}{\partial y_1} \, dy_1 \cdot \frac{\partial \mathbf{R}}{\partial y_1} \, dy_1 = G_{11}(dy_1)^2. \tag{5.4.4}$$

Similarly, the original length is given by

$$|pa|^2 = \mathbf{i}_1 \, dy_1 \cdot \mathbf{i}_1 \, dy_1 = (dy_1)^2. \tag{5.4.5}$$

With the substitution of Eqs. (5.4.4) and (5.4.5), Eq. (5.4.3) becomes

$$E_1 = \sqrt{G_{11}} - 1. \tag{5.4.6}$$

By definition (see Eq. 5.3.2) the component of the strain tensor γ_{11} is the one related to G_{11},

$$\gamma_{11} = \tfrac{1}{2}(G_{11} - 1). \tag{5.4.7}$$

The replacement of G_{11} in Eq. (5.4.6) by its equivalent from Eq. (5.4.7) yields the desired result,

$$E_1 = \frac{dS_1}{ds_1} - 1 = \sqrt{1 + 2\gamma_{11}} - 1. \tag{5.4.8}$$

---

* In the Eulerian formulation, the reference length would be the final length dS_1, which would then appear in the denominator.

Similarly, the relative elongations of the other two edges can be obtained:

$$E_2 = \frac{dS_2}{ds_2} - 1 = \sqrt{1 + 2\gamma_{22}} - 1, \qquad (5.4.9)$$

$$E_3 = \frac{dS_3}{ds_3} - 1 = \sqrt{1 + 2\gamma_{33}} - 1. \qquad (5.4.10)$$

The changes in length of the line elements along the coordinate directions $\mathbf{i}_1$, $\mathbf{i}_2$, $\mathbf{i}_3$ are thus seen to be related to the components of the strain tensor γ_{11}, γ_{22}, and γ_{33}, respectively. These are generally referred to as the *extensional components* of strain.

Next, the changes in angles will be calculated in terms of the strain tensor. Since the angle between edges pa and pb is a right angle, it is convenient and customary to denote the angle between PA and PB as $(\pi/2 - \phi_{12})$. Thus ϕ_{12} is the change which the originally right angle undergoes during deformation. The definition of the scalar product yields for this angle:

$$\cos\left(\frac{\pi}{2} - \phi_{12}\right) = \frac{[(\partial\mathbf{R}/\partial y_1)\, dy_1] \cdot [(\partial\mathbf{R}/\partial y_2)\, dy_2]}{(dS_1)(dS_2)}, \qquad (5.4.11)$$

wherein use has again been made of the fact that $(\partial\mathbf{R}/\partial y_1)\, dy_1$ and $(\partial\mathbf{R}/\partial y_2)\, dy_2$ are the vector representations of the edges PA and PB. The numerator of Eq. (5.4.11) is included in Eq. (5.2.17) as

$$G_{12} = \frac{\partial\mathbf{R}}{\partial y_1} \cdot \frac{\partial\mathbf{R}}{\partial y_2}, \qquad (5.4.12)$$

and the denominator contains (see Eqs. 5.4.8 and 5.4.9)

$$dS_1 = \sqrt{1 + 2\gamma_{11}}\, dy_1 = (1 + E_1)\, dy_1, \qquad (5.4.13)$$

$$dS_2 = \sqrt{1 + 2\gamma_{22}}\, dy_2 = (1 + E_2)\, dy_2. \qquad (5.4.14)$$

If these are substituted in Eq. (5.4.11), it becomes

$$\cos\left(\frac{\pi}{2} - \phi_{12}\right) = \sin\phi_{12} = \frac{2\gamma_{12}}{(1 + E_1)(1 + E_2)}, \qquad (5.4.15)$$

where (see Eq. 5.3.2)

$$\gamma_{12} = \tfrac{1}{2}G_{12}. \qquad (5.4.16)$$

Similarly, the changes which angles bpc and cpa undergo during deformation are given by

$$\sin\phi_{23} = \frac{2\gamma_{23}}{(1 + E_2)(1 + E_3)}, \qquad (5.4.17)$$

$$\sin\phi_{31} = \frac{2\gamma_{31}}{(1 + E_3)(1 + E_1)}. \qquad (5.4.18)$$

Thus the angular changes between adjacent edges of the rectangular parallelepiped are related to the components γ_{12}, γ_{23}, and γ_{31}, as well as to the relative elongations. These components of the strain tensor are referred to as the *shearing components*.

In the preceding few paragraphs, the components of the strain tensor have been given geometrical substance which is certainly not evident in the definition, Eq. (5.3.1). It has been shown that those components which have both subscripts identical, such as γ_{11}, γ_{22}, γ_{33}, are related to the changes in length along the coordinate axes. Those components which have two different subscripts, such as γ_{12}, γ_{23}, γ_{31}, are related to the changes in angle between line segments which were originally oriented along the coordinate axes. The geometrical interpretations which have been presented may not be entirely satisfying, since the relative elongations E_n are not linearly related to the strain tensor. Also, the angular interpretations involve a trigonometric function as well as the relative elongations. A more explicit and hence more satisfying geometrical picture will result when the magnitudes of the changes in length and angle are suitably restricted, as will be done in the following section. At this stage of the development, however, no restrictions have been placed on the strain tensor. It is equally applicable to materials such as steel or rubber, which differ widely in response to forces.

5.5 SMALL STRAIN

The magnitudes of the relative elongations E_1, E_2, and E_3, as well as the changes in angles ϕ_{12}, ϕ_{23}, and ϕ_{31}, encountered in engineering practice are generally very small. For example, the largest relative elongation in the wing of a transport airplane even in very turbulent weather will be less than 0.006. Its largest value in a steel railroad bridge under a locomotive will be less than 0.001. Even if rupture of the wing or of a steel girder in the bridge were to take place, the largest relative elongations would still be small compared to unity. The changes in angle are of a correspondingly small order of magnitude. Thus the strains encountered are usually small in structures of high-strength materials such as metals. Obviously, the relative elongations in a rubber balloon will be much larger and, indeed, are of three or four orders of magnitude greater. However, for the usual engineering materials under nonfailure conditions the strains encountered in the usual structure are on the order of 10^{-3}, and such strains certainly merit the appellation of "small" strains. With this restriction placed on the magnitude, tremendous simplifications may be made in strain analysis.

The simplification and the resulting analytical benefits arise from the direct geometrical meanings which can be assigned to the components of the strain tensor. Previously [see Eqs. (5.4.8), (5.4.9), (5.4.10), (5.4.15), (5.4.17), and (5.4.18)], the components γ_{mm} (no sum) were related in not too simple a fashion to the relative elongations, E_m; and the components, γ_{mn} ($m \neq n$) were related in a comparatively complex fashion to the changes in angles, ϕ_{mn}. If Eq. (5.4.8) is rearranged into the form

$$E_1(E_1 + 2) = 2\gamma_{11},\tag{5.5.1}$$

then under the assumption of small strains,

$$E_1 \ll 1 < 2,\tag{5.5.2}$$

and therefore

$$\gamma_{11} \cong E_1.\tag{5.5.3}$$

Similar results are obtained for γ_{22} and γ_{33}:

$$\gamma_{22} \cong E_2,\tag{5.5.4}$$

$$\gamma_{33} \cong E_3.\tag{5.5.5}$$

Thus the assumption of small strains means that the components of the Cartesian strain tensor, γ_{11}, γ_{22}, and γ_{33}, are the relative elongations E_1, E_2, E_3, respectively:

$$\gamma_{mm} \cong \frac{dS_m}{ds_m} - 1 \quad \text{(no sum)}.\tag{5.5.6}$$

In Eq. (5.4.15), the sine of the angle can be replaced by the angle itself for small strain. Also, the denominator on the right-hand side of the equation can be replaced by unity, because both E_1 and E_2 are small compared to unity. There results for Eq. (5.4.15) the simpler expression

$$\gamma_{12} \cong \tfrac{1}{2}\phi_{12}.\tag{5.5.7}$$

Similarly,

$$\gamma_{23} \cong \tfrac{1}{2}\phi_{23},\tag{5.5.8}$$

$$\gamma_{31} \cong \tfrac{1}{2}\phi_{31}.\tag{5.5.9}$$

Thus under the assumption of small strains, the components of the strain tensor, γ_{12}, γ_{23}, and γ_{31}, are one-half of the changes in angles ϕ_{12}, ϕ_{23}, and ϕ_{31}, respectively.

The results of the "small strain" assumption are summarized in pictorial form in Fig. 5.4. This figure is similar to Fig. 5.3, except that the vectors have been omitted. In their place the scalar quantities which are the lengths and angles of the deformed element of volume under the restriction of small strain have been written.

A few remarks should be made at this point concerning semantics. As originally introduced and without the "smallness" restriction, γ_{mn} was called the *strain tensor*. In the ordinary usage of engineering, the term "strain" is used to

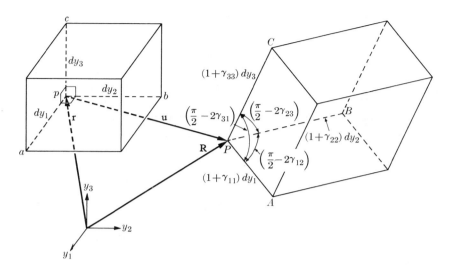

Fɪɢ. 5.4   Pictorial representation of components of small strain.

describe the nondimensional quantities and has been denoted herein by E_m and ϕ_{mn}. That is, the term "strain" refers to the relative changes in length and/or the changes which orthogonal directions in the unstrained body undergo. Thus it is customary to discuss the extensional or normal strains (which herein are E_1, E_2, E_3) and the shear strains (which herein are ϕ_{12}, ϕ_{23}, ϕ_{31}). The major part of engineering can restrict itself to a study of small strains and within this restriction, γ_{11}, γ_{22}, and γ_{33} are called the *extensional* or *normal strains*, and γ_{12}, γ_{23}, and γ_{31} are called the *shear strains*.

It appears worth while to mention the difference between the so-called engineering definition of strain components and the definition of strain tensor adopted in the present text. In most classical treatments relating to the theory of elasticity and plasticity published in the United States (for example, Refs. 1, 2, and 3), x, y, and z are often used as the rectangular Cartesian coordinates. For small strains, the normal strains are defined in the same way as the relative elongations of the elements along the x-, y-, and z-directions. They are usually denoted by ϵ_x, ϵ_y, and ϵ_z, respectively. The shear strains, in the engineering notation, are defined as the changes in angles between lines which are along the x-, y-, and z-axes prior to the deformation. They are denoted by γ_{xy}, γ_{yz}, and γ_{xz}, respectively. It is seen that if the x-, y-, and z-axes coincide with the y_1-, y_2-, and y_3-axes, respectively, the following relations exist:

$$\epsilon_x \equiv \gamma_{11}, \qquad \gamma_{xy} \equiv 2\gamma_{12},$$
$$\epsilon_y \equiv \gamma_{22}, \qquad \gamma_{yz} \equiv 2\gamma_{23}, \qquad (5.5.10)$$
$$\epsilon_z \equiv \gamma_{33}, \qquad \gamma_{xz} \equiv 2\gamma_{31}.$$

5.6 THE STRAIN TRANSFORMATION LAWS

The components of the strain tensor have been defined by means of Eq. (5.3.1) in a rectangular Cartesian coordinate frame of reference y_1, y_2, y_3. It is of great practical importance, as well as of theoretical interest, to ascertain the manner in which the strain tensor changes under a coordinate transformation. The line elements PQ and pq (see Fig. 5.2) in the deformed and undeformed body remain the same irrespective of the coordinate system used. In other words, these are invariant quantities with respect to a transformation of coordinates. Consequently in a new coordinate system, $\tilde{y}_1$, $\tilde{y}_2$, $\tilde{y}_3$, Eq. (5.3.1) becomes

$$2\tilde{\gamma}_{mn}(\tilde{y}_1, \tilde{y}_2, \tilde{y}_3)\, d\tilde{y}_m\, d\tilde{y}_n = (dS)^2 - (ds)^2, \tag{5.6.1}$$

and therefore the left-hand side of Eq. (5.6.1) is equal to the left-hand side of Eq. (5.3.1):

$$\tilde{\gamma}_{mn}(\tilde{y}_1, \tilde{y}_2, \tilde{y}_3)\, d\tilde{y}_m\, d\tilde{y}_n = \gamma_{mn}(y_1, y_2, y_3)\, dy_m\, dy_n. \tag{5.6.2}$$

The differentials, dy_m, are tensors of order one, and transform according to Eq. (A.6.2):

$$dy_m = \frac{\partial y_m}{\partial \tilde{y}_r}\, d\tilde{y}_r. \tag{5.6.3}$$

Equation (5.6.3) may also be obtained directly from the functional relations between the original and new coordinates,

$$y_m = y_m(\tilde{y}_1, \tilde{y}_2, \tilde{y}_3). \tag{5.6.4}$$

By virtue of the formula for the total derivative of a function of many variables, we have for the total differential of Eq. (5.6.4) the result given by Eq. (5.6.3). (See Ref. 5.4, Chapter 14.) Since both y_m and $\tilde{y}_r$ are rectangular Cartesian coordinates, it should be recognized that the partial derivatives $\partial y_m/\partial \tilde{y}_r$ are the direction cosines (see Appendix, Section A.4):

$$\frac{\partial y_m}{\partial \tilde{y}_r} = l_{m\tilde{r}}. \tag{5.6.5}$$

The substitution of Eqs. (5.6.4) and (5.6.2), and a minor bit of rearrangement give

$$(\gamma_{mn} l_{m\tilde{r}} l_{n\tilde{p}} - \tilde{\gamma}_{rp})\, d\tilde{y}_r\, d\tilde{y}_p = 0. \tag{5.6.6}$$

In obtaining Eq. (5.6.6) it was necessary to relabel the dummy indices in the term $\tilde{\gamma}_{mn}\, d\tilde{y}_m\, d\tilde{y}_n$ with r and p. Since $d\tilde{y}_1$, $d\tilde{y}_2$, and $d\tilde{y}_3$ are independent quantities, Eq. (5.6.6) then yields the desired strain transformation law:

$$\tilde{\gamma}_{rp}(\tilde{y}_1, \tilde{y}_2, \tilde{y}_3) = \gamma_{mn}(y_1, y_2, y_3) l_{m\tilde{r}} l_{n\tilde{p}}. \tag{5.6.7}$$

The preceding equation, which relates the components of the strain tensor in one coordinate system to the components in another coordinate system should be written out in full by the reader in order for him to fully appreciate the meaning of a second-order tensor. Observe that, in general, each component $\tilde{\gamma}_{rp}$ is related to all six components γ_{mn}, but that there are nine terms on the right-hand side of Eq. (5.6.7). Geometrical representations of special cases of Eq. (5.6.7) have been developed. One of these is known as *Mohr's circle* for strain, and is discussed in a problem at the end of the chapter.

5.7 PRINCIPAL STRAINS AND PRINCIPAL DIRECTIONS

It was stated at the beginning of this chapter that the goal of strain analysis is to specify the changes in length of all the line segments which originate at a point in the body, and the changes in angles subtended by all pairs of line segments. All this information is contained in the strain tensor, γ_{mn}. There are values of the strain tensor which merit special consideration. The points p and q (see Fig. 5.3) in the unstrained body have been arbitrarily chosen. If point p is held fixed and point q is varied such that the line segment pq assumes different directions in the body, then the quantity $(dS)^2 - (ds)^2$ will generally change in value.

Thus it is germane to seek those directions for which the relative elongation becomes an extremum, i.e., becomes a maximum and/or minimum (see Appendix, Section A.16). It will be shown that there are three such directions which are mutually orthogonal, and along which the shear components of strain are zero. These directions are called the *principal directions*, and the corresponding values of γ_{mn} are called the *principal strains*.

Let us assume that the six components of strain are known relative to the coordinates y_n. Then the component $\tilde{\gamma}_{11}$ relative to another coordinate set $\tilde{y}_n$ is given by (see Eq. 5.6.7)

$$\tilde{\gamma}_{11} = l_{\bar{1}m}l_{\bar{1}n}\gamma_{nm}. \tag{5.7.1}$$

Each admissible set of direction cosines $l_{\bar{1}m}$ corresponds to a different direction at point p of the body (see Fig. 5.3), and for each set, $\tilde{\gamma}_{11}$ will take on different values. Thus Eq. (5.7.1) can be viewed as an equation for $\tilde{\gamma}_{11}$ in which the independent variables are the three direction cosines. For emphasis of this aspect and also for ease of writing, we set

$$Q(\lambda) = \tilde{\gamma}_{11} = \lambda_m\lambda_n\gamma_{nm}, \tag{5.7.2}$$

where the direction cosines have been replaced by

$$\lambda_m = l_{\bar{1}m}. \tag{5.7.3}$$

It is observed that the trio $\lambda_1, \lambda_2, \lambda_3$ locates the orientation of the line segment pq relative to the axes y_1, y_2, and y_3, and this trio must satisfy the condition

$$(\lambda_1)^2 + (\lambda_2)^2 + (\lambda_3)^2 = \delta_{mn}\lambda_m\lambda_n = 1. \tag{5.7.4}$$

The relative elongation (see Eq. 5.4.8) in the $\tilde{y}_1$-direction is

$$\widetilde{E}_1 = \sqrt{1 + 2\tilde{\gamma}_{11}} - 1. \tag{5.7.5}$$

We see from this expression that $\widetilde{E}_1$ will be an extremum when the quantity $Q(\lambda)$ is an extremum.

The extreme-value problem to be solved can now be stated in terms of $Q(\lambda)$ because the minimal and maximal values of $\tilde{\gamma}_{11}$ are directly related to the extreme value of $\widetilde{E}_1$. It is desired to find the values of $\lambda_1, \lambda_2, \lambda_3$ for which $Q(\lambda)$ assumes its largest and/or smallest value. However, not all values of $\lambda_1, \lambda_2, \lambda_3$ are admissible because Eq. (5.7.4) must be satisfied. The usual procedure for handling problems of this kind can be followed if Eq. (5.7.4) is used to eliminate one of the direction cosines from Eq. (5.7.2). Then the resulting equation will contain only two direction cosines, and the application of the necessary conditions, $\partial Q/\partial \lambda_m = 0$, will lead to a pair of simultaneous equations. However, there is a more convenient and elegant manner by which we may deal with the condition given by Eq. (5.7.4), which is known as a *constraint*. The determination of the extreme values of a quantity subject to a constraint is easily accomplished by Lagrange's method of undetermined multipliers (see Appendix, Section A.16). This method preserves the symmetries in the mathematics, in that no preference is given to any one of the variables $\lambda_1, \lambda_2,$ or λ_3. The constraint, Eq. (5.7.4), is put into the form

$$P(\lambda) = \delta_{mn}\lambda_m\lambda_n - 1 = 0. \tag{5.7.6}$$

A parameter α called the *Lagrange multiplier*, which is independent of the direction cosines, is introduced as an unknown factor.

It is shown in the Appendix that the necessary conditions for $Q(\lambda)$ to have an extreme value are given by the equations

$$\frac{\partial Q}{\partial \lambda_m} - \alpha\frac{\partial P}{\partial \lambda_m} = 0. \tag{5.7.7}$$

Substitution of Eqs. (5.7.2) and (5.7.6) yields the set of homogeneous algebraic equations

$$(\gamma_{mn} - \alpha\delta_{mn})\lambda_n = 0. \tag{5.7.8}$$

Equations (5.7.8) constitute three simultaneous equations for $\lambda_1, \lambda_2,$ and λ_3, and because these are homogeneous, a nontrivial solution can exist if and only if the determinant of the coefficients is zero:

$$\begin{vmatrix} \gamma_{11} - \alpha & \gamma_{12} & \gamma_{13} \\ \gamma_{21} & \gamma_{22} - \alpha & \gamma_{23} \\ \gamma_{31} & \gamma_{32} & \gamma_{33} - \alpha \end{vmatrix} = 0. \tag{5.7.9}$$

The expansion of the determinant yields a cubic equation in α, and because the determinant is symmetric, there are three real principal values. The expanded determinant can be written as

$$-(\alpha)^3 + J_1(\alpha)^2 - J_2\alpha + J_3 = 0, \qquad (5.7.10)$$

in which J_1, J_2, J_3 are called the first, second, and third strain invariants, respectively. That J_1, J_2, J_3 are invariants will be subsequently demonstrated. It can be verified by the expansion of Eq. (5.7.9) that

$$J_1 = \gamma_{11} + \gamma_{22} + \gamma_{33}, \qquad (5.7.11)$$

$$J_2 = \begin{vmatrix} \gamma_{11} & \gamma_{12} \\ \gamma_{21} & \gamma_{22} \end{vmatrix} + \begin{vmatrix} \gamma_{11} & \gamma_{13} \\ \gamma_{31} & \gamma_{33} \end{vmatrix} + \begin{vmatrix} \gamma_{22} & \gamma_{23} \\ \gamma_{32} & \gamma_{33} \end{vmatrix}, \qquad (5.7.12)$$

$$J_3 = \begin{vmatrix} \gamma_{11} & \gamma_{12} & \gamma_{13} \\ \gamma_{21} & \gamma_{22} & \gamma_{23} \\ \gamma_{31} & \gamma_{32} & \gamma_{33} \end{vmatrix}. \qquad (5.7.13)$$

The three values of α which are the roots of the cubic equation (5.7.10) will be labeled with Roman numeral subscripts as α_I, α_{II}, and α_{III}. (We will presently demonstrate that α_I, α_{II}, and α_{III} are the sought-after principal strains.) In order to find the corresponding principal directions, each of the roots is substituted individually back into Eqs. (5.7.8). The principal directions will be identified by a Roman numeral superscript. Thus $\lambda_1^{(I)}$, $\lambda_2^{(I)}$, $\lambda_3^{(I)}$ are the direction cosines of the direction along which the root α_I (principal strain) occurs. To find this triplet of numbers, α_I is put into Eqs. (5.7.8) yielding

$$(\gamma_{11} - \alpha_I)\lambda_1^{(I)} + \gamma_{12}\lambda_2^{(I)} + \gamma_{13}\lambda_3^{(I)} = 0,$$
$$\gamma_{21}\lambda_1^{(I)} + (\gamma_{22} - \alpha_I)\lambda_2^{(I)} + \gamma_{23}\lambda_3^{(I)} = 0, \qquad (5.7.14)$$
$$\gamma_{31}\lambda_1^{(I)} + \gamma_{32}\lambda_2^{(I)} + (\gamma_{33} - \alpha_I)\lambda_3^{(I)} = 0.$$

However, these represent only two independent equations for $\lambda_m^{(I)}$, because the determinant of the coefficients of $\lambda_m^{(I)}$ is zero. The third independent equation is the constraint, Eq. (5.7.4), which now becomes

$$[\lambda_1^{(I)}]^2 + [\lambda_2^{(I)}]^2 + [\lambda_3^{(I)}]^2 = 1. \qquad (5.7.15)$$

The actual computational procedure would be to solve for $\lambda_1^{(I)}$ and $\lambda_2^{(I)}$ in terms of $\lambda_3^{(I)}$ with any pair of Eqs. (5.7.14), and then to use Eq. (5.7.15) to calculate an explicit value for $\lambda_3^{(I)}$. The value of $\lambda_3^{(I)}$ would then be substituted back into Eqs. (5.7.14) to calculate explicit values for $\lambda_1^{(I)}$ and $\lambda_2^{(I)}$.

Once the direction cosines $\lambda_m^{(I)}$ have been determined, then the principal strain γ_I, which is $Q(\lambda_m^{(I)})$, is calculated from

$$\gamma_I = \gamma_{mn}\lambda_m^{(I)}\lambda_n^{(I)}. \tag{5.7.16}$$

There remains the task of demonstrating that the Lagrange multiplier α_I is the principal strain γ_I. Equation (5.7.8) is true for α_I and $\lambda_n^{(I)}$,

$$(\gamma_{mn} - \alpha_I\,\delta_{mn})\lambda_n^{(I)} = 0. \tag{5.7.17}$$

We multiply Eq. (5.7.17) by $\lambda_m^{(I)}$ and sum on m to obtain

$$\gamma_{mn}\lambda_n^{(I)}\lambda_m^{(I)} = \alpha_I\,\delta_{mn}\lambda_n^{(I)}\lambda_m^{(I)}. \tag{5.7.18}$$

The left-hand side of Eq. (5.7.18) is γ_I as indicated by Eq. (5.7.16), and the right-hand side by virtue of the properties of the direction cosines (see Eq. 5.7.4) reduces to α_I. Therefore Eq. (5.7.18) is the desired proof that

$$\gamma_I \equiv \alpha_I. \tag{5.7.19}$$

In a similar manner the equivalence of γ_{II} and α_{II}, as well as of γ_{III} and α_{III}, can be demonstrated. The three values are usually arranged such that

$$\gamma_I > \gamma_{II} > \gamma_{III}. \tag{5.7.20}$$

In factored form, the cubic equation (5.7.10) becomes

$$-(\gamma - \gamma_I)(\gamma - \gamma_{II})(\gamma - \gamma_{III}) = 0. \tag{5.7.21}$$

The strain invariants can then be expressed in terms of the principal strains:

$$J_1 = \gamma_I + \gamma_{II} + \gamma_{III}, \tag{5.7.22}$$

$$J_2 = \gamma_I\gamma_{II} + \gamma_{II}\gamma_{III} + \gamma_{III}\gamma_I, \tag{5.7.23}$$

$$J_3 = \gamma_I\gamma_{II}\gamma_{III}. \tag{5.7.24}$$

The fact that J_1, J_2, and J_3 are invariants should now be apparent. It is evident that the magnitudes of the principal strains γ_I, γ_{II}, γ_{III} are independent of the coordinate frame which is chosen, and therefore J_1, J_2, and J_3, the coefficients of Eq. (5.7.10), must also be independent of the frame of reference.

An important property of the three principal directions is that they are mutually orthogonal. This can be demonstrated as follows: First, Eqs. (5.7.8) are written for $\lambda_n^{(I)}$ and $\lambda_n^{(II)}$,

$$(\gamma_{mn} - \gamma_I\,\delta_{mn})\lambda_n^{(I)} = 0, \tag{5.7.25}$$

$$(\gamma_{mn} - \gamma_{II}\,\delta_{mn})\lambda_n^{(II)} = 0. \tag{5.7.26}$$

Second, Eq. (5.7.25) is multiplied by $\lambda_m^{(II)}$, and Eq. (5.7.26) is multiplied by $\lambda_m^{(I)}$:

$$(\gamma_{mn} - \gamma_I \, \delta_{mn})\lambda_n^{(I)} \, \lambda_m^{(II)} = 0, \tag{5.7.27}$$

$$(\gamma_{mn} - \gamma_{II} \, \delta_{mn})\lambda_n^{(II)}\lambda_m^{(I)} = 0. \tag{5.7.28}$$

Third, Eq. (5.7.27) is subtracted from Eq. (5.7.28) to yield

$$(\gamma_I - \gamma_{II}) \, \delta_{mn}\lambda_m^{(I)}\lambda_n^{(II)} = 0. \tag{5.7.29}$$

The symmetry of the strain tensor and of the Kronecker delta has permitted a rearrangement of the dummy indices according to the scheme:

$$\gamma_{mn}\lambda_n^{(II)}\lambda_m^{(I)} = \gamma_{mn}\lambda_m^{(II)}\lambda_n^{(I)},$$

$$\delta_{mn}\lambda_m^{(II)}\lambda_n^{(I)} = \delta_{mn}\lambda_m^{(I)}\lambda_n^{(II)}.$$

Finally, if the principal strains γ_I and γ_{II} are unequal, then Eq. (5.7.29) can be satisfied only if

$$\delta_{mn}\lambda_m^{(I)}\lambda_n^{(II)} = 0, \tag{5.7.30}$$

which can be recognized as the condition that the principal directions $\lambda_m^{(I)}$ and $\lambda_n^{(II)}$ are orthogonal (see Eq. A.5.12). A similar series of algebraic manipulations will demonstrate the orthogonalities between $\lambda_m^{(III)}$ and $\lambda_n^{(I)}$, and between $\lambda_m^{(III)}$ and $\lambda_n^{(II)}$.

Another important attribute of the principal directions is that the components of shear strain belonging to the orthogonal axes given by $\lambda_m^{(I)}$, $\lambda_m^{(II)}$, and $\lambda_m^{(III)}$ are all identically zero. Equation (5.7.25) can be rearranged with the aid of Eq. (5.7.3) to read

$$\gamma_{mn}l_{\bar{1}n} = \gamma_I \, \delta_{mn}l_{\bar{1}n}. \tag{5.7.31}$$

The multiplication of Eq. (5.7.31) by $l_{\bar{s}m}$ leads to

$$\gamma_{mn}l_{\bar{1}n}l_{\bar{s}m} = \gamma_I\delta_{mn}l_{\bar{1}n}l_{\bar{s}m}. \tag{5.7.32}$$

We make the observation that the left-hand side is the transformation law for the components represented by $\tilde{\gamma}_{1s}$ (see Eq. 5.6.2) while the right-hand side reduces to $\gamma_I\delta_{1s}$, that is,

$$\tilde{\gamma}_{1s} = \gamma_I \, \delta_{1s}, \tag{5.7.33}$$

where the subscripts 1 and s refer to the principal axes. Clearly, $\tilde{\gamma}_{12}$ and $\tilde{\gamma}_{13}$ (which we may also cumbersomely write as $\gamma_{I,II}$ and $\gamma_{I,III}$, respectively) are zero. In a similar manner the shear component, $\gamma_{II,III}$, can be shown to be zero.

The vanishing of the shear strains can also be shown by choosing the principal directions as the reference rectangular Cartesian frame (see Fig. 5.5) in which the coordinate axes y_I, y_{II}, and y_{III} represent the principal direction at point p.

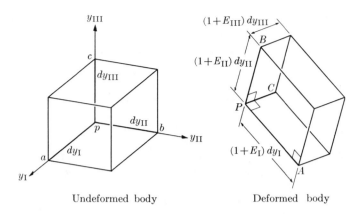

Undeformed body   Deformed body

FIG. 5.5   Absence of shear strains for principal axes.

This is permissible since the principal directions are orthogonal. Roman numerals will be used to denote this special set of coordinates, and the corresponding components of the strain tensor will be denoted by $\gamma_{I,I}$, $\gamma_{I,II}$, $\gamma_{I,III}$, etc. However, a comparison of the two sets of expressions [Eqs. (5.7.22) through (5.7.24) and Eqs. (5.7.11) through (5.7.13)] for the strain invariants shows that the only nonzero components of the strain tensor, which is defined with respect to the principal directions, are $\gamma_{I,I}$, $\gamma_{II,II}$, and $\gamma_{III,III}$. More specifically, if Eqs. (5.7.13) and (5.7.24), which define the strain invariants, are compared, it is evident that

$$J_3 = \gamma_I \gamma_{II} \gamma_{III} = \begin{vmatrix} \gamma_I & 0 & 0 \\ 0 & \gamma_{II} & 0 \\ 0 & 0 & \gamma_{III} \end{vmatrix}. \tag{5.7.34}$$

This result means that the components which would be labeled $\gamma_{I,II}$, $\gamma_{II,III}$, $\gamma_{III,I}$ are zero, and that

$$\gamma_I \equiv \gamma_{I,I}, \qquad \gamma_{II} \equiv \gamma_{II,II}, \qquad \gamma_{III} \equiv \gamma_{III,III} \tag{5.7.35}$$

are the only nonzero components of strain with respect to the principal directions. Thus it has been shown that the strain tensor, if referred to the principal directions, consists only of the three extensional components.

Furthermore, we have found (see Eq. 5.4.15) that

$$\sin \phi_{I,II} = \frac{2\gamma_{I,II}}{(1 + E_I)(1 + E_{II})}, \tag{5.7.36}$$

where $\phi_{I,II}$ is defined as the change which the originally right angle apb undergoes during deformation. Since $\gamma_{I,II}$ is zero, then $\phi_{I,II}$ is also zero. Similar conclusions can be drawn for $\phi_{II,III}$ and $\phi_{III,I}$. This represents another way in which the principal directions may be interpreted; i.e., an element of volume defined in the undeformed body by means of the principal directions does not undergo any shearing deformations. It follows, then, that the element of volume in the deformed state is also a rectangular parallelepiped (see Fig. 5.5).

5.8 THE STRAIN-DISPLACEMENT RELATIONS

It is desirable to introduce at this stage the displacement vector $\mathbf{u}$, which connects the initial (undeformed) position p of the material particle and its final (deformed) position P, by the vector equation (cf. Fig. 5.2):

$$\mathbf{R}(y_1, y_2, y_3) = \mathbf{r}(y_1, y_2, y_3) + \mathbf{u}(y_1, y_2, y_3). \qquad (5.8.1)$$

The component form of $\mathbf{u}$ is

$$\mathbf{u} = u_m(y_1, y_2, y_3)\mathbf{i}_m. \qquad (5.8.2)$$

At this point the reader should clearly understand what is meant by the position vector $\mathbf{R}(y_1, y_2, y_3)$. In the undeformed body, the position vector $\mathbf{r}(y_1, c_2, c_3)$ will trace out a line which is parallel to the y_1-axis if y_1 is varied while y_2 and y_3 are set equal to constants c_2 and c_3 (see Fig. 5.6). However, the position vector $\mathbf{R}$ is dependent on the displacement vector $\mathbf{u}$,

$$\mathbf{R}(y_1, c_2, c_3) = \mathbf{r}(y_1, c_2, c_3) + \mathbf{u}(y_1, c_2, c_3). \qquad (5.8.3)$$

An examination of Eq. (5.8.3) shows that $\mathbf{u}(y_1, c_2, c_3)$ can still have components in all three directions, and therefore $\mathbf{R}(y_1, c_2, c_3)$ will trace out a

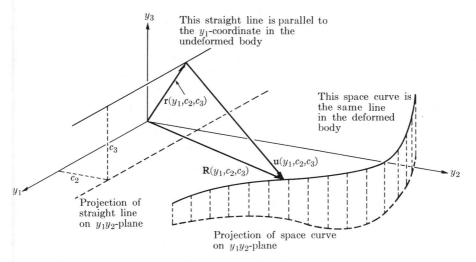

FIG. 5.6 Coordinate line in undeformed and deformed body.

curve in space. Thus, whereas points of the coordinate y_1 in the undeformed body lie on a straight line, the end-points of the position vector $\mathbf{R}(y_1, c_2, c_3)$ which identifies similar points in the deformed body, usually trace out a space curve.

Using Eq. (5.8.1) we find that the partial derivatives of the position vector $\mathbf{R}$ are

$$\frac{\partial \mathbf{R}}{\partial y_m} = \frac{\partial \mathbf{r}}{\partial y_m} + \frac{\partial \mathbf{u}}{\partial y_m} = \mathbf{1}_m + \frac{\partial u_n}{\partial y_m} \mathbf{i}_n, \tag{5.8.4}$$

where with the use of Eq. (5.8.2), we get

$$\frac{\partial \mathbf{u}}{\partial y_k} = \frac{\partial u_r}{\partial y_k} \mathbf{i}_r. \tag{5.8.5}$$

In detail, Eq. (5.8.4) becomes

$$\frac{\partial \mathbf{R}}{\partial y_1} = \left(1 + \frac{\partial u_1}{\partial y_1}\right)\mathbf{i}_1 + \frac{\partial u_2}{\partial y_1}\mathbf{i}_2 + \frac{\partial u_3}{\partial y_1}\mathbf{i}_3,$$

$$\frac{\partial \mathbf{R}}{\partial y_2} = \frac{\partial u_1}{\partial y_2}\mathbf{i}_1 + \left(1 + \frac{\partial u_2}{\partial y_2}\right)\mathbf{i}_2 + \frac{\partial u_3}{\partial y_2}\mathbf{i}_3, \tag{5.8.6}$$

$$\frac{\partial \mathbf{R}}{\partial y_3} = \frac{\partial u_1}{\partial y_3}\mathbf{i}_1 + \frac{\partial u_2}{\partial y_3}\mathbf{i}_2 + \left(1 + \frac{\partial u_3}{\partial y_3}\right)\mathbf{i}_3.$$

The fundamental metric tensor of the deformed body, G_{mn}, now can be expressed in terms of the displacement vector as

$$G_{mn} = \delta_{mn} + \frac{\partial \mathbf{u}}{\partial y_m} \cdot \mathbf{i}_n + \frac{\partial \mathbf{u}}{\partial y_n} \cdot \mathbf{i}_m + \frac{\partial \mathbf{u}}{\partial y_m} \cdot \frac{\partial \mathbf{u}}{\partial y_n}. \tag{5.8.7}$$

In order to further clarify the displacements, Fig. 5.7 is drawn with the deformed volume element moved by pure translation until points p and P coincide. The resulting figure shows the displacements of points a, b, c, relative to point p, and pictures more clearly the rotations of the line segments pa, pb, and pc as these are changed by the process of deformation into PA, PB, and PC, respectively. It can be seen that Fig. 5.7, in actuality, schematically represents the three vector equations (5.8.6). The dashed lines which are parallel to the y_n-axes are the orthogonal components of the vectors

$$\frac{\partial \mathbf{u}}{\partial y_1} dy_1, \qquad \frac{\partial \mathbf{u}}{\partial y_2} dy_2, \qquad \text{and} \qquad \frac{\partial \mathbf{u}}{\partial y_3} dy_3,$$

which connect points a to A, b to B, and c to C, respectively.

By introducing Eq. (5.8.7) we express the strain tensor in terms of the displacement vector as

$$\gamma_{mn} = \frac{1}{2}\left[\frac{\partial \mathbf{u}}{\partial y_m} \cdot \mathbf{i}_n + \frac{\partial \mathbf{u}}{\partial y_n} \cdot \mathbf{i}_m + \frac{\partial \mathbf{u}}{\partial y_m} \cdot \frac{\partial \mathbf{u}}{\partial y_n}\right], \tag{5.8.8}$$

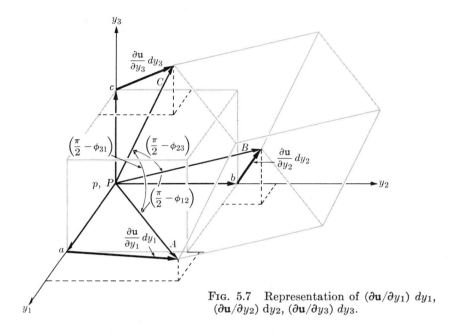

FIG. 5.7   Representation of $(\partial\mathbf{u}/\partial y_1)\,dy_1$, $(\partial\mathbf{u}/\partial y_2)\,dy_2$, $(\partial\mathbf{u}/\partial y_3)\,dy_3$.

which becomes, with the introduction of Eq. (5.8.5),

$$\gamma_{mn} = \frac{1}{2}\left(\frac{\partial u_m}{\partial y_n} + \frac{\partial u_n}{\partial y_m} + \frac{\partial u_r}{\partial y_m}\frac{\partial u_s}{\partial y_n}\delta_{rs}\right). \tag{5.8.9}$$

(See Eqs. (5.14.3b) for all six components of the strain-displacement relations given by Eq. 5.8.9)

It should be observed that the strain tensor is composed of a part which is linear in the derivatives of the displacements and a part which, since it involves the squares and products of the derivatives, is nonlinear. The linear portion is attributable to the first two terms of Eq. (5.8.8) or Eq. (5.8.9), and the non-linear to the last term.

5.9   LINEAR STRAIN

There is one further restriction which is made in the analysis of strain, and this simplification is generally referred to as *infinitesimal strain* although the term *linear strain* is more connotative of the true implication of the simplification. The linear strain case encompasses the vast majority of the literature in all aspects of elasticity (and even plasticity), not only because the mathematical aspects are less complex, but also because the linear strain case does correctly describe the behavior of most structural forms with sufficient accuracy for engineering purposes. However, it should be pointed out that whenever stability of a structural form is in question, the case of linear strain is no longer applicable.

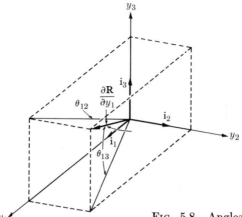

FIG. 5.8   Angles of rotation of $\partial\mathbf{R}/\partial y_1$.

The linear strain simplification neglects the nonlinear terms in the strain-displacement relations given by Eqs. (5.8.9). These are the terms involving the squares and products of the first derivatives of the components of the displacement vector $\mathbf{u}$. It can be seen that these nonlinear terms arise from the term formed by the scalar product $(\partial\mathbf{u}/\partial y_m) \cdot (\partial\mathbf{u}/\partial y_n)$ of Eq. (5.8.8); hence in the linear-strain case this term is omitted. Since linear strain requires basic modifications to the strain-displacement relations, a new symbol, ϵ_{mn}, will be used to denote the linear strain tensor. The definition for ϵ_{mn} is

$$\epsilon_{mn} = \frac{1}{2}\left(\frac{\partial\mathbf{u}}{\partial y_n}\cdot\mathbf{i}_m + \frac{\partial\mathbf{u}}{\partial y_m}\cdot\mathbf{i}_n\right), \qquad (5.9.1)$$

or

$$\epsilon_{mn} = \frac{1}{2}\left(\frac{\partial u_m}{\partial y_n} + \frac{\partial u_n}{\partial y_m}\right). \qquad (5.9.2)$$

There is a geometrical justification for neglecting the nonlinear terms. This can be shown by examining Fig. 5.8, which is a condensation of Fig. 5.3. In this figure, the vector which lies along the edge PA of the deformed element of volume has been translated so that the P-end coincides with the corner p of the undeformed element of volume. The two angles θ_{12} and θ_{13} represent the rotation of vector $\partial\mathbf{R}/\partial y_1$ out of the planes located by $\mathbf{i}_2$ and $\mathbf{i}_3$, respectively. Using the scalar product, we obtain

$$\left|\frac{\partial\mathbf{R}}{\partial y_1}\right|\cos\left(\frac{\pi}{2}-\theta_{12}\right) = \frac{\partial\mathbf{R}}{\partial y_1}\cdot\mathbf{i}_2, \qquad (5.9.3)$$

$$\left|\frac{\partial\mathbf{R}}{\partial y_1}\right|\cos\left(\frac{\pi}{2}-\theta_{13}\right) = \frac{\partial\mathbf{R}}{\partial y_1}\cdot\mathbf{i}_3. \qquad (5.9.4)$$

With the aid of Eqs. (5.2.17), (5.4.6) and (5.8.6), we can express the preceding two equations as

$$\sin \theta_{12} = \frac{1}{E_1 + 1} \frac{\partial u_2}{\partial y_1},$$ (5.9.5)

$$\sin \theta_{13} = \frac{1}{E_1 + 1} \frac{\partial u_3}{\partial y_1}.$$ (5.9.6)

If we now stipulate that the magnitude of the elongation E_1 and the angles θ_{12} and θ_{13} be small, then we can write

$$\theta_{12} \approx \frac{\partial u_2}{\partial y_1},$$ (5.9.7)

$$\theta_{13} \approx \frac{\partial u_3}{\partial y_1}.$$ (5.9.8)

In a similar manner we can derive the equation

$$\theta_{mn} \approx \frac{\partial u_n}{\partial y_m} \quad \text{for } m \neq n.$$ (5.9.9)

The case of linear strain, as has already been stated, assumes that squares and products of θ_{mn} are negligible compared to θ_{mn} itself. However, it has been demonstrated that this premise is based upon angles of rotation which are suitably small.

It may prove informative to pursue this discussion from a slightly different viewpoint. The linearization of the strain-displacement relations is usually not discussed. Instead, the analysis of strain proceeds from the assumption that the derivatives of the components of displacement are infinitesimals. (We have shown in the preceding paragraph that these need be only suitably small.) Thus the base vectors of the deformed body (see Eqs. 5.8.6) can be simplified to

$$\frac{\partial \mathbf{R}}{\partial y_1} \cong \mathbf{i}_1 + \frac{\partial u_2}{\partial y_1} \mathbf{i}_2 + \frac{\partial u_3}{\partial y_1} \mathbf{i}_3,$$

$$\frac{\partial \mathbf{R}}{\partial y_2} \cong \frac{\partial u_1}{\partial y_2} \mathbf{i}_1 + \mathbf{i}_2 + \frac{\partial u_3}{\partial y_2} \mathbf{i}_3,$$ (5.9.10)

$$\frac{\partial \mathbf{R}}{\partial y_3} \cong \frac{\partial u_1}{\partial y_3} \mathbf{i}_1 + \frac{\partial u_2}{\partial y_3} \mathbf{i}_2 + \mathbf{i}_3,$$

because $\partial u_m / \partial y_n$, being an infinitesimal, can be neglected with respect to unity. Also the edges $(\partial \mathbf{R}/\partial y_m)\, dy_m$ (no sum) of the deformed element of volume have directions which are only infinitesimally different from the directions of the corresponding edges, $dy_m \mathbf{i}_m$ (no sum), of the undeformed element of volume (see Fig. 5.3). In other words, line segments (such as pa, pb, pc) undergo only infinitesimal rotations as the body becomes deformed. This can be visualized by projecting points A, B, a, and b of Fig. 5.4 onto the $y_1 y_2$-plane for this case

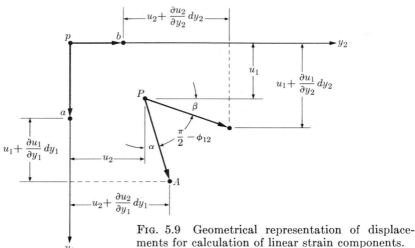

FIG. 5.9 Geometrical representation of displacements for calculation of linear strain components.

of infinitesimal deformation. In Fig. 5.9, the true lengths of lines PA and PB are shown to within an order of magnitude of an infinitesimal even though PA and PB do not lie in a plane parallel to the $y_1 y_2$-plane.

For the case at hand, the displacements and angles have been magnified for purposes of presentation. If the infinitesimal nature of the derivatives is considered, then as is done in the traditional manner, the strain in the y_1-direction is defined as

$$\epsilon_{11} = \frac{|PA| - |pa|}{|pa|} = \frac{(1 + (\partial u_1/\partial y_1))\, dy_1 - dy_1}{dy_1} = \frac{\partial u_1}{\partial y_1}, \quad (5.9.11)$$

a result which was previously derived (see Eqs. 5.9.2) in a more rigorous manner as a simplification of an exact expression. Similarly, the change in angle ϕ_{12} (see Fig. 5.9) can be defined as

$$\phi_{12} = \alpha + \beta = \frac{(\partial u_2/\partial y_1)\, dy_1}{dy_1} + \frac{(\partial u_1/\partial y_2)\, dy_2}{dy_2}, \quad (5.9.12)$$

$$\phi_{12} = \frac{\partial u_2}{\partial y_1} + \frac{\partial u_1}{\partial y_2},$$

providing the infinitesimal nature of derivatives is kept in mind. This latter result is related to the linear strain tensor ϵ_{mn} by

$$\phi_{12} = 2\epsilon_{12}. \quad (5.9.13)$$

(See Eqs. 5.5.7 and 5.9.2.) Obviously, a figure such as 5.9 can also be drawn for the $y_2 y_3$- and $y_3 y_1$-planes.

We have presented two different classifications of strain tensors: small versus large strains, and linear versus nonlinear strain-displacement relations. In general, when the relative elongations (E_n) and/or angles of shearing (ϕ_{nm})

are large, the nonlinear strain-displacement relation must be used. Thus linear strain must also imply small strain. However, there are cases in which the elongations and shearing angles are extremely small but for which the rotations of the elements are considerable. In these cases it is still necessary to use the nonlinear strain-displacement relations. It should be clear that linear and small strains are not two mutually exclusive classes of strains. The generally used term of "infinitesimal strain" is then not an appropriate one for "linear strain" because it creates the erroneous impression that "linear strain" is even smaller than the so-called "small strain."

Whether or not the linear, small, or even the large components of a strain should be used depends on (1) the materials of construction, (2) the structural configuration, and (3) the mode of structural behavior. For example, if rubber is used in an inflated structure, then clearly the strains will be large. If the structure is a diving board constructed of oak, then the theory of small strains is sufficiently accurate. If the structure is a floor beam constructed of steel, it is generally adequate to use linear strain-displacement relations to describe the behavior. For flat sheets of metal which are subjected to in-plane compressive forces, it is necessary to use nonlinear strain-displacement relations in order to account for the possibility of buckling.

5.10 THE CHANGE IN VOLUME

The element of volume in the undeformed body is the volume of the rectangular parallelepiped whose edges are shown as $dy_1\mathbf{i}_1$, $dy_2\mathbf{i}_2$, and $dy_3\mathbf{i}_3$ in Fig. 5.3, and is given by

$$dV^{(0)} = dy_1\, dy_2\, dy_3. \tag{5.10.1}$$

The element of volume in the deformed body in the limit is the parallelepiped with edges shown as $(\partial\mathbf{R}/\partial y_1)\, dy_1$, $(\partial\mathbf{R}/\partial y_2)\, dy_2$, and $(\partial\mathbf{R}/\partial y_3)\, dy_3$ (see Fig. 5.3). Using the triple scalar product, we find the deformed element of volume to be

$$dV = \left(\frac{\partial\mathbf{R}}{\partial y_1}\, dy_1\right) \cdot \left(\frac{\partial\mathbf{R}}{\partial y_2}\, dy_2\right) \times \left(\frac{\partial\mathbf{R}}{\partial y_3}\, dy_3\right). \tag{5.10.2}$$

A much more useful form of Eq. (5.10.2) will be obtained if the components of strain are introduced by means of the following development: First, the vector formula, Eq. (A.14.3) is applied to Eq. (5.10.2) to yield

$$(dV)^2 = \begin{vmatrix} \dfrac{\partial\mathbf{R}}{\partial y_1}\cdot\dfrac{\partial\mathbf{R}}{\partial y_1} & \dfrac{\partial\mathbf{R}}{\partial y_1}\cdot\dfrac{\partial\mathbf{R}}{\partial y_2} & \dfrac{\partial\mathbf{R}}{\partial y_1}\cdot\dfrac{\partial\mathbf{R}}{\partial y_3} \\[2ex] \dfrac{\partial\mathbf{R}}{\partial y_2}\cdot\dfrac{\partial\mathbf{R}}{\partial y_1} & \dfrac{\partial\mathbf{R}}{\partial y_2}\cdot\dfrac{\partial\mathbf{R}}{\partial y_2} & \dfrac{\partial\mathbf{R}}{\partial y_2}\cdot\dfrac{\partial\mathbf{R}}{\partial y_3} \\[2ex] \dfrac{\partial\mathbf{R}}{\partial y_3}\cdot\dfrac{\partial\mathbf{R}}{\partial y_1} & \dfrac{\partial\mathbf{R}}{\partial y_3}\cdot\dfrac{\partial\mathbf{R}}{\partial y_2} & \dfrac{\partial\mathbf{R}}{\partial y_3}\cdot\dfrac{\partial\mathbf{R}}{\partial y_3} \end{vmatrix} (dy_1\, dy_2\, dy_3)^2. \tag{5.10.3}$$

Second, the fundamental metric tensor, G_{mn}, of the deformed body is used to replace the scalar products in Eq. (5.10.3). (See Eq. 5.2.17):

$$(dV)^2 = \begin{vmatrix} G_{11} & G_{12} & G_{13} \\ G_{21} & G_{22} & G_{23} \\ G_{31} & G_{32} & G_{33} \end{vmatrix} (dy_1 \, dy_2 \, dy_3)^2. \tag{5.10.4}$$

Third, the determinant is set equal to G,

$$G = \begin{vmatrix} G_{11} & G_{12} & G_{13} \\ G_{21} & G_{22} & G_{23} \\ G_{31} & G_{32} & G_{33} \end{vmatrix}, \tag{5.10.5}$$

with the result that

$$dV = \sqrt{G} \, dy_1 \, dy_2 \, dy_3. \tag{5.10.6}$$

Fourth, the strain tensor is made to appear in the determinant G by means of Eq. (5.3.2). This may be expressed in the determinant form

$$G = \begin{vmatrix} 1 + 2\gamma_{11} & 2\gamma_{12} & 2\gamma_{13} \\ 2\gamma_{21} & 1 + 2\gamma_{22} & 2\gamma_{23} \\ 2\gamma_{31} & 2\gamma_{32} & 1 + 2\gamma_{33} \end{vmatrix}. \tag{5.10.7}$$

Finally, Eq. (5.10.7) is expanded and the terms are grouped accordingly so that

$$G = 1 + 2J_1 + 4J_2 + 8J_3, \tag{5.10.8}$$

where J_1, J_2, J_3 are the previously defined strain invariants [see Eqs. (5.7.11), (5.7.12), and (5.7.13)]. The ratio of the original volume to the deformed volume is thus seen to be

$$\frac{dV}{dV^{(0)}} = \sqrt{1 + 2J_1 + 4J_2 + 8J_3}. \tag{5.10.9}$$

As for elongations, it is sometimes more informative to work with a ratio of the change in the volume to the original volume. This ratio is called the *cubical dilatation* or just simply the dilatation, and is expressed as

$$\frac{dV - dV^{(0)}}{dV^{(0)}} = \frac{dV}{dV^{(0)}} - 1 = \sqrt{1 + 2J_1 + 4J_2 + 8J_3} - 1. \tag{5.10.10}$$

For small strains, the volume-change ratio becomes simplified because the strain invariants J_2 and J_3 are of the order $(E_1)^2$ and $(E_1)^3$, respectively, and can be neglected with respect to J_1, which is of the order E_1 (see Section 5.4). Further, if the binominal expansion is used on $1 + 2J_1$, and if only the lowest-order terms are retained, then

$$\frac{dV}{dV^{(0)}} - 1 \cong J_1 = \gamma_{11} + \gamma_{22} + \gamma_{33}. \qquad (5.10.11)$$

Under the assumption of small strains, and hence linear strains, it is seen that the first strain invariant is the dilatation.

5.11 TWO SIMPLE EXAMPLES OF STRAIN

(a) Uniform strain. This example of strain is one of uniform extension in which the displacement vector $\mathbf{u}$ has the components given by

$$\mathbf{u} = c_1 y_1 \mathbf{i}_1 + c_2 y_2 \mathbf{i}_2 + c_3 y_3 \mathbf{i}_3, \qquad (5.11.1)$$

where c_1, c_2, and c_3 are constants. Since the position vector to points in the deformed body is

$$\mathbf{R} = \mathbf{r} + \mathbf{u} = (y_1 + c_1 y_1)\mathbf{i}_1 + (y_2 + c_2 y_2)\mathbf{i}_2 + (y_3 + c_3 y_3)\mathbf{i}_3, \qquad (5.11.2)$$

then

$$\frac{\partial \mathbf{R}}{\partial y_1} = (1 + c_1)\mathbf{i}_1, \qquad \frac{\partial \mathbf{R}}{\partial y_2} = (1 + c_2)\mathbf{i}_2, \qquad \frac{\partial \mathbf{R}}{\partial y_3} = (1 + c_3)\mathbf{i}_3. \qquad (5.11.3)$$

The fundamental metric tensor of the deformed body becomes

$$(G_{mn}) = \begin{pmatrix} (1 + c_1)^2 & 0 & 0 \\ 0 & (1 + c_2)^2 & 0 \\ 0 & 0 & (1 + c_3)^2 \end{pmatrix} \qquad (5.11.4)$$

and the determinant becomes

$$G = (1 + c_1)^2 (1 + c_2)^2 (1 + c_3)^2. \qquad (5.11.5)$$

The components of the strain tensor are then

$$(\gamma_{mn}) = \begin{pmatrix} \frac{1}{2}[(1 + c_1)^2 - 1] & 0 & 0 \\ 0 & \frac{1}{2}[(1 + c_2)^2 - 1] & 0 \\ 0 & 0 & \frac{1}{2}[(1 + c_3)^2 - 1] \end{pmatrix} \cdot \qquad (5.11.6)$$

It should be observed that the axes y_1, y_2, y_3 also coincide with the principal axes, since the strain tensor does not have any shearing components. The three strain invariants are

$$J_1 = \tfrac{1}{2}\{[(1 + c_1)^2 - 1] + [(1 + c_2)^2 - 1] + [(1 + c_3)^2 - 1]\},$$

$$J_2 = \tfrac{1}{4}\{[(1 + c_1)^2 - 1][(1 + c_2)^2 - 1] + [(1 + c_2)^2 - 1][(1 + c_3)^2 - 1]$$
$$+ [(1 + c_3)^2 - 1][(1 + c_1)^2 - 1]\}, \qquad (5.11.7)$$

$$J_3 = \tfrac{1}{8}[(1 + c_1)^2 - 1][(1 + c_2)^2 - 1][(1 + c_3)^2 - 1].$$

If the constants c_1, c_2, and c_3 are so small that their squares and products are negligible, then the strain tensor γ_{mn} can be replaced by the linear strain tensor ϵ_{mn}. Its components are

$$\epsilon_{mn} = \begin{pmatrix} c_1 & 0 & 0 \\ 0 & c_2 & 0 \\ 0 & 0 & c_3 \end{pmatrix}. \qquad (5.11.8)$$

Also the dilatation becomes simplified (see Eq. 5.10.10),

$$\frac{dV}{dV^{(0)}} - 1 \approx \epsilon_{11} + \epsilon_{22} + \epsilon_{33} = c_1 + c_2 + c_3. \qquad (5.11.9)$$

(b) Simple shear. This example is one of simple shear in which the displacement vector $\mathbf{u}$ has the components given by

$$\mathbf{u} = cy_2 \mathbf{i}_1, \qquad (5.11.10)$$

where c is a constant. Each point in the undeformed body moves parallel to the y_1-axis by an amount which is proportional to its y_2-coordinate. Thus rectangles drawn on the body on the $y_1 y_2$-plane before deformation become parallelograms after deformation (see Fig. 5.10).

The position vector, $\mathbf{R}$, to points in the deformed body is

$$\mathbf{R} = (y_1 + cy_2)\mathbf{i}_1 + y_2 \mathbf{i}_2 + y_3 \mathbf{i}_3, \qquad (5.11.11)$$

and the base vectors of the deformed body are

$$\frac{\partial \mathbf{R}}{\partial y_1} = \mathbf{i}_1, \qquad \frac{\partial \mathbf{R}}{\partial y_2} = c\mathbf{i}_1 + \mathbf{i}_2, \qquad \frac{\partial \mathbf{R}}{\partial y_3} = \mathbf{i}_3. \qquad (5.11.12)$$

Hence the fundamental metric tensor of the deformed body has the form

$$(G_{mn}) = \begin{pmatrix} 1 & c & 0 \\ c & 1 + c^2 & 0 \\ 0 & 0 & 1 \end{pmatrix} \qquad (5.11.13)$$

The determinant, G, has the value

$$G = 1, \qquad (5.11.14)$$

and the components of the strain tensor are given by

$$(\gamma_{mn}) = \begin{pmatrix} 0 & \tfrac{1}{2}c & 0 \\ \tfrac{1}{2}c & \tfrac{1}{2}c^2 & 0 \\ 0 & 0 & 0 \end{pmatrix}. \qquad (5.11.15)$$

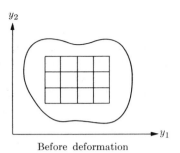

Before deformation

Therefore, the three strain invariants have the value

$$J_1 = \tfrac{1}{2}c^2, \qquad J_2 = -\tfrac{1}{4}c^2, \qquad J_3 = 0. \qquad (5.11.16)$$

Again if c is sufficiently small, then γ_{mn} can be replaced by ϵ_{mn}:

$$\epsilon_{mn} = \begin{pmatrix} 0 & \tfrac{1}{2}c & 0 \\ \tfrac{1}{2}c & 0 & 0 \\ 0 & 0 & 0 \end{pmatrix}. \qquad (5.11.17)$$

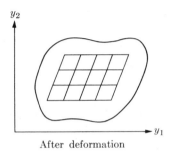

After deformation

FIG. 5.10   Simple shear strain.

Note that the dilatation for this case of linear simple shear is zero:

$$\frac{dV}{dV^{(0)}} - 1 = \epsilon_{11} + \epsilon_{22} + \epsilon_{33} = 0. \qquad (5.11.18)$$

5.12   THE DEVIATOR AND SPHERICAL STRAIN TENSORS

It will often be found convenient to consider the small strain tensor as the sum of two other strain tensors, called the *deviator* and *spherical strain tensors*. This is accomplished by defining

$$\gamma_{mn} = \hat{\gamma}_{mn} + \hat{\hat{\gamma}}_{mn}, \qquad (5.12.1)$$

where

$$\hat{\gamma}_{mn} = \gamma_{mn} - \tfrac{1}{3}\delta_{mn}\gamma_{rr}, \qquad (5.12.2)$$

$$\hat{\hat{\gamma}}_{mn} = \tfrac{1}{3}\delta_{mn}\gamma_{rr}. \qquad (5.12.3)$$

Here the symbols $\wedge$ and $\hat{\wedge}$ are used to indicate a resolution of γ_{mn}, rather than a transformation to other coordinates. If Eqs. (5.12.2) and (5.12.3) are substituted in Eq. (5.12.1), an identity results. The tensor designated by $\hat{\gamma}_{mn}$ is called the deviator strain tensor and the one designated by $\hat{\hat{\gamma}}_{mn}$ is called the spherical strain tensor.

It should be observed that the extensional component of the deviator strain tensor is merely the strain tensor minus one-third of the first strain invariant. The extensional components of the spherical strain tensor are all identical, and are equal to one-third of the first strain invariant.

The concept of the strain invariants can be applied to both the deviator and the spherical strain tensors. These can be expressed in terms of the strain invariants of the strain tensor. There results

$$\hat{J}_1 = 0, \tag{5.12.4}$$

$$\hat{J}_2 = J_2 - \tfrac{1}{3}(J_1)^2, \tag{5.12.5}$$

$$\hat{J}_3 = J_3 - \tfrac{1}{3}J_1J_2 + \tfrac{2}{27}(J_1)^3, \tag{5.12.6}$$

$$\mathring{J}_1 = J_1, \tag{5.12.7}$$

$$\mathring{J}_2 = \tfrac{1}{3}(J_1)^2, \tag{5.12.8}$$

$$\mathring{J}_3 = \tfrac{1}{27}(J_1)^3. \tag{5.12.9}$$

The motivation for and the resulting convenience from splitting the strain tensor into two parts can be seen from an examination of the first strain invariants, $\hat{J}_1$ and $\mathring{J}_1$. We recall that the dilatation in the special case of small strain is equal to the first strain invariant. Since the first strain invariant $\hat{J}_1$ of the deviator strain tensor is zero, there is no volume change associated with the deviator strains. Consequently, all the volume change is attributable to the spherical strains. These results will be found to be of greater significance in plasticity than in elasticity.

5.13 COMPATIBILITY RELATIONS FOR LINEAR STRAINS

The linear strain tensor has been specified in terms of six components, ϵ_{mn}. However, the six components of strain are themselves functions of the three components of displacement u_m:

$$\epsilon_{mn} = \frac{1}{2}\left(\frac{\partial u_m}{\partial y_n} + \frac{\partial u_n}{\partial y_m}\right). \tag{5.13.1}$$

These strain-displacement relations can be considered as six partial differential equations for the determination of the three components of displacement. Obviously, there is a redundancy. Therefore the components of the strain tensor are not completely independent, and certain conditions, called the *compatibility relations*, must exist between the components of the strain.

The conditions of compatibility can be interpreted physically by examining the deformed body. In continuum elasticity, it is assumed that the process of deformation does not create cracks or holes in the body, i.e., the body remains continuous. The assumption that the deformed body is continuous can be formulated in terms of the difference in displacement between two points P and R in the deformed body.

Let Δ_1, Δ_2, and Δ_3 represent respectively the i_1-, i_2-, and i_3-components of the difference in displacements between points P and R (see Fig. 5.11). This can be calculated by

$$\Delta_n = \oint_P^R du_n, \tag{5.13.2}$$

where $\mathscr{S}_P^R$ represents a line integral. Since the deformed body is continuous, Δ_n should be independent of the path of integration, that is, Δ_n should have the same value regardless of whether the integration occurs along the path labeled (1), (2), or (3) (see Fig. 5.11) or any other path. Thus it will be shown that the mathematical requirement on the integral of Eq. (5.13.2) will lead to the compatibility relations.

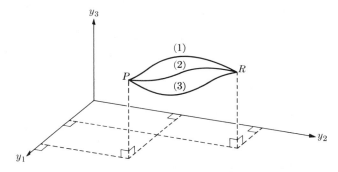

FIG. 5.11   Alternate paths of integration.

The derivative of Eq. (5.8.2) yields

$$d\mathbf{u} = \frac{\partial u_n}{\partial y_m} \mathbf{i}_n \, dy_m, \tag{5.13.3}$$

which can be rearranged to read

$$d\mathbf{u} = \left\{ \frac{1}{2} \left(\frac{\partial u_n}{\partial y_m} + \frac{\partial u_m}{\partial y_n} \right) + \frac{1}{2} \left(\frac{\partial u_n}{\partial y_m} - \frac{\partial u_m}{\partial y_n} \right) \right\} \mathbf{i}_n \, dy_m. \tag{5.13.4}$$

By introducing a new tensor, which is antisymmetric,*

$$\omega_{nm} = \frac{1}{2} \left(\frac{\partial u_n}{\partial y_m} - \frac{\partial u_m}{\partial y_n} \right), \tag{5.13.5}$$

we can write Eq. (5.13.4) in the form

$$d\mathbf{u} = (\epsilon_{nm} + \omega_{nm}) \mathbf{i}_n \, dy_m. \tag{5.13.6}$$

---

* An antisymmetric tensor has the property $\omega_{nm} = -\omega_{mn}$.

It is evident upon examining Eq. (5.13.6) that

$$du_n = (\epsilon_{nm} + \omega_{nm}) \, dy_m, \tag{5.13.7}$$

and hence Eq. (5.13.2) becomes

$$\Delta_n = \oint_P^R (\epsilon_{nm} + \omega_{nm}) \, dy_m. \tag{5.13.8}$$

An integration by parts of the second term in Eq. (5.13.8) leads to

$$\Delta_n = y_m \omega_{nm} \Big|_P^R + \oint_P^R \left[\epsilon_{nk} - y_m \frac{\partial \omega_{nm}}{\partial y_k} \right] dy_k, \tag{5.13.9}$$

where a dummy index on ϵ_{nm} has been relabeled.

It is easily verified by differentiating Eqs. (5.13.1) and (5.13.5) that

$$\frac{\partial \omega_{nm}}{\partial y_k} = \frac{\partial \epsilon_{nk}}{\partial y_m} - \frac{\partial \epsilon_{mk}}{\partial y_n}. \tag{5.13.10}$$

Combination of Eqs. (5.13.9) and (5.13.10) yields

$$\Delta_n = y_m \omega_{nm} \Big|_P^R + \oint_P^R \left[\epsilon_{nk} - y_m \left(\frac{\partial \epsilon_{nk}}{\partial y_m} - \frac{\partial \epsilon_{mk}}{\partial y_n} \right) \right] dy_k. \tag{5.13.11}$$

The first term in Eq. (5.13.11) depends only on the end points P and R, and hence is independent of the path of integration. Therefore, in order to fulfill the requirement only the integral term need be considered. It is shown in the theory of line integrals that the necessary and sufficient condition for the line integral to be independent of the path of integration (Ref. 5.4) is for the following equations to be satisfied:

$$\frac{\partial \xi_{nk}}{\partial y_l} = \frac{\partial \xi_{nl}}{\partial y_k}, \tag{5.13.12}$$

where for convenience, we introduce for the integrand,

$$\xi_{nk} = \epsilon_{nk} - y_m \left(\frac{\partial \epsilon_{nk}}{\partial y_m} - \frac{\partial \epsilon_{mk}}{\partial y_n} \right). \tag{5.13.13}$$

Introducing Eq. (5.13.13) into (5.13.12), we obtain

$$\begin{aligned}
\frac{\partial \xi_{nk}}{\partial y_l} &= \frac{\partial \epsilon_{nk}}{\partial y_l} - y_m \left[\frac{\partial^2 \epsilon_{nk}}{\partial y_m \, \partial y_l} - \frac{\partial^2 \epsilon_{mk}}{\partial y_n \, \partial y_l} \right] - \delta_{ml} \left[\frac{\partial \epsilon_{nk}}{\partial y_m} - \frac{\partial \epsilon_{mk}}{\partial y_n} \right], \\
\frac{\partial \xi_{nl}}{\partial y_k} &= \frac{\partial \epsilon_{nl}}{\partial y_k} - y_m \left[\frac{\partial^2 \epsilon_{nl}}{\partial y_m \, \partial y_k} - \frac{\partial^2 \epsilon_{ml}}{\partial y_n \, \partial y_k} \right] - \delta_{mk} \left[\frac{\partial \epsilon_{nl}}{\partial y_m} - \frac{\partial \epsilon_{ml}}{\partial y_n} \right],
\end{aligned} \tag{5.13.14}$$

where

$$\frac{\partial y_m}{\partial y_l} = \delta_{ml}. \tag{5.13.15}$$

If we substitute Eqs. (5.13.14) in Eq. (5.13.12), we find that

$$\frac{\partial \xi_{nk}}{\partial y_l} - \frac{\partial \xi_{nl}}{\partial y_k} = y_m \left[\frac{\partial^2 \epsilon_{nk}}{\partial y_m \partial y_l} + \frac{\partial^2 \epsilon_{ml}}{\partial y_n \partial y_k} - \frac{\partial^2 \epsilon_{nl}}{\partial y_m \partial y_k} - \frac{\partial^2 \epsilon_{mk}}{\partial y_n \partial y_l} \right]. \tag{5.13.16}$$

Since the y_m are independent, the necessary and sufficient conditions that Δ_n be independent of the path of integration thus become

$$\frac{\partial^2 \epsilon_{nk}}{\partial y_m \partial y_l} + \frac{\partial^2 \epsilon_{ml}}{\partial y_n \partial y_k} - \frac{\partial^2 \epsilon_{nl}}{\partial y_m \partial y_k} - \frac{\partial^2 \epsilon_{mk}}{\partial y_n \partial y_l} = 0. \tag{5.13.17}$$

These are the sought-after *compatibility relations*.

The independent compatibility relations can be shown to result from an incompatibility tensor (which is symmetric) defined as

$$S_{mn} = e_{mpq} e_{nrs} R_{pqrs}, \tag{5.13.18}$$

where R_{pqrs} is called the *Riemann tensor:*

$$R_{pqrs} = \frac{\partial^2 \epsilon_{ps}}{\partial y_q \partial y_r} + \frac{\partial^2 \epsilon_{qr}}{\partial y_p \partial y_s} - \frac{\partial^2 \epsilon_{pr}}{\partial y_q \partial y_s} - \frac{\partial^2 \epsilon_{qs}}{\partial y_p \partial y_r}. \tag{5.13.19}$$

Then the independent relations, Eqs. (5.13.18), can be obtained by setting S_{11}, S_{12}, etc., equal to zero.

The development of the compatibility relations presented in this section has connected the requirements on the strain tensor to the assumption that the process of deformation has not created any cracks or holes. A more direct approach devoid of any physical meaning can be used. Thus, since

$$\epsilon_{11} = \frac{\partial u_1}{\partial y_1}, \qquad \epsilon_{22} = \frac{\partial u_2}{\partial y_2},$$

$$\epsilon_{12} = \frac{1}{2} \left(\frac{\partial u_1}{\partial y_2} + \frac{\partial u_2}{\partial y_1} \right), \tag{5.13.20}$$

and

$$\frac{\partial^2 \epsilon_{11}}{\partial y_2 \partial y_2} = \frac{\partial^3 u_1}{\partial y_1 \partial y_2 \partial y_2},$$

$$\frac{\partial^2 \epsilon_{22}}{\partial y_1 \partial y_1} = \frac{\partial^3 u_2}{\partial y_2 \partial y_1 \partial y_1}, \tag{5.13.21}$$

$$\frac{\partial^2 \epsilon_{12}}{\partial y_1 \partial y_2} = \frac{1}{2} \left(\frac{\partial^3 u_1}{\partial y_2 \partial y_1 \partial y_2} + \frac{\partial^3 u_2}{\partial y_1 \partial y_1 \partial y_2} \right),$$

the first of Eqs. (5.13.17), namely,

$$\frac{\partial^2 \epsilon_{11}}{\partial y_2 \, \partial y_2} + \frac{\partial^2 \epsilon_{22}}{\partial y_1 \, \partial y_1} - 2 \frac{\partial^2 \epsilon_{12}}{\partial y_1 \, \partial y_2} = 0,$$

is seen to be merely an expression of a mathematical identity, because the order of partial differentiation is not important. In a similar manner, the other five compatibility relations, which may be obtained by expanding Eqs. (5.13.17), can be shown to be mathematical identities. In general, the compatibility relations provide differential equations relating the displacement components, and are useful in obtaining solutions to problems in solid mechanics.

5.14 SUMMARY

The primary objective of this chapter has been to develop the equations of consistent deformations for a general strained-solid continuum. If the displacements u_m are continuous functions, the strain components γ_{mn} represented by the strain displacement relations will satisfy the conditions of consistent deformations. The conditions may also be represented by the six strain-compatibility equations which are derived from the strain-displacement relations.

The equations of consistent deformation are one of three sets of equations required for a complete solution for the internal forces and strains in a body under a general system of external loads and supports. The other two sets, the equilibrium equations and the constitutive equations, will be discussed in the following three chapters. The equations of consistent deformation and the equations of equilibrium are completely independent of the material properties of the body.

The expanded forms of some of the expressions and relations developed in this chapter are summarized below in terms of a rectangular Cartesian coordinate system.

Pythagorean relations for line elements

Undeformed body:

$$(ds)^2 = \delta_{mn} \, dy_m \, dy_n$$
$$= (dy_1)^2 + (dy_2)^2 + (dy_3)^2; \tag{5.14.1a}$$

Deformed body:

$$(dS)^2 = G_{mn} \, dy_m \, dy_n,$$
$$(dS)^2 = G_{11}(dy_1)^2 + G_{22}(dy_2)^2 + G_{33}(dy_3)^2 + 2G_{12} \, dy_1 \, dy_2$$
$$+ 2G_{23} \, dy_2 \, dy_3 + 2G_{31} \, dy_3 \, dy_1. \tag{5.14.1b}$$

Definition of strain tensor

$$2\gamma_{mn}\, dy_m\, dy_n = (dS)^2 - (ds)^2; \tag{5.14.2a}$$

or

$$\gamma_{mn} = \tfrac{1}{2}(G_{mn} - \delta_{mn}),$$

$$\gamma_{11} = \tfrac{1}{2}(G_{11} - 1), \qquad \gamma_{12} = \tfrac{1}{2}G_{12},$$

$$\gamma_{22} = \tfrac{1}{2}(G_{22} - 1), \qquad \gamma_{23} = \tfrac{1}{2}G_{23}, \tag{5.14.2b}$$

$$\gamma_{33} = \tfrac{1}{2}(G_{33} - 1), \qquad \gamma_{31} = \tfrac{1}{2}G_{31}.$$

Strain-displacement relations for exact and small strains

$$\gamma_{mn} = \frac{1}{2}\left[\frac{\partial u_m}{\partial y_n} + \frac{\partial u_n}{\partial y_m} + \frac{\partial u_r}{\partial y_m}\frac{\partial u_s}{\partial y_n}\,\delta_{rs}\right]; \tag{5.14.3a}$$

$$\gamma_{11} = \frac{\partial u_1}{\partial y_1} + \frac{1}{2}\left[\left(\frac{\partial u_1}{\partial y_1}\right)^2 + \left(\frac{\partial u_2}{\partial y_1}\right)^2 + \left(\frac{\partial u_3}{\partial y_1}\right)^2\right],$$

$$\gamma_{22} = \frac{\partial u_2}{\partial y_2} + \frac{1}{2}\left[\left(\frac{\partial u_1}{\partial y_2}\right)^2 + \left(\frac{\partial u_2}{\partial y_2}\right)^2 + \left(\frac{\partial u_3}{\partial y_2}\right)^2\right],$$

$$\gamma_{33} = \frac{\partial u_3}{\partial y_3} + \frac{1}{2}\left[\left(\frac{\partial u_1}{\partial y_3}\right)^2 + \left(\frac{\partial u_2}{\partial y_3}\right)^2 + \left(\frac{\partial u_3}{\partial y_3}\right)^2\right], \tag{5.14.3b}$$

$$\gamma_{12} = \frac{1}{2}\left[\frac{\partial u_1}{\partial y_2} + \frac{\partial u_2}{\partial y_1} + \frac{\partial u_1}{\partial y_1}\frac{\partial u_1}{\partial y_2} + \frac{\partial u_2}{\partial y_1}\frac{\partial u_2}{\partial y_2} + \frac{\partial u_3}{\partial y_1}\frac{\partial u_3}{\partial y_2}\right],$$

$$\gamma_{23} = \frac{1}{2}\left[\frac{\partial u_2}{\partial y_3} + \frac{\partial u_3}{\partial y_2} + \frac{\partial u_1}{\partial y_2}\frac{\partial u_1}{\partial y_3} + \frac{\partial u_2}{\partial y_2}\frac{\partial u_2}{\partial y_3} + \frac{\partial u_3}{\partial y_2}\frac{\partial u_3}{\partial y_3}\right],$$

$$\gamma_{31} = \frac{1}{2}\left[\frac{\partial u_3}{\partial y_1} + \frac{\partial u_1}{\partial y_3} + \frac{\partial u_1}{\partial y_3}\frac{\partial u_1}{\partial y_1} + \frac{\partial u_2}{\partial y_3}\frac{\partial u_2}{\partial y_1} + \frac{\partial u_3}{\partial y_3}\frac{\partial u_3}{\partial y_1}\right].$$

Relations between elongations, angle changes, dilation and strain components

$$E_1 = \sqrt{1 + 2\gamma_{11}} - 1, \qquad \sin\phi_{12} = \frac{2\gamma_{12}}{(1 + E_1)(1 + E_2)},$$

$$E_2 = \sqrt{1 + 2\gamma_{22}} - 1, \qquad \sin\phi_{23} = \frac{2\gamma_{23}}{(1 + E_2)(1 + E_3)},$$

$$E_3 = \sqrt{1 + 2\gamma_{23}} - 1, \qquad \sin\phi_{31} = \frac{2\gamma_{31}}{(1 + E_3)(1 + E_1)},$$

$$\frac{dV}{dV^{(0)}} - 1 = \sqrt{1 + 2J_1 + 4J_2 + 8J_3} - 1. \tag{5.14.4}$$

Strain transformation law

$$\tilde{\gamma}_{mn} = \gamma_{rs}l_{r\tilde{m}}l_{s\tilde{n}};$$ (5.14.5a)

$$\begin{aligned}
\tilde{\gamma}_{11} = {}& (l_{1\tilde{1}})^2\gamma_{11} + (l_{2\tilde{1}})^2\gamma_{22} + (l_{3\tilde{1}})^2\gamma_{33} \\
& + 2(l_{1\tilde{1}}l_{2\tilde{1}}\gamma_{12} + l_{2\tilde{1}}l_{3\tilde{1}}\gamma_{23} + l_{3\tilde{1}}l_{1\tilde{1}}\gamma_{31}),
\end{aligned}$$

$$\begin{aligned}
\tilde{\gamma}_{22} = {}& (l_{1\tilde{2}})^2\gamma_{11} + (l_{2\tilde{2}})^2\gamma_{22} + (l_{3\tilde{2}})^2\gamma_{33} \\
& + 2(l_{1\tilde{2}}l_{2\tilde{2}}\gamma_{12} + l_{2\tilde{2}}l_{3\tilde{2}}\gamma_{23} + l_{3\tilde{2}}l_{1\tilde{2}}\gamma_{31}),
\end{aligned}$$

$$\begin{aligned}
\tilde{\gamma}_{33} = {}& (l_{1\tilde{3}})^2\gamma_{11} + (l_{2\tilde{3}})^2\gamma_{22} + (l_{3\tilde{3}})^2\gamma_{33} \\
& + 2(l_{1\tilde{3}}l_{2\tilde{3}}\gamma_{12} + l_{2\tilde{3}}l_{3\tilde{3}}\gamma_{23} + l_{3\tilde{3}}l_{1\tilde{3}}\gamma_{31}),
\end{aligned}$$

$$\begin{aligned}
\tilde{\gamma}_{12} = {}& l_{1\tilde{1}}l_{1\tilde{2}}\gamma_{11} + l_{2\tilde{1}}l_{2\tilde{2}}\gamma_{22} + l_{3\tilde{1}}l_{3\tilde{2}}\gamma_{33} \\
& + (l_{1\tilde{1}}l_{2\tilde{2}} + l_{2\tilde{1}}l_{1\tilde{2}})\gamma_{12} + (l_{2\tilde{1}}l_{3\tilde{2}} + l_{3\tilde{1}}l_{2\tilde{2}})\gamma_{23} \\
& + (l_{3\tilde{1}}l_{1\tilde{2}} + l_{1\tilde{1}}l_{3\tilde{2}})\gamma_{31},
\end{aligned}$$

(5.14.5b)

$$\begin{aligned}
\tilde{\gamma}_{23} = {}& l_{1\tilde{2}}l_{1\tilde{3}}\gamma_{11} + l_{2\tilde{2}}l_{2\tilde{3}}\gamma_{22} + l_{3\tilde{2}}l_{3\tilde{3}}\gamma_{33} \\
& + (l_{1\tilde{2}}l_{2\tilde{3}} + l_{2\tilde{2}}l_{1\tilde{3}})\gamma_{12} + (l_{2\tilde{2}}l_{3\tilde{3}} + l_{3\tilde{2}}l_{2\tilde{3}})\gamma_{23} \\
& + (l_{3\tilde{2}}l_{1\tilde{3}} + l_{1\tilde{2}}l_{3\tilde{3}})\gamma_{31},
\end{aligned}$$

$$\begin{aligned}
\tilde{\gamma}_{31} = {}& l_{1\tilde{3}}l_{1\tilde{1}}\gamma_{11} + l_{2\tilde{3}}l_{2\tilde{1}}\gamma_{22} + l_{3\tilde{3}}l_{3\tilde{1}}\gamma_{33} \\
& + (l_{1\tilde{3}}l_{2\tilde{1}} + l_{2\tilde{3}}l_{1\tilde{1}})\gamma_{12} + (l_{2\tilde{3}}l_{3\tilde{1}} + l_{3\tilde{3}}l_{2\tilde{1}})\gamma_{23} \\
& + (l_{3\tilde{3}}l_{1\tilde{1}} + l_{1\tilde{3}}l_{3\tilde{1}})\gamma_{31}.
\end{aligned}$$

Characteristic equation for principal strains

$$-(\gamma)^3 + J_1(\gamma)^2 - J_2(\gamma) + J_3 = 0.$$ (5.14.6)

Strain invariants

$$J_1 = \gamma_{11} + \gamma_{22} + \gamma_{33},$$

$$J_2 = \begin{vmatrix} \gamma_{11} & \gamma_{12} \\ \gamma_{21} & \gamma_{22} \end{vmatrix} + \begin{vmatrix} \gamma_{11} & \gamma_{13} \\ \gamma_{31} & \gamma_{33} \end{vmatrix} + \begin{vmatrix} \gamma_{22} & \gamma_{23} \\ \gamma_{32} & \gamma_{33} \end{vmatrix},$$ (5.14.7)

$$J_3 = \begin{vmatrix} \gamma_{11} & \gamma_{12} & \gamma_{13} \\ \gamma_{21} & \gamma_{22} & \gamma_{23} \\ \gamma_{31} & \gamma_{32} & \gamma_{33} \end{vmatrix}.$$

Strain invariants in terms of principal strains

$$\begin{aligned}
J_1 &= \gamma_{\text{I}} + \gamma_{\text{II}} + \gamma_{\text{III}}, \\
J_2 &= \gamma_{\text{I}}\gamma_{\text{II}} + \gamma_{\text{II}}\gamma_{\text{III}} + \gamma_{\text{I}}\gamma_{\text{III}}, \\
J_3 &= \gamma_{\text{I}}\gamma_{\text{II}}\gamma_{\text{III}}.
\end{aligned}$$ (5.14.8)

Equations for direction cosines of principal directions

$$[\gamma_{mn} - \gamma_{\mathrm{I}} \, \delta_{mn}]\lambda_n^{(\mathrm{I})} = 0,$$

$$[\gamma_{mn} - \gamma_{\mathrm{II}} \, \delta_{mn}]\lambda_n^{(\mathrm{II})} = 0, \qquad (5.14.9)$$

$$[\gamma_{mn} - \gamma_{\mathrm{III}} \, \delta_{mn}]\lambda_n^{(\mathrm{III})} = 0.$$

Small strain relations to elongations, angle changes and dilatation

$$E_1 \cong \gamma_{11}, \qquad \phi_{12} \cong 2\gamma_{12},$$

$$E_2 \cong \gamma_{22}, \qquad \phi_{23} \cong 2\gamma_{23}, \qquad \frac{dV}{dV^{(0)}} - 1 \cong J_1(\gamma_{mn}) \quad \text{(small strain)}.$$

$$E_3 \cong \gamma_{33}, \qquad \phi_{31} \cong 2\gamma_{31}, \qquad\qquad\qquad\qquad (5.14.10)$$

Since linear strain, by definition, is always a small strain, we also have

$$E_1 = \epsilon_{11}, \qquad \phi_{12} = 2\epsilon_{12},$$

$$E_2 = \epsilon_{22}, \qquad \phi_{23} = 2\epsilon_{23}, \qquad \frac{dV}{dV^{(0)}} - 1 = J_1(\epsilon_{mn}) \quad \text{(linear strain)}.$$

$$E_3 = \epsilon_{33}, \qquad \phi_{31} = 2\epsilon_{31}, \qquad\qquad\qquad\qquad (5.14.11)$$

Strain-displacement relations for linear strain

$$\epsilon_{mn} = \frac{1}{2}\left(\frac{\partial u_m}{\partial y_n} + \frac{\partial u_n}{\partial y_m}\right); \qquad (5.14.12a)$$

$$\epsilon_{11} = \frac{\partial u_1}{\partial y_1}, \qquad \epsilon_{12} = \frac{1}{2}\left(\frac{\partial u_1}{\partial y_2} + \frac{\partial u_2}{\partial y_1}\right),$$

$$\epsilon_{22} = \frac{\partial u_2}{\partial y_2}, \qquad \epsilon_{23} = \frac{1}{2}\left(\frac{\partial u_2}{\partial y_3} + \frac{\partial u_3}{\partial y_2}\right), \qquad (5.14.12b)$$

$$\epsilon_{33} = \frac{\partial u_3}{\partial y_3}, \qquad \epsilon_{31} = \frac{1}{2}\left(\frac{\partial u_3}{\partial y_1} + \frac{\partial u_1}{\partial y_3}\right).$$

Resolution of small strain tensor into deviator and spherical strain tensors

$$\gamma_{mn} = \hat{\gamma}_{mn} + \mathring{\gamma}_{mn}, \qquad (5.14.13)$$

where $\hat{\gamma}_{mn} =$ deviator strain tensor, and $\mathring{\gamma}_{mn} =$ spherical strain tensor.

Components of the deviator strain tensor

$$\hat{\gamma}_{mn} = \gamma_{mn} - \tfrac{1}{3}\delta_{mn}\gamma_{rr}; \qquad (5.14.14a)$$

$$\hat{\gamma}_{11} = \gamma_{11} - \tfrac{1}{3}(\gamma_{11} + \gamma_{22} + \gamma_{33}), \qquad \hat{\gamma}_{12} = \gamma_{12},$$

$$\hat{\gamma}_{22} = \gamma_{22} - \tfrac{1}{3}(\gamma_{11} + \gamma_{22} + \gamma_{33}), \qquad \hat{\gamma}_{23} = \gamma_{23}, \qquad (5.14.14b)$$

$$\hat{\gamma}_{33} = \gamma_{33} - \tfrac{1}{3}(\gamma_{11} + \gamma_{22} + \gamma_{33}), \qquad \hat{\gamma}_{31} = \gamma_{31}.$$

Components of the spherical strain tensor

$$\hat{\gamma}_{mn} = \tfrac{1}{3}\delta_{mn}\gamma_{rr};\tag{5.14.15a}$$

$$\hat{\gamma}_{11} = \tfrac{1}{3}(\gamma_{11} + \gamma_{22} + \gamma_{33}), \qquad \hat{\gamma}_{12} = 0,$$

$$\hat{\gamma}_{22} = \tfrac{1}{3}(\gamma_{11} + \gamma_{22} + \gamma_{33}), \qquad \hat{\gamma}_{23} = 0, \tag{5.14.15b}$$

$$\hat{\gamma}_{33} = \tfrac{1}{3}(\gamma_{11} + \gamma_{22} + \gamma_{33}), \qquad \hat{\gamma}_{31} = 0.$$

Linear strain compatibility conditions

$$\frac{\partial^2 \epsilon_{nk}}{\partial y_m\, \partial y_l} + \frac{\partial^2 \epsilon_{ml}}{\partial y_n\, \partial y_k} + \frac{\partial^2 \epsilon_{nl}}{\partial y_m\, \partial y_k} - \frac{\partial^2 \epsilon_{mk}}{\partial y_n\, \partial y_l} = 0; \tag{5.14.16a}$$

$$\frac{\partial^2 \epsilon_{11}}{\partial y_2\, \partial y_2} + \frac{\partial^2 \epsilon_{22}}{\partial y_1\, \partial y_1} - 2\frac{\partial^2 \epsilon_{12}}{\partial y_1\, \partial y_2} = 0,$$

$$\frac{\partial^2 \epsilon_{22}}{\partial y_3\, \partial y_3} + \frac{\partial^2 \epsilon_{33}}{\partial y_2\, \partial y_2} - 2\frac{\partial^2 \epsilon_{23}}{\partial y_2\, \partial y_3} = 0,$$

$$\frac{\partial^2 \epsilon_{33}}{\partial y_1\, \partial y_1} + \frac{\partial^2 \epsilon_{11}}{\partial y_3\, \partial y_3} - 2\frac{\partial^2 \epsilon_{31}}{\partial y_3 \partial\, y_1} = 0,$$

$$\tag{5.14.16b}$$

$$\frac{\partial^2 \epsilon_{11}}{\partial y_2\, \partial y_3} + \frac{\partial^2 \epsilon_{23}}{\partial y_1\, \partial y_1} - \frac{\partial^2 \epsilon_{31}}{\partial y_1\, \partial y_2} - \frac{\partial^2 \epsilon_{12}}{\partial y_1\, \partial y_3} = 0,$$

$$\frac{\partial^2 \epsilon_{22}}{\partial y_3\, \partial y_1} + \frac{\partial^2 \epsilon_{31}}{\partial y_2\, \partial y_2} - \frac{\partial^2 \epsilon_{12}}{\partial y_2\, \partial y_3} - \frac{\partial^2 \epsilon_{23}}{\partial y_2\, \partial y_1} = 0,$$

$$\frac{\partial^2 \epsilon_{33}}{\partial y_1\, \partial y_2} + \frac{\partial^2 \epsilon_{12}}{\partial y_3\, \partial y_3} - \frac{\partial^2 \epsilon_{23}}{\partial y_3\, \partial y_1} - \frac{\partial^2 \epsilon_{31}}{\partial y_3\, \partial y_2} = 0.$$

PROBLEMS

5.1   Expand Eq. (5.8.9) into six expressions for the individual strain components γ_{11}, γ_{22}, γ_{33}, γ_{12}, γ_{23}, and γ_{31}.

5.2   As a simple exercise in the application of the Lagrange multiplier method, find the extreme values of the function $f(x, y) = 4 - x^2 - y^2$, subject to the constraint $x^2 - y^2 = 9$. (Refer to the Appendix, Section A.16, for similar examples.)

5.3   Compute the direction cosines $l_{\tilde{1}m}$ of the axis $\tilde{y}_1$ which is perpendicular to the face ABC of the tetrahedron shown in Fig. P.5.3.

5.4   In the y_1-, y_2-, y_3-coordinate system the components of the strain tensor are shown by the matrix

$$[\gamma_{mn}] = \begin{bmatrix} 0.01 & -0.005 & 0 \\ -0.005 & +0.02 & +0.01 \\ 0 & +0.01 & -0.02 \end{bmatrix}.$$

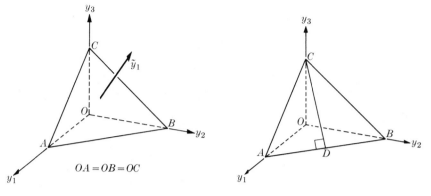

FIGURE P.5.3   FIGURE P.5.4

On the tetrahedron shown in Fig. P.5.4, where $OA = OB = OC$ (see Problem 5.3), what is the percentage elongation of the line AB, and what is the change in angle ADC which was 90° in the unstrained position?

5.5 At a point in a body the shear strain components γ_{13} and γ_{23} are zero with respect to the y_1-, y_2-, and y_3-coordinate system. In a new coordinate system $\tilde{y}_1$, $\tilde{y}_2$, and $\tilde{y}_3$ where $\tilde{y}_3$ and y_3 coincide (Fig. P.5.5), show that the following strain transformation relations hold:

$$\tilde{\gamma}_{11} = \frac{\gamma_{11} + \gamma_{22}}{2} + \frac{\gamma_{11} - \gamma_{22}}{2} \cos 2\theta + \gamma_{12} \sin 2\theta,$$

$$\tilde{\gamma}_{22} = \frac{\gamma_{11} + \gamma_{22}}{2} - \frac{\gamma_{11} - \gamma_{22}}{2} \cos 2\theta - \gamma_{12} \sin 2\theta,$$

$$\tilde{\gamma}_{12} = - \frac{\gamma_{11} - \gamma_{22}}{2} \sin 2\theta + \gamma_{12} \cos 2\theta,$$

$$\tilde{\gamma}_{23} = \tilde{\gamma}_{13} = 0,$$

where θ is the angle between the y_1-axis and the $\tilde{y}_1$-axis. *Note:* One can conclude that y_3 (hence also $\tilde{y}_3$) is one of the principal axes because the shear strains are zero along the principal directions.

5.6 The coordinates y_1, y_2, y_3 are located such that y_3 coincides with one of the principal axes, say y_{III}, and the angle from y_I to y_1 is θ, as shown in Fig. P.5.6(a). (Note that y_1, y_2, y_I, and y_{II} are all on the same plane.) The two principal axes are assigned such that $\gamma_I > \gamma_{II}$. Using the result in the preceding problem show that γ_{11} and γ_{12} can be determined graphically according to the following procedure. A circle (Fig. P.5.6b) called *Mohr's circle* is constructed with its center O located at $((\gamma_I + \gamma_{II})/2, O)$ and its radius equal to $(\gamma_I - \gamma_{II})/2$. A line OP is then drawn at an angle 2θ from the horizontal axis. [Both θ in Fig. 5.6(a) and 2θ in Fig. P.5.6(b) are measured counterclockwise.] The strain components γ_{11} and γ_{12} are then represented by the coordinates of point P. The abscissa of P is the normal strain γ_{11}, and the ordinate is the shear strain γ_{12}. Here γ_{11} is positive when P is located to the

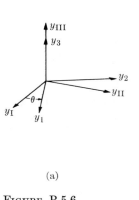

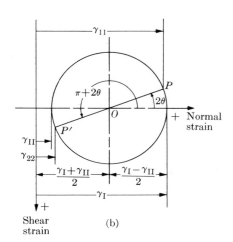

(a)

FIGURE P.5.6

Shear
strain

(b)

right of the vertical axis, and γ_{12} is positive when P is located below the horizontal axis. Thus in Fig. P.5.6(b) γ_{11} is positive and γ_{12} is negative. Note that γ_{22} is the normal stress with respect to an axis at an angle $\pi/2 + \theta$ from y_I, and hence is represented by the abscissa of point P'. How do you interpret the fact that P' represents a shear strain equal in magnitude to γ_{12} but of opposite sign?

For a certain coordinate system y_1, y_2, y_3, the following strain components are given:

$$\gamma_{23} = \gamma_{13} = 0, \qquad \gamma_{11} = 0.005, \qquad \gamma_{22} = -0.001, \qquad \text{and} \qquad \gamma_{12} = 0.003.$$

Construct the corresponding Mohr circle. Determine γ_I and γ_{II}, and locate the two principal directions y_I and y_{II} with respect to the y_1- and y_2-axes.

5.7   Draw Mohr's circles for the following general states of strain:

(a) $\gamma_{11} > \gamma_{22} > 0, \qquad \gamma_{12} < 0;$

(b) $\gamma_{22} > 0, \qquad \gamma_{11} < 0, \qquad \gamma_{12} > 0;$

(c) $\gamma_{11} = \gamma_{22} > 0, \qquad \gamma_{12} < 0.$

5.8   Let y_1 and y_2 be tangent to the surface of a solid and y_3, which is then perpendicular to the surface, be a principal direction. (In later chapters we shall show that for isotropic materials the normal direction to a surface which is free of stress is always a principal direction.) (a) Show that the strain components γ_{11}, γ_{12}, and γ_{22} can be calculated by measuring the normal strains E_A, E_B, E_C along any three different directions on the surface. (b) Let the three directions be given by $\theta_A = 0°$, $\theta_B = 45°$, $\theta_C = 90°$. Determine the principal strains γ_I and γ_{II} in terms of the three normal strain readings E_A, E_B, and E_C. What is the angle between y_1 and y_I? (c) Let the three directions be given by $\theta_A = 0°$, $\theta_B = 60°$, $\theta_C = 120°$. Determine γ_I and γ_{II} in terms of the three normal strain readings.

A device for measuring the normal strain on a surface is called a *strain gage*. The arrangement of three or more strain gages to determine the state of strain on a surface

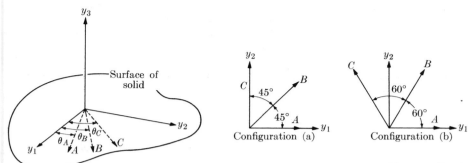

FIGURE P.5.8

is called a *strain rosette*. In Fig. P.5.8, Configuration (a) is called a *rectangular or 45° rosette;* Configuration (b) is called a *delta rosette.*

5.9 By means of the Mohr circle show that the shear strain γ_{12} is given by

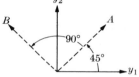

$$\gamma_{12} = \frac{E_A - E_B}{2},$$

where E_A and E_B are the normal strains along two mutually perpendicular directions A and B.

FIGURE P.5.9

5.10 (a) At a point O on the surface of a deformable body three strain gages were mounted as shown in the sketch. The angles between the y_1-axes and the $\tilde{y}_1$- and $\tilde{\tilde{y}}_1$-axes are 60°. Under a certain loading condition, the percentage elongations in these gages are as follows:

$$E_1 = -0.005,$$
$$\tilde{E}_1 = +0.005,$$
$$\tilde{\tilde{E}}_1 = +0.010.$$

Determine the strain components γ_{11}, γ_{12}, and γ_{22}.

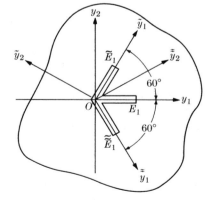

(b) Consider now a deformable body made of rubber. Under a given loading condition the percentage elongations obtained by these strain gages are as follows:

$$E_1 = -0.5,$$
$$\tilde{E}_1 = +0.5,$$
$$\tilde{\tilde{E}}_1 = +1.0.$$

FIGURE P.5.10

What are the corresponding strain components γ_{11}, γ_{12}, and γ_{22}? Suppose that prior to loading, a 90° cross were inscribed to coincide with the y_1- and y_2-axes. What would the angle between the two lines be after deformation?

Note that in (a) E_1, $\frac{1}{2}\phi_{12}$, E_2 and $\tilde{E}_1$, $\frac{1}{2}\tilde{\phi}_{12}$, $\tilde{E}_2$ are related by the tensor transformation law, while in (b) they are no longer related by the tensor transformation law. Why?

5.11 An initial rectangular block 1 in. $\times$ 2 in. $\times$ 1 in. (see Fig. P.5.11) is deformed such that the displacement **u** of every point can be represented in the terms of the following equation

$$\mathbf{u}(y_1, y_2, y_3) = (3 - 0.5\, y_3)\mathbf{i}_2 - 0.5\, y_3 \mathbf{i}_3.$$

(a) Sketch the deformed block. (b) Determine the strains γ_{mn} (use strain-displacement relations). (c) Determine the relative elongation E_m and angle changes.

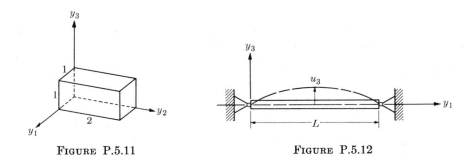

FIGURE P.5.11   FIGURE P.5.12

5.12 The beam shown in the sketch is deformed such that points lying on the undeformed axis of the beam do not experience any horizontal displacement (that is, $u_1 = 0$), but do experience a lateral displacement given by

$$u_3 = A\frac{y_1}{L}\left(1 - \frac{y_1}{L}\right),$$

where A is a constant. Determine the distribution of the normal strain γ_{11} along the axis of the beam. What are the consequences of employing the linear strain-displacement relations in the solution of the problem?

5.13 In the previous problem, assume that the normal strain γ_{11} is constant along the deformed axis, i.e., the displacement $u_1(y_1, 0)$ is not zero. Determine $u_1(y_1, 0)$ and $\gamma_{11}(y_1, 0)$.

5.14 Consider the displacement vector specified by

$$\mathbf{u}(y_1, y_2, y_3) = c(y_2 + y_3)\mathbf{i}_1.$$

(a) Sketch the final position and shape of a unit cube which has its nearest corner located at (1, 1, 1) and which was originally oriented with its edges parallel to the y_1-, y_2-, y_3-coordinates in the undeformed body.

(b) Determine the components of the strain tensor and the strain invariants in terms of the components of displacement.

(c) Find the components of the strain tensor for coordinates $\tilde{y}_1$, $\tilde{y}_2$, and $\tilde{y}_3$ which are related to the y_1-, y_2-, y_3-coordinates as follows:

$$y_1 = \tilde{y}_1, \qquad y_2 = \tilde{y}_3 \cos\frac{\pi}{4} + \tilde{y}_2 \cos\frac{\pi}{4}, \qquad y_3 = \tilde{y}_3 \cos\frac{\pi}{4} - \tilde{y}_2 \cos\frac{\pi}{4}.$$

5.15 Use the *Eulerian* representation, coordinates Y_1, Y_2, Y_3 (Fig. P.5.15), to determine the following: (a) the base vectors in a deformed and an undeformed system; (b) relations connecting the base vectors of the undeformed system to those of the deformed system, and the components of displacement $u_m(Y_1, Y_2, Y_3)$; (c) the fundamental metric tensor of the deformed system H_{mn} and of the undeformed system h_{mn}; and (d) the strain-displacement relations for the Eulerian strain tensor which is defined as $\eta_{mn} = \frac{1}{2}(H_{mn} - h_{mn})$.

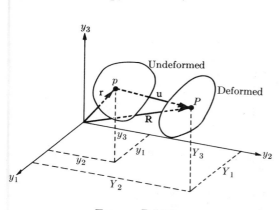

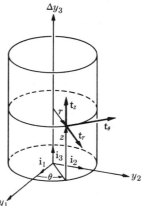

FIGURE P.5.15   FIGURE P.5.16

5.16 Strain problems for cylindrically shaped bodies are in general tackled most conveniently in terms of *cylindrical polar coordinates*. A cylindrical polar-coordinate system is shown in relation to the rectangular Cartesian system in Fig. P.5.16. The coordinate transformation between the two systems is given by

$$y_1 = r \cos\theta, \qquad y_2 = r \sin\theta, \qquad y_3 = z.$$

The set of unit tangent vectors (see Fig. P.5.16) associated with the cylindrical coordinates is related to the unit vectors $\mathbf{i}_1$, $\mathbf{i}_2$, $\mathbf{i}_3$:

$$\mathbf{t}_r = \cos\theta\, \mathbf{i}_1 + \sin\theta\, \mathbf{i}_2, \qquad \mathbf{t}_\theta = -\sin\theta\, \mathbf{i}_1 + \cos\theta\, \mathbf{i}_2, \qquad \mathbf{t}_z = \mathbf{i}_3.$$

We can write the displacement vector as

$$\mathbf{u}(r, \theta, z) = u\mathbf{t}_r + v\mathbf{t}_\theta + w\mathbf{t}_z.$$

For cylindrical coordinates, the components of strain can be defined as follows (cf. Eq. 5.3.1):

$$\frac{1}{2}[(dS)^2 - (ds)^2] = \gamma_{rr}(dr)^2 + \gamma_{\theta\theta}(r\,d\theta)^2 + \gamma_{zz}(dz)^2$$
$$+ 2\gamma_{\theta r}r\,d\theta\,dr + 2\gamma_{\theta z}r\,d\theta\,dz + 2\gamma_{zr}\,dz\,dr.$$

Verify the following relations:

$$d\mathbf{R} = d\mathbf{r} + d\mathbf{u},$$

$$d\mathbf{r} = \mathbf{t}_r\,dr + r\mathbf{t}_\theta\,d\theta + \mathbf{t}_z\,dz,$$

$$d\mathbf{u} = \frac{\partial\mathbf{u}}{\partial r}\,dr + \frac{\partial\mathbf{u}}{\partial\theta}\,d\theta + \frac{\partial\mathbf{u}}{\partial z}\,dz,$$

where $\mathbf{R}$ and $\mathbf{r}$ are the position vectors to points in the deformed and undeformed body, respectively (see Section 5.8).

Next show that

$$\tfrac{1}{2}[(dS)^2 - (ds)^2] = d\mathbf{r}\cdot d\mathbf{u} + \tfrac{1}{2}d\mathbf{u}\cdot d\mathbf{u}$$

and (see Eq. 5.8.8)

$$\gamma_{rr} = \mathbf{t}_r\cdot\frac{\partial\mathbf{u}}{\partial r} + \frac{1}{2}\frac{\partial\mathbf{u}}{\partial r}\cdot\frac{\partial\mathbf{u}}{\partial r},$$

$$\gamma_{\theta\theta} = \frac{1}{r^2}\left\{r\mathbf{t}_\theta\cdot\frac{\partial\mathbf{u}}{\partial\theta} + \frac{1}{2}\frac{\partial\mathbf{u}}{\partial\theta}\cdot\frac{\partial\mathbf{u}}{\partial\theta}\right\},$$

$$\gamma_{zz} = \mathbf{t}_z\cdot\frac{\partial\mathbf{u}}{\partial z} + \frac{1}{2}\frac{\partial\mathbf{u}}{\partial z}\cdot\frac{\partial\mathbf{u}}{\partial z},$$

$$\gamma_{r\theta} = \frac{1}{2r}\left\{\mathbf{t}_r\cdot\frac{\partial\mathbf{u}}{\partial\theta} + r\mathbf{t}_\theta\cdot\frac{\partial\mathbf{u}}{\partial r} + \frac{\partial\mathbf{u}}{\partial r}\cdot\frac{\partial\mathbf{u}}{\partial\theta}\right\},$$

$$\gamma_{\theta z} = \frac{1}{2r}\left\{r\mathbf{t}_\theta\cdot\frac{\partial\mathbf{u}}{\partial z} + \mathbf{t}_z\cdot\frac{\partial\mathbf{u}}{\partial\theta} + \frac{\partial\mathbf{u}}{\partial\theta}\cdot\frac{\partial\mathbf{u}}{\partial z}\right\},$$

$$\gamma_{zr} = \frac{1}{2}\left\{\mathbf{t}_r\cdot\frac{\partial\mathbf{u}}{\partial z} + \mathbf{t}_z\cdot\frac{\partial\mathbf{u}}{\partial r} + \frac{\partial\mathbf{u}}{\partial z}\cdot\frac{\partial\mathbf{u}}{\partial r}\right\}.$$

Finally, verify the strain-displacement relations (compare with Eqs. 5.14.3b):

$$\gamma_{rr} = \frac{\partial u}{\partial r} + \frac{1}{2}\left[\left(\frac{\partial u}{\partial r}\right)^2 + \left(\frac{\partial v}{\partial r}\right)^2 + \left(\frac{\partial w}{\partial r}\right)^2\right],$$

$$\gamma_{\theta\theta} = \frac{1}{r}\frac{\partial v}{\partial\theta} + \frac{u}{r} + \frac{1}{2}\left[\left(\frac{1}{r}\frac{\partial u}{\partial\theta} - \frac{v}{r}\right)^2 + \left(\frac{1}{r}\frac{\partial v}{\partial\theta} + \frac{u}{r}\right)^2 + \left(\frac{1}{r}\frac{\partial w}{\partial\theta}\right)^2\right],$$

$$\gamma_{zz} = \frac{\partial u}{\partial z} + \frac{1}{2}\left[\left(\frac{\partial u}{\partial z}\right)^2 + \left(\frac{\partial v}{\partial z}\right)^2 + \left(\frac{\partial w}{\partial z}\right)^2\right],$$

$$\gamma_{r\theta} = \frac{1}{2}\left\{\frac{1}{r}\frac{\partial u}{\partial\theta} - \frac{v}{r} + \frac{\partial v}{\partial r} + \frac{\partial u}{\partial r}\left(\frac{1}{r}\frac{\partial u}{\partial\theta} - \frac{v}{r}\right) + \frac{\partial v}{\partial r}\left(\frac{1}{r}\frac{\partial v}{\partial\theta} + \frac{u}{r}\right) + \frac{\partial w}{\partial r}\frac{1}{r}\frac{\partial w}{\partial\theta}\right\},$$

$$\gamma_{\theta z} = \frac{1}{2}\left\{\frac{\partial v}{\partial z} + \frac{1}{r}\frac{\partial w}{\partial\theta} + \frac{\partial u}{\partial z}\left(\frac{1}{r}\frac{\partial u}{\partial\theta} - \frac{v}{r}\right) + \frac{\partial v}{\partial z}\left(\frac{1}{r}\frac{\partial v}{\partial\theta} + \frac{u}{r}\right) + \frac{\partial w}{\partial z}\frac{1}{r}\frac{\partial w}{\partial\theta}\right\},$$

$$\gamma_{zr} = \frac{1}{2}\left\{\frac{\partial u}{\partial z} + \frac{\partial w}{\partial r} + \frac{\partial u}{\partial r}\frac{\partial u}{\partial z} + \frac{\partial v}{\partial r}\frac{\partial v}{\partial z} + \frac{\partial w}{\partial r}\frac{\partial w}{\partial z}\right\}.$$

Note: The components of strain denoted in this problem as γ_{rr} $\gamma_{r\theta}$, $\gamma_{\theta\theta}$, etc., are dimensionless and hence are referred to as the *physical* components of the strain tensor.   There are tensor methods of obtaining these results more elegant than the method of this problem.   However, tensors in general curvilinear coordinates are beyond the scope of this introductory text.   The reader who is interested can find such a treatment in Green and Zerna, *Theoretical Elasticity*.   See Bibliography 5.4.

5.17.   Strain problems for spherically shaped bodies are usually solved most easily in terms of *spherical coordinates*.   A spherical coordinate system is shown in relation to the rectangular Cartesian system in Fig. P.5.17.   The coordinate transformation between the two systems is given by

$$y_1 = \rho \sin \phi \cos \theta,$$
$$y_2 = \rho \sin \phi \sin \theta,$$
$$y_3 = \rho \cos \phi.$$

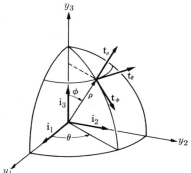

The set of unit tangent vectors (see Fig. P.5.17) associated with the spherical coordinates is related to the unit vectors $\mathbf{i}_1$, $\mathbf{i}_2$, $\mathbf{i}_3$:

$$\mathbf{t}_\rho = \sin\phi\cos\theta\mathbf{i}_1 + \sin\phi\sin\theta\mathbf{i}_2 + \cos\phi\mathbf{i}_3,$$
$$\mathbf{t}_\phi = \cos\phi\cos\theta\mathbf{i}_1 + \cos\phi\sin\theta\mathbf{i}_2 - \sin\phi\mathbf{i}_3,$$
$$\mathbf{t}_\theta = -\sin\theta\mathbf{i}_1 + \cos\theta\mathbf{i}_2.$$

FIGURE P.5.17

We can write the displacement vector in terms of components parallel to $\mathbf{t}_\rho$, $\mathbf{t}_\phi$, and $\mathbf{t}_\theta$, respectively:

$$\mathbf{u}(\rho, \phi, \theta) = w\mathbf{t}_\rho + u\mathbf{t}_\phi + v\mathbf{t}_\theta.$$

For spherical coordinates, the components of strain can be defined as follows (cf. Eq. 5.3.1):

$$\tfrac{1}{2}[(dS)^2 - (ds)^2] = \gamma_{\rho\rho}(d\rho)^2 + \gamma_{\phi\phi}(\rho d\phi)^2 + \gamma_{\theta\theta}(\rho \sin\phi \, d\theta)^2$$
$$+ 2\gamma_{\rho\phi} \, d\rho \, \rho d\phi + 2\gamma_{\phi\theta} \, \rho d\phi \, \rho \sin\phi \, d\theta$$
$$+ 2\gamma_{\theta\rho} \, \rho \sin\phi \, d\theta \, d\rho.$$

Verify the following relations:

$$d\mathbf{R} = d\mathbf{r} + d\mathbf{u},$$
$$d\mathbf{r} = \mathbf{t}_\rho \, d\rho + \mathbf{t}_\phi \, \rho \, d\phi + \mathbf{t}_\theta \, \rho \sin\phi \, d\theta,$$
$$d\mathbf{u} = \frac{\partial \mathbf{u}}{\partial \rho} \, d\rho + \frac{\partial \mathbf{u}}{\partial \phi} \, d\phi + \frac{\partial \mathbf{u}}{\partial \theta} \, d\theta,$$

where $\mathbf{R}$ and $\mathbf{r}$ are the position vectors to points in the deformed and undeformed body, respectively (see Section 5.8).

Next show that

$$\tfrac{1}{2}[(dS)^2 - (ds)^2] = d\mathbf{r} \cdot d\mathbf{u} + \tfrac{1}{2}\,d\mathbf{u} \cdot d\mathbf{u}$$

and that (see Eq. 5.8.8)

$$\gamma_{\rho\rho} = \mathbf{t}_\rho \cdot \frac{\partial \mathbf{u}}{\partial \rho} + \frac{1}{2}\frac{\partial \mathbf{u}}{\partial \rho} \cdot \frac{\partial \mathbf{u}}{\partial \rho},$$

$$\gamma_{\phi\phi} = \frac{1}{\rho^2}\left\{\rho \mathbf{t}_\phi \cdot \frac{\partial \mathbf{u}}{\partial \phi} + \frac{1}{2}\frac{\partial \mathbf{u}}{\partial \phi} \cdot \frac{\partial \mathbf{u}}{\partial \phi}\right\},$$

$$\gamma_{\theta\theta} = \frac{1}{(\rho \sin \phi)^2}\left\{\rho \sin \phi\, \mathbf{t}_\theta \cdot \frac{\partial \mathbf{u}}{\partial \theta} + \frac{1}{2}\frac{\partial \mathbf{u}}{\partial \theta} \cdot \frac{\partial \mathbf{u}}{\partial \theta}\right\},$$

$$\gamma_{\rho\phi} = \frac{1}{2\rho}\left\{\mathbf{t}_\rho \cdot \frac{\partial \mathbf{u}}{\partial \phi} + \rho\, \mathbf{t}_\phi \cdot \frac{\partial \mathbf{u}}{\partial \rho} + \frac{\partial \mathbf{u}}{\partial \rho} \cdot \frac{\partial \mathbf{u}}{\partial \phi}\right\},$$

$$\gamma_{\phi\theta} = \frac{1}{2\rho^2 \sin \phi}\left\{\rho \mathbf{t}_\phi \cdot \frac{\partial \mathbf{u}}{\partial \theta} + \rho \sin \phi\, \mathbf{t}_\theta \cdot \frac{\partial \mathbf{u}}{\partial \phi} + \frac{\partial \mathbf{u}}{\partial \phi} \cdot \frac{\partial \mathbf{u}}{\partial \theta}\right\},$$

$$\gamma_{\theta\rho} = \frac{1}{2\rho \sin \phi}\left\{\rho \sin \phi\, \mathbf{t}_\theta \cdot \frac{\partial \mathbf{u}}{\partial \rho} + \mathbf{t}_\rho \cdot \frac{\partial \mathbf{u}}{\partial \theta} + \frac{\partial \mathbf{u}}{\partial \theta} \cdot \frac{\partial \mathbf{u}}{\partial \rho}\right\},$$

and also

$$\frac{\partial \mathbf{u}}{\partial \rho} = \frac{\partial w}{\partial \rho}\mathbf{t}_\rho + \frac{\partial u}{\partial \rho}\mathbf{t}_\phi + \frac{\partial v}{\partial \rho}\mathbf{t}_\theta,$$

$$\frac{\partial \mathbf{u}}{\partial \phi} = \left[\frac{\partial w}{\partial \phi} - u\right]\mathbf{t}_\rho + \left[\frac{\partial u}{\partial \phi} + w\right]\mathbf{t}_\phi + \left[\frac{\partial v}{\partial \phi}\right]\mathbf{t}_\theta,$$

$$\frac{\partial \mathbf{u}}{\partial \theta} = \left[\frac{\partial w}{\partial \theta} - v \sin \phi\right]\mathbf{t}_\rho + \left[\frac{\partial u}{\partial \theta} - v \cos \phi\right]\mathbf{t}_\phi + \left[\frac{\partial v}{\partial \theta} + w \sin \phi + u \cos \phi\right]\mathbf{t}_\theta.$$

Finally, verify the following strain-displacement relations (cf. Eq. 5.14.3b):

$$\gamma_{\rho\rho} = \frac{\partial w}{\partial \rho} + \frac{1}{2}\left[\left(\frac{\partial w}{\partial \rho}\right)^2 + \left(\frac{\partial u}{\partial \rho}\right)^2 + \left(\frac{\partial v}{\partial \rho}\right)^2\right],$$

$$\gamma_{\phi\phi} = \frac{1}{\rho}\frac{\partial u}{\partial \phi} + \frac{w}{\rho} + \frac{1}{2}\left[\left(\frac{1}{\rho}\frac{\partial w}{\partial \phi} - \frac{u}{\rho}\right)^2 + \left(\frac{1}{\rho}\frac{\partial u}{\partial \phi} + \frac{w}{\rho}\right)^2 + \left(\frac{1}{\rho}\frac{\partial v}{\partial \phi}\right)^2\right],$$

$$\gamma_{\theta\theta} = \frac{1}{\rho \sin \phi}\frac{\partial v}{\partial \theta} + \frac{w}{\rho} + \frac{u}{\rho}\cot \phi$$
$$+ \frac{1}{\sin^2 \phi}\left[\left(\frac{1}{\rho}\frac{\partial w}{\partial \theta} - \frac{v}{\rho}\sin \phi\right)^2 + \left(\frac{1}{\rho}\frac{\partial u}{\partial \theta} - \frac{v}{\rho}\cos \phi\right)^2\right.$$
$$\left. + \left(\frac{1}{\rho}\frac{\partial v}{\partial \theta} + \frac{w}{\rho}\sin \phi + \frac{u}{\rho}\cos \phi\right)^2\right],$$

$$\gamma_{\rho\phi} = \frac{1}{2}\left\{\frac{1}{\rho}\frac{\partial w}{\partial \phi} - \frac{u}{\rho} + \frac{\partial u}{\partial \rho} + \left[\frac{\partial w}{\partial \rho}\left(\frac{1}{\rho}\frac{\partial w}{\partial \phi} - \frac{u}{\rho}\right) + \frac{\partial u}{\partial \rho}\left(\frac{1}{\rho}\frac{\partial u}{\partial \phi} + \frac{w}{\rho}\right) + \frac{\partial v}{\partial \rho}\frac{1}{\rho}\frac{\partial v}{\partial \phi}\right]\right\},$$

$$\gamma_{\phi\theta} = \frac{1}{2}\left\{\frac{1}{\rho\sin\phi}\frac{\partial u}{\partial\theta} - \cot\phi\frac{v}{\rho} + \frac{1}{\rho}\frac{\partial v}{\partial\phi} + \frac{1}{\sin\phi}\left[\left(\frac{1}{\rho}\frac{\partial w}{\partial\phi} - \frac{u}{\rho}\right)\left(\frac{1}{\rho}\frac{\partial w}{\partial\theta} - \sin\phi\frac{v}{\rho}\right)\right.\right.$$

$$\left.\left. + \left(\frac{1}{\rho}\frac{\partial u}{\partial\phi} + \frac{w}{\rho}\right)\left(\frac{1}{\rho}\frac{\partial u}{\partial\theta} - \cot\phi\frac{v}{\rho}\right) + \left(\frac{1}{\rho}\frac{\partial v}{\partial\phi}\right)\left(\frac{1}{\rho}\frac{\partial v}{\partial\theta} + \sin\phi\frac{w}{\rho} + \cos\phi\frac{u}{\rho}\right)\right]\right\},$$

$$\gamma_{\theta\rho} = \frac{1}{2}\left\{\frac{\partial v}{\partial\rho} + \frac{1}{\rho\sin\phi}\frac{\partial w}{\partial\theta} - \frac{v}{\rho} + \frac{1}{\sin\phi}\left[\left(\frac{\partial w}{\partial\rho}\right)\left(\frac{1}{\rho}\frac{\partial w}{\partial\theta} - \frac{v}{\rho}\sin\phi\right)\right.\right.$$

$$\left.\left. + \left(\frac{\partial u}{\partial\rho}\right)\left(\frac{1}{\rho}\frac{\partial u}{\partial\theta} - \frac{v}{\rho}\cos\phi\right) + \left(\frac{\partial v}{\partial\rho}\right)\left(\frac{1}{\rho}\frac{\partial v}{\partial\theta} + \frac{w}{\rho}\sin\phi + \frac{u}{\rho}\cos\phi\right)\right]\right\}.$$

[See note at end of Problem 5.16.]

5.18   Determine the dilatations (relative volume changes) for the following two states of strain:

(a)
$$[\gamma_{mn}] = \begin{bmatrix} -0.5 & 1 & 0 \\ 1 & 2 & 0.5 \\ 0 & 0.5 & 0 \end{bmatrix},$$

(b)
$$[\gamma_{mn}] = \begin{bmatrix} -0.005 & 0.01 & 0 \\ 0.01 & 0.02 & 0.005 \\ 0 & 0.005 & 0 \end{bmatrix}.$$

5.19   The body shown in Fig. P.5.19 is assumed to be in a state of strain given by:

$$\epsilon_{11} = 0.01(y_2)^2,$$
$$\epsilon_{12} = -0.01\, y_1 y_2,$$
all other $\epsilon_{mn} = 0.$

Using the compatibility relations check to see whether this strain state is possible for a continuous body. If so, find the displacements u_m corresponding to these strains. *Hint:* Note the boundary conditions on the edge $y_1 = 0$. Sketch the displacement components y_1, y_2 and y_3 (exaggerate the scale if necessary).

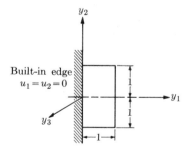

FIGURE P.5.19

5.20   With the aid of the ideas developed in Sections 5.4 and 5.5, prove that the components of strain γ_{rr}, $\gamma_{\theta\theta}$, γ_{zz}, and $\gamma_{r\theta}$, $\gamma_{\theta z}$, γ_{zr} derived in Problem 5.16 are the relative elongations and changes in angles for small strains. What are the strain-displacement relations for linear strain?

5.21   With the aid of the ideas developed in Sections 5.4 and 5.5, prove that the components of strain γ_{rr}, $\gamma_{\phi\phi}$, $\gamma_{\theta\theta}$ and $\gamma_{r\phi}$, $\gamma_{\phi\theta}$, $\gamma_{\theta r}$ derived in Problem 5.17 are the relative elongations and changes in angles for small strains. What are the strain-displacement relations for linear strain?

REFERENCES

(1) S. TIMOSHENKO and J. N. GOODIER. *Theory of Elasticity*, 2nd Ed., McGraw-Hill, New York, 1951.
(2) E. E. SECHLER, *Elasticity in Engineering*, John Wiley and Sons, New York, 1952.
(3) W. PRAGER and P. G. HODGE, Jr., *Theory of Perfectly Plastic Solids*, John Wiley and Sons, New York, 1951.
(4) G. B. THOMAS, *Calculus and Analytic Geometry*, Addison-Wesley, Reading, Mass., 1961.

Analysis of Stress

6.1 INTRODUCTION

The analysis of stress is concerned with a study of the cohesive forces, those which hold the structure together as an integral unit. On the scale of the individual atoms, these are the covalent, ionic, metallic, and van der Waals forces that bind the atoms together. In principle, stress analysis can proceed from such a viewpoint, but even the solid-state physicist is forced to idealize the discrete nature of the solid to a continuum and to adopt the ideas of continuum mechanics. The notion of stress which is adopted, therefore, is of an integrated effect of the atomic forces as seen from a macroscopic point of view. It is interesting to note that this continuum approach yields useful results down to a scale of dimension on the order of the spacing between atoms.

A structure or body is said to be "stress free" if the only internal forces present are those necessary to give the structure or body its desired shape in the absence of all external influences including gravitational attraction. Also specifically excluded are internal forces that might be generated during the process of manufacturing the material or of fabricating the specific configuration of the structure or body. Thus the stress-free structure is one in which all the atoms are arranged in their equilibrium interatomic positions. In the usage of solid mechanics, "stresses" will arise if the equilibrium interatomic spacings are disturbed. Thus the analysis of stress is actually concerned with the changes in atomic forces rather than with the absolute level of the atomic forces.

6.2 CONCEPT OF STRESS AT A POINT

The concept of stress at a point p in the interior of a body is introduced by visualizing an imaginary cut or slit of area ΔA surrounding point p (see Fig. 6.1). A pair of matching faces (i.e., surfaces) would be produced by the cut. According to the ideas of internal forces developed in Chapter 3, the detailed action of one face on the other can be represented by an equipollent

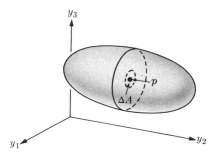

FIG. 6.1 Body cut by imaginary plane.

141

system consisting of a force vector $\mathbf{F}$ and moment vector $\mathbf{M}$. These are shown on the left face produced by the cut in Fig. 6.2 on which the orientation of the element of area ΔA is specified by the outward unit normal vector $\mathbf{n}$. The element of area ΔA is permitted to shrink in size toward zero, but in a manner such that point p always remains inside and $\mathbf{n}$ remains the normal vector. Then, from physical considerations, it is assumed that the following limits occur:

$$\lim_{\Delta A \to 0} \left(\frac{\Delta \mathbf{F}}{\Delta A} \right) = \frac{d\mathbf{F}}{dA} = \boldsymbol{\sigma}, \tag{6.2.1}$$

$$\lim_{\Delta A \to 0} \left(\frac{\Delta \mathbf{M}}{\Delta A} \right) = 0.^* \tag{6.2.2}$$

The vector denoted by $\boldsymbol{\sigma}$ is called the *stress vector* at point p with respect to an infinitesimal element of area dA which has an orientation specified by a unit normal, $\mathbf{n}$, as illustrated in Fig. 6.3. Note that the dimensions of $\boldsymbol{\sigma}$ are force per unit area.

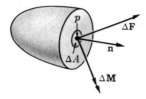

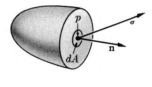

Fɪɢ. 6.2 Equipollent internal force system acting on Δa of isolated left half of body.

Fɪɢ. 6.3 Stress vector acting on dA of isolated left half of body.

6.3 THE STRESS TENSOR

If a structure is subjected to external loads, the stress-free shape of the structure will be changed. In the previous chapter, the deformations were analyzed in terms of the strain tensor and the displacement vector. As has been stated, stresses arise because the interatomic spacings are disturbed. If an element of volume in the form of a rectangular parallelepiped in the undeformed body is figuratively isolated (as in Fig. 5.3), then in the present context, there are no stresses acting on the exposed surfaces. There will be, however, stresses acting on the exposed surfaces of the deformed element of volume. The effect is illustrated in Fig. 6.4, which is based on Fig. 5.3.

In Fig. 6.4, the stresses that act on the surfaces of the deformed element of volume are summarized in the form of the stress vectors $\boldsymbol{\sigma}_1$, $\boldsymbol{\sigma}_2$, $\boldsymbol{\sigma}_3$ and their differential changes $d\boldsymbol{\sigma}_1$, $d\boldsymbol{\sigma}_2$, $d\boldsymbol{\sigma}_3$. The stress vectors have dimensions of force

---

* There also is a branch of elasticity which assumes this limit to be finite. The resulting quantity is called a *couple-stress vector*. See Cosserat [1] and Mindlin [2].

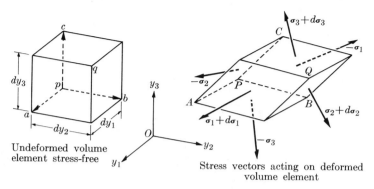

FIG. 6.4   Stress vectors acting on deformed volume element.

per unit area, and are assumed to act at the respective centers of each of the faces. Note carefully that Fig. 6.4 schematically represents the stress vectors, which act on the deformed element of volume, and not the force vectors. The latter are obtained by multiplying each stress vector by the appropriate surface area. It should also be observed that a stress vector is not, in general, oriented perpendicular to the surface on which it acts.

Fundamentally, the analysis of stress should proceed on the basis of the stress vectors acting on the deformed element of volume. This method would lead to certain geometrical difficulties best avoided at this stage in solid mechanics. Briefly, these difficulties arise because the element of volume in the deformed body (see Fig. 6.4) is not a rectangular parallelepiped. Its shape can only be described in terms of the displacements or strain tensor, and this means that the analysis of stress becomes inextricably entwined with the analysis of strain. In many of the modern problems confronting the men in structural research such an intertwining of stress and strain is unavoidable and, indeed, is the challenge. There is a restriction, however, which can be made to circumvent these difficulties. This restriction is embodied in the assumption of linear strains (see Section 5.9). It will be convenient in our discussion to adopt the assumption of linear strain for the analysis of stress. We will see later, however, that some of the results we are developing can also be applied to the general case of small strain.

It will be recalled that the assumption of linear strain implies that the edges of the deformed element of volume have undergone negligible rotations from their corresponding positions in the undeformed body (see Section 5.9). Additionally, the lengths of the edges in the deformed state were found to differ only by a negligible amount from their original lengths. Hence the surface areas of the faces differ only infinitesimally from the corresponding areas of the undeformed element of volume. All these factors contrive to make the deformed volume element indistinguishable from the undeformed volume element insofar as the analysis of stress is concerned. Thus the analysis of stress for linear

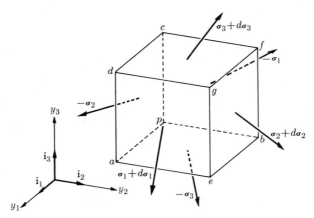

Fig. 6.5   Stress vectors neglecting body deformation.

strains can be based on an element of volume which is a rectangular parallelepiped and which can be assumed to be taken from the undeformed body (see Fig. 6.5).

The stress vectors σ_m are identified by a single subscript which denotes the direction of the unit normal vector of the surface on which the stress vector acts. For example, σ_2 acts on face *padc* which is perpendicular to the unit vector i_2. Similarly, σ_1 acts on a face perpendicular to i_1, and σ_3 acts on a face perpendicular to i_3. The stress vectors are written in component form as

$$\sigma_1 = \sigma_{1m}i_m = \sigma_{11}i_1 + \sigma_{12}i_2 + \sigma_{13}i_3, \tag{6.3.1}$$

$$\sigma_2 = \sigma_{2m}i_m = \sigma_{21}i_1 + \sigma_{22}i_2 + \sigma_{23}i_3, \tag{6.3.2}$$

$$\sigma_3 = \sigma_{3m}i_m = \sigma_{31}i_1 + \sigma_{32}i_2 + \sigma_{33}i_3, \tag{6.3.3}$$

or in the compact tensor form as

$$\sigma_n = \sigma_{nm}i_m. \tag{6.3.4}$$

The stress tensor comprises the nine components σ_{nm} (it will presently be shown that the stress tensor has only six independent components). The one described above is the Cartesian stress tensor, since the definition is based on the use of rectangular Cartesian coordinates. It is observed that the first subscript in σ_{nm} coincides with the subscript in the stress vector σ_n, and that the second subscript coincides with the unit vector i_m. Thus the first subscript locates the plane on which σ_{nm} is acting, and the second subscript gives its direction. In this special case of rectangular Cartesian coordinates, the components of the stress tensor admit to a fairly simple interpretation. The components σ_{11}, σ_{22}, σ_{33} which have identical subscripts act perpendicular to the surface, and are usually referred to as the normal or extensional stresses. If

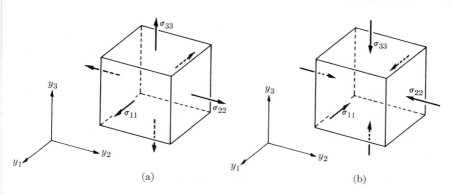

FIG. 6.6   Normal stress sign convention: (a) normal stresses in tension (positive); (b) normal stresses in compression (negative).

the sense of the normal stress is directed outward from the element of volume, the normal stress is referred to as *tensile* (or tension) and is assigned a positive number (see Fig. 6.6a). If the stress is directed inward toward the center of the element of volume, the normal stress is *compressive* (or compression) and is negative (see Fig. 6.6b).

The other six components which have different subscripts lie in the plane of the surface, and are called the *shear components*. It is a bit more involved to assign a positive or negative sign to the shear component than to the normal component. The explanation will be aided by referring to Fig. 6.7, which shows the shear components (in this figure the $d\boldsymbol{\sigma}_m$ have been set equal to zero for purposes of clarity). Two figures have been drawn. On both of these, the shear stresses as depicted are all positive. Negative shear stresses on any of the six faces of the rectangular parallelepiped point in directions opposite to those shown in Fig. 6.7. The first subscript in the shear stress, σ_{nm} $(m \neq n)$, coincides

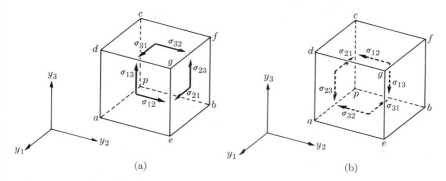

FIG. 6.7   Positive shear stress convention: (a) shear stress components on faces having positive, outward normals point in positive axis directions; (b) shear stress components on faces having negative, outward normals point in negative axis directions.

with the subscripts of the unit normal, i_n, which is perpendicular to the surface on which the shear stress acts. The second subscript coincides with the direction in which the line segment representing the shear stress is aligned. Thus the shear stress σ_{23} acts on surface *begf* whose normal is i_2, and is aligned in a direction parallel to the y_3-axis. Similarly, the shear stress σ_{32} acts on surface *cdgf* whose normal is i_3, and is aligned parallel to the y_2-axis. Each of the six faces of the rectangular parallelepiped can be identified by specifying the outward directed unit normal. Thus i_1 designates the face *aegd*, whereas the opposite face *pbfc* is designated by $-i_1$. The shear stress is positive if either (a) it acts on a face which has a positively directed outward normal and has a sense in the positive direction of the axis designated by the second subscript (Fig. 6.7a), or (b) it acts on a face which has a negatively directed outward normal and has a sense in the negative direction of the axis designated by the second subscript (Fig. 6.7b).

6.4   THE TRANSFORMATION OF STRESS

The Cartesian stress tensor has been defined with respect to an arbitrarily selected coordinate system, y_n (cf. Fig. 6.5). It is important to ascertain the components of the stress tensor if another coordinate system is chosen. This process is called the *transformation of stress*. Figure 6.8 shows a coordinate system, y_n, and another coordinate system, $\tilde{y}_n$. Both coordinate systems can be used to describe the state of stress at point p.

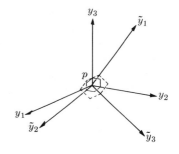

FIGURE 6.8

The law of transformation is arrived at by considering the equilibrium of a tetrahedron *pabc* formed by slicing the parallelepiped of Fig. 6.5 by a fourth face perpendicular to $\tilde{y}_1$ (see Fig. 6.9). Face *abc*, which is canted to the axes y_n, is located by the unit normal vector j_1. The system of unit vectors j_n associated with the axes $\tilde{y}_n$ are fixed in the face *abc*, with j_1 perpendicular to, and j_2, j_3 lying in the canted face, *abc*. The vectors j_n can be expressed in terms of components relative to the axes y_n:

$$j_m = l_{\tilde{m}n} i_n, \qquad (6.4.1)$$

where the numbers designated by $l_{\tilde{m}n}$ are merely the direction cosines which connect the axes $\tilde{y}_n$ with y_n (see Appendix, Section A.12).

The stress vectors which act on the tetrahedron *pabc* are schematically represented in Fig. 6.10. The stress vectors σ_1, σ_2, σ_3 are the same as the ones shown in Fig. 6.5; σ_1 is the stress vector acting on the canted face *abc*. The vector $\mathbf{F}$ shown in Fig. 6.10 is a body force vector with dimensions of force per unit volume. The total body force for this element is $\frac{1}{2}\mathbf{F} \, dV$, and is assumed to act at the center of mass of the tetrahedron. Since the tetrahedron is assumed

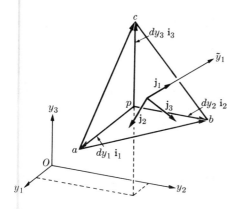

Fɪɢ. 6.9 System of unit vectors on canted face of infinitesimal tetrahedron at point p.

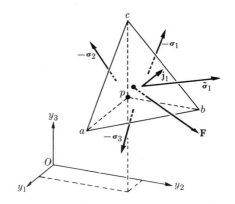

Fɪɢ. 6.10 Stresses and body force per unit volume acting on infinitesimal tetrahedron at point p.

to be in equilibrium, the following force equilibrium equation must be true:

$$\tfrac{1}{2}\tilde{\sigma}_1 \, d\tilde{A}_1 - \tfrac{1}{2}\sigma_1 \, dA_1 - \tfrac{1}{2}\sigma_2 \, dA_2 - \tfrac{1}{2}\sigma_3 \, dA_3 + \tfrac{1}{2}\mathbf{F} \, dV = 0, \qquad (6.4.2)$$

where $\tfrac{1}{2} \, dA_1$ is the area of face pac which is a triangle, $\tfrac{1}{2} \, dA_2$ is the area of triangle pbc, $\tfrac{1}{2} \, dA_3$ is the area of triangle pab, $\tfrac{1}{2} \, d\tilde{A}_1$ is the area of triangle abc, and $\tfrac{1}{2} \, dV$ is the volume of the tetrahedron. The term containing the body force vector is of higher order and hence can be neglected. We have, then,

$$\tilde{\sigma}_1 \, d\tilde{A}_1 \;=\; \sigma_m \, dA_m. \qquad (6.4.3)$$

Before any further progress can be made, we must develop an expression connecting $d\tilde{A}_1$ and dA_m. We recall from vector analysis that the cross product of two vectors yields a vector whose magnitude is equal to the area of the parallelogram formed by the two vectors. Thus, with reference to Fig. 6.9,

$$\mathbf{ab} \times \mathbf{ac} = d\tilde{A}_1 \mathbf{j}_1, \qquad (6.4.4)$$

where $\mathbf{ab}$ and $\mathbf{ac}$ are vectors representing the edges of the canted face. It is easy to verify from Fig. 6.9 that

$$\mathbf{ab} = \mathbf{pb} - \mathbf{pa}, \qquad (6.4.5)$$

and

$$\mathbf{ac} = \mathbf{pc} - \mathbf{pa}, \qquad (6.4.6)$$

where $\mathbf{pa}$, $\mathbf{pb}$, and $\mathbf{pc}$ represent the edges of the tetrahedron. If we substitute these in Eq. (6.4.4), we find

$$\mathbf{pb} \times \mathbf{pc} - \mathbf{pa} \times \mathbf{pc} - \mathbf{pb} \times \mathbf{pa} = d\tilde{A}_1 \mathbf{j}_1. \qquad (6.4.7)$$

We make use of the following relations (see Fig. 6.9):

$$\mathbf{pb} \times \mathbf{pc} = \quad dA_1\mathbf{i}_1, \tag{6.4.8}$$

$$\mathbf{pa} \times \mathbf{pc} = -dA_2\mathbf{i}_2, \tag{6.4.9}$$

$$\mathbf{pb} \times \mathbf{pa} = -dA_3\mathbf{i}_3. \tag{6.4.10}$$

When we substitute them in Eq. (6.4.7) we get

$$dA_1\mathbf{i}_1 + dA_2\mathbf{i}_2 + dA_3\mathbf{i}_3 = d\tilde{A}_1\mathbf{j}_1, \tag{6.4.11}$$

and hence the unit normal to the canted face has the component form [cf. Eqs. (6.4.7) and (6.4.1)]:

$$\mathbf{j}_1 = \frac{dA_m}{d\tilde{A}_1}\,\mathbf{i}_m. \tag{6.4.12}$$

If this latter expression is compared with Eq. (6.4.1), we see that

$$l_{\bar{1}m} = \frac{dA_m}{d\tilde{A}_1}, \tag{6.4.13}$$

that is, the ratios of the areas of the orthogonal faces of the tetrahedron to the area of the fourth face are also the direction cosines which the unit normal vector $\mathbf{j}_1$ makes with the y_n-axes. Thus Eq. (6.4.3) becomes

$$\tilde{\sigma}_1 = l_{\bar{1}m}\sigma_m. \tag{6.4.14}$$

In the coordinate system $\tilde{y}_n$, which is fixed in the canted face, the component form of $\tilde{\sigma}_1$ is written as

$$\tilde{\sigma}_1 = \tilde{\sigma}_{11}\mathbf{j}_1 + \tilde{\sigma}_{12}\mathbf{j}_2 + \tilde{\sigma}_{13}\mathbf{j}_3 = \tilde{\sigma}_{1m}\mathbf{j}_m, \tag{6.4.15}$$

where the components $\tilde{\sigma}_{nm}$ $(n = 1)$ in Eq. (6.4.15) are the sought-after representation of stress at p with respect to $\tilde{y}_n$. By substituting Eq. (6.4.15) into the left-hand side of Eq. (6.4.14), and Eq. (6.3.4) into the right-hand side, we find that the equilibrium equation of the tetrahedron becomes

$$\tilde{\sigma}_{1m}\mathbf{j}_m = l_{\bar{1}m}\sigma_{mn}\mathbf{i}_n. \tag{6.4.16}$$

This is a vector equation from which we can obtain three scalar equations by forming the scalar product with $\mathbf{j}_1$, $\mathbf{j}_2$, and $\mathbf{j}_3$, respectively. We note that

$$\mathbf{j}_m \cdot \mathbf{i}_n = l_{\tilde{m}n}. \tag{6.4.17}$$

The scalar equations of equilibrium, which are the desired equations relating σ_{mn} to $\tilde{\sigma}_{mn}$, are summarized by

$$\tilde{\sigma}_{1r} = \sigma_{mn}l_{\bar{1}m}l_{\tilde{r}n}. \tag{6.4.18}$$

These represent three of the transformation equations relating $\tilde{\sigma}_{mn}$ to σ_{mn}. The remaining equations are obtained by forming two other tetrahedrons in which the fourth faces are perpendicular to $\tilde{y}_2$ and $\tilde{y}_3$, respectively. By cyclically changing the 1 to a 2 and then to a 3, it is clear that the entire set of transformation equations is contained in the following:

$$\tilde{\sigma}_{rs}(\tilde{y}_1, \tilde{y}_2, \tilde{y}_3) = \sigma_{mn}(y_1, y_2, y_3) l_{\tilde{r}m} l_{\tilde{s}n}. \tag{6.4.19}$$

It should be carefully noted again that the stress tensor σ_{mn} exists at a point, for example, at the point p with coordinates y_1, y_2, y_3. Limiting processes, whereby the element of volume is shrunk in size, have been tacitly assumed to occur; thus, in the limit, these results refer to a point in the body. Note that the transformation equation for the Cartesian stress tensor (Eq. 6.4.19) is of the same mathematical form as the transformation equation for the Cartesian strain tensor (Eq. 5.6.7).

6.5 THE SYMMETRY OF THE STRESS TENSOR: MOMENT EQUILIBRIUM

The stress tensor σ_{mn} is symmetric, that is,

$$\sigma_{mn} = \sigma_{nm}. \tag{6.5.1}$$

This property of the stress tensor is a consequence of the moment equilibrium of the element of volume.† We can reproduce Fig. 6.5 for this special purpose, and to avoid cluttering the picture, show only two force vectors in Fig. 6.11.

First, let us recall that the moment of a force about a point p (see Chapter 3 and Appendix) can be calculated with the help of the vector product. Applying this process to the pair of force vectors shown in Fig. 6.11, we obtain for the

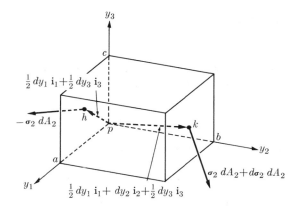

FIGURE 6.11

† It also depends on the absence of couple stresses, i.e., on the assumption made in Eq. (6.2.2). See footnote on p. 142.

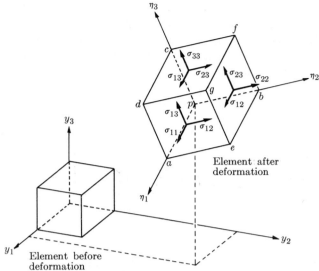

FIGURE 6.12

moment of these forces about point p,

$$- [\tfrac{1}{2} dy_1 \mathbf{i}_1 + \tfrac{1}{2} dy_3 \mathbf{i}_3] \times [\boldsymbol{\sigma}_2 \, dA_2]$$

$$+ [dy_2 \mathbf{i}_2 + \tfrac{1}{2} dy_1 \mathbf{i}_1 + \tfrac{1}{2} dy_3 \mathbf{i}_3] \times \left[\boldsymbol{\sigma}_2 \, dA_2 + \frac{\partial \boldsymbol{\sigma}_2}{\partial y_2} \, dA_2 \, dy_2 \right]$$

$$= (dy_2 \mathbf{i}_2) \times (\boldsymbol{\sigma}_2 \, dA_2) + \text{higher-order terms.} \qquad (6.5.2)$$

Similar equations can be obtained for the remaining two pairs of forces for which the subscripts are 1 and 3 respectively. Since the element of volume is in equilibrium, the sum of all contributing moments must be zero:

$$(dy_2 \mathbf{i}_2) \times (\boldsymbol{\sigma}_2 \, dA_2) + (dy_3 \mathbf{i}_3) \times (\boldsymbol{\sigma}_3 \, dA_3) + (dy_1 \mathbf{i}_1) \times (\boldsymbol{\sigma}_1 dA_1) = 0.$$

$$(6.5.3)$$

The body force vector does not contribute to the moment equilibrium equation since it contributes only a higher-order term. By canceling the common factor $dy_1 \, dy_2 \, dy_3$, Eq. (6.5.3) has the form

$$\mathbf{i}_m \times \boldsymbol{\sigma}_m = 0. \qquad (6.5.4)$$

This in turn becomes, with the introduction of the stress tensor,

$$\mathbf{i}_m \times (\sigma_{mn} \mathbf{i}_n) = 0, \qquad \text{or} \qquad \sigma_{mn} e_{mnr} \mathbf{i}_r = 0. \qquad (6.5.5)$$

The properties of the vector product of the unit orthogonal vectors reduce Eq. (6.5.5) to

$$(\sigma_{12} - \sigma_{21})\mathbf{i}_3 + (\sigma_{31} - \sigma_{13})\mathbf{i}_2 + (\sigma_{23} - \sigma_{32})\mathbf{i}_1 = 0. \qquad (6.5.6)$$

In order for the vector equation (6.5.6) to be satisfied, each component must independently vanish, that is, $\sigma_{12} = \sigma_{21}, \sigma_{31} = \sigma_{13}, \sigma_{23} = \sigma_{32}$ or, compactly,

$$\sigma_{mn} = \sigma_{nm}. \tag{6.5.7}$$

This result, then, demonstrates that the stress tensor is symmetric, and hence the stress tensor, σ_{mn}, consists of six independent components.

At this point it is worth while to extend the definition of the stress tensor to include the situation in which the strains are small but the rotations are not. In Fig. 6.12 $pabcdefg$ represents the new location of an element that was originally a parallelepiped with all edges parallel to the Cartesian coordinates y_1, y_2, and y_3. Under the condition of finite rotation the edges pa, pb, and pc will not be parallel to the y_1-, y_2-, and y_3-axes, respectively. However, under the assumption of small strain, the changes in angles ϕ_{nm} (see Section 5.5) are negligible to the extent that the edges pa, pb, and pc may still be considered mutually orthogonal. Thus at the point p, another set of rectangular Cartesian coordinates η_1, η_2, and η_3 can be located as shown. This set can be used to define the components of the stress tensors σ_{mn} that are acting on the corresponding faces of the element in its deformed position. It is clear that the transformation laws and the symmetric properties of the stress tensor which have been derived are also applicable to the present extended definition.

6.6 THE DIFFERENTIAL EQUATIONS OF EQUILIBRIUM

The moment equilibrium of the element of volume is assured by the symmetry of the stress tensor. There is still the force equilibrium to be considered. By vectorially adding up all the forces which result from the stress vectors shown in Fig. 6.5 and a body force vector $\mathbf{F}$, we find that the force equilibrium equation is

$$dA_1\,d\boldsymbol{\sigma}_1 + dA_2\,d\boldsymbol{\sigma}_2 + dA_3\,d\boldsymbol{\sigma}_3 + \mathbf{F}\,dV = 0. \tag{6.6.1}$$

Since the $d\boldsymbol{\sigma}_n$ shown in Fig. 6.5 represent changes only in the corresponding y_n-direction, then

$$d\boldsymbol{\sigma}_n = \frac{\partial \boldsymbol{\sigma}_n}{\partial y_n}\,dy_n \qquad \text{(no sum on } n\text{)}, \tag{6.6.2}$$

and the equation of force equilibrium becomes

$$\left(\frac{\partial \boldsymbol{\sigma}_1}{\partial y_1} + \frac{\partial \boldsymbol{\sigma}_2}{\partial y_2} + \frac{\partial \boldsymbol{\sigma}_3}{\partial y_3} + \mathbf{F}\right) dy_1\,dy_2\,dy_3 = 0. \tag{6.6.3}$$

If the stress tensor is introduced, and if the body force vector is written in component form as

$$\mathbf{F} = F_n \mathbf{i}_n, \tag{6.6.4}$$

then Eq. (6.6.3) assumes the form

$$\left(\frac{\partial \sigma_{mn}}{\partial y_m} + F_n\right) \mathbf{i}_n = 0, \tag{6.6.5}$$

where the common factor $dy_1 \, dy_2 \, dy_3$ has been canceled.

The scalar equations of equilibrium are obtained by forming the appropriate scalar products. Thus the scalar equations of equilibrium in the directions of the y_1-, y_2-, y_3-axes are obtained by forming the scalar product of Eq. (6.6.5) with $\mathbf{i}_1$, $\mathbf{i}_2$, and $\mathbf{i}_3$, respectively:

$$\frac{\partial \sigma_{11}}{\partial y_1} + \frac{\partial \sigma_{21}}{\partial y_2} + \frac{\partial \sigma_{31}}{\partial y_3} + F_1 = 0, \tag{6.6.6}$$

$$\frac{\partial \sigma_{12}}{\partial y_1} + \frac{\partial \sigma_{22}}{\partial y_2} + \frac{\partial \sigma_{32}}{\partial y_3} + F_2 = 0, \tag{6.6.7}$$

$$\frac{\partial \sigma_{13}}{\partial y_1} + \frac{\partial \sigma_{23}}{\partial y_2} + \frac{\partial \sigma_{33}}{\partial y_3} + F_3 = 0. \tag{6.6.8}$$

It is well to remember at this stage that the above set of equations of equilibrium has been derived on the assumption of linear strains. For the case of small strain but finite rotations, the stress components σ_{mn} are defined with respect to the rotated system of axes η_1, η_2, and η_3, and hence Eqs. (6.6.6), (6.6.7), and (6.6.8) are no longer the proper equations of equilibrium.

6.7   THE EQUATIONS OF EQUILIBRIUM ON THE SURFACE OF A BODY

The equations of equilibrium derived in the previous section are those which must hold in the interior of a body. There will be surface loads applied to some portion of the surface of a body. The intensity, i.e., the force per unit area, will be denoted as $\boldsymbol{\sigma}^*$ where the asterisk indicates that this is a prescribed quantity. Consider the body shown in Fig. 6.13. We imagine that the distribution of $\boldsymbol{\sigma}^*$ over some region of the surface is known. In particular, let us focus our attention on a triangular-shaped infinitesimal element of area at a point q. The orientation of the element of area is specified by the outward unit normal $\mathbf{n}$. Let us figuratively carve an infinitesimal tetrahedron out of the body at point q in such a manner that the element of area is the fourth face (see Fig. 6.14).

The three orthogonal faces of the tetrahedron are interior surfaces of the body and hence have the internal stress vectors $\boldsymbol{\sigma}_m$ acting on them. Figures 6.14 and 6.10 are identical except that one is located at the surface of the body and the other in the interior. The equilibrium of the tetrahedron of Fig. 6.14 is assumed, since the internal stresses at point q satisfy the relation (cf. Eq. 6.4.3),

$$\boldsymbol{\sigma}_m \, dA_m = \boldsymbol{\sigma}^* \, dA_s, \tag{6.7.1}$$

where dA_s is the triangular area on the surface. Let the unit normal **n** have components given by

$$\mathbf{n} = n_m \mathbf{i}_m. \qquad (6.7.2)$$

Then according to Eq. (6.4.13)

$$n_m = \frac{dA_m}{dA_s}. \qquad (6.7.3)$$

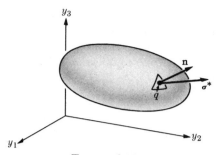

FIGURE 6.13

By introducing the stress tensor (Eq. 6.3.4) as well as Eq. (6.7.3) into Eq. (6.7.1) we obtain the desired equation of equilibrium at the surface of a body,

$$\sigma_{mn} n_m \mathbf{i}_n = \boldsymbol{\sigma}^*. \qquad (6.7.4)$$

The result summarized in Eq. (6.7.4) applies at the surface of a body; hence it is also the boundary condition for the stress tensor σ_{mn}. It relates the boundary values of σ_{mn}, that is, the values at the bounding surfaces, to the prescribed surface stresses $\boldsymbol{\sigma}^*$.

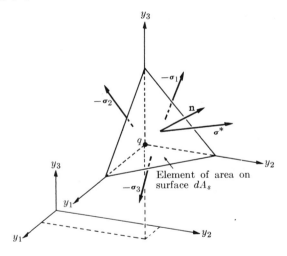

FIGURE 6.14

6.8   PRINCIPAL STRESSES AND PRINCIPAL DIRECTIONS

There are two questions which can be asked regarding the stresses at point p:

(1) At what orientation of plane abc (see Fig. 6.9) does the stress vector $\boldsymbol{\sigma}$ have no shear components; that is, when does $\boldsymbol{\sigma}$ consist only of a component normal to the plane abc?

(2) At what orientation of plane abc (see Fig. 6.9) is the normal component of $\boldsymbol{\sigma}$ a maximum or a minimum?

It will be demonstrated that the same set of planes answers both questions. These planes are called the *principal planes,* and the corresponding components of stress are called the *principal stresses.*

The answer to question (1) can be formulated by writing

$$\boldsymbol{\sigma} = \tau^* \mathbf{n}, \tag{6.8.1}$$

where τ^* is the magnitude of the normal component of $\boldsymbol{\sigma}$, and $\mathbf{n}$ is the unit normal to the plane on which τ^* acts (see Fig. 6.15).

If we combine Eqs. (6.8.1), (6.3.4), and (6.4.1), in which $\mathbf{n}$ takes on the role of $\mathbf{j}_1$, and n_m the role of $l_{\bar{1}m}$, with Eq. (6.4.14), where $\boldsymbol{\sigma}$ takes on the role of $\tilde{\boldsymbol{\sigma}}_1$, and n_m the role of $l_{\bar{1}m}$, we get the following result:

$$\sigma_{mn} \mathbf{i}_n n_m = \tau^* n_m \mathbf{i}_m, \tag{6.8.2}$$

wherein $\boldsymbol{\sigma}_m$ has been expressed in its component form (see Eq. 6.3.4). The scalar product of Eq. (6.8.2) with $\mathbf{i}_r$ yields

$$(\sigma_{mr} - \tau^* \delta_{mr}) n_m = 0. \tag{6.8.3}$$

Before analyzing Eq. (6.8.3), let us proceed to question (2).

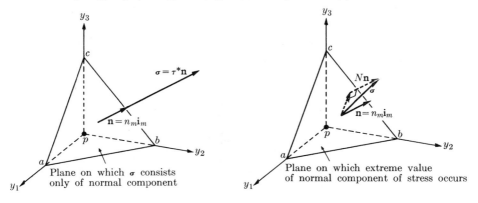

FIGURE 6.15   FIGURE 6.16

Let the normal component of $\boldsymbol{\sigma}$ be denoted by N (see Fig. 6.16). Then

$$N = \mathbf{n} \cdot \boldsymbol{\sigma}. \tag{6.8.4}$$

If we substitute Eqs. (6.3.4), (6.4.1), and (6.4.14), as we did in the development of Eq. (6.8.2), into Eq. (6.8.1), we may express the normal component N in terms of the stress tensor:

$$N = (n_r \mathbf{i}_r) \cdot (\sigma_{mn} \mathbf{i}_n n_m) = \sigma_{mn} n_m n_n. \tag{6.8.5}$$

Thus, given the stress tensor σ_{mn}, and the components n_m of the unit normal **n** which locates the plane on which $\boldsymbol{\sigma}$ acts, Eq. (6.8.5) is an explicit relation between the direction cosines of the plane abc and the normal component of the stress vector acting on abc. The maximum value of N is desired, and also the orientation of the plane on which it acts. This is again an extreme-value problem with a constraint (see Section 5.7 and Appendix, Section A.16), because the direction cosines n_m must satisfy the condition specified by

$$\delta_{mr}n_m n_r = 1. \tag{6.8.6}$$

If this equation of constraint is rewritten as

$$M = \delta_{mr}n_m n_r - 1 = 0, \tag{6.8.7}$$

then the necessary conditions for the existence of extreme values of N are embodied in the following set of partial differential equations (see Appendix, Section A.16):

$$\frac{\partial N}{\partial n_m} - \tau\frac{\partial M}{\partial n_m} = 0, \qquad \frac{\partial N}{\partial \tau} - M = 0, \tag{6.8.8}$$

where τ is the Lagrange multiplier. The first three of the necessary conditions result in a set of three simultaneous, homogeneous equations:

$$(\sigma_{mr} - \tau\delta_{mr})n_r = 0, \tag{6.8.9}$$

and the last necessary condition reduces to the equation of constraint (6.8.7). Equations (6.8.9) have exactly the same form as Eqs. (6.8.3), and hence

$$\tau^* = \tau. \tag{6.8.10}$$

Thus the Lagrange multiplier is the normal component of stress, and the planes on which the maximum value of the normal component N occurs are also the planes on which the shearing components of $\boldsymbol{\sigma}$ are zero.

The necessary and sufficient condition that the homogeneous equations (6.8.9) have a solution is that the determinant formed by the coefficients of n_m vanish:

$$\begin{vmatrix} \sigma_{11} - \tau & \sigma_{12} & \sigma_{13} \\ \sigma_{21} & \sigma_{22} - \tau & \sigma_{23} \\ \sigma_{31} & \sigma_{32} & \sigma_{33} - \tau \end{vmatrix} = 0. \tag{6.8.11}$$

At this stage, the reader should observe the mathematical similarities between the search for principal strains and directions (see Section 5.7), and the search for principal stresses and directions.

The expansion of the determinant yields a cubic equation in τ. This can be written as

$$-(\tau)^3 + I_1(\tau)^2 - I_2(\tau) + I_3 = 0, \tag{6.8.12}$$

where the coefficients I_1, I_2, I_3 are called the first, second, and third stress invariants, respectively. When expressed in terms of the stress tensor with respect to y_n they can be expressed by

$$I_1 = \sigma_{11} + \sigma_{22} + \sigma_{33}, \tag{6.8.13}$$

$$I_2 = \begin{vmatrix} \sigma_{11} & \sigma_{12} \\ \sigma_{21} & \sigma_{22} \end{vmatrix} + \begin{vmatrix} \sigma_{22} & \sigma_{23} \\ \sigma_{32} & \sigma_{33} \end{vmatrix} + \begin{vmatrix} \sigma_{11} & \sigma_{13} \\ \sigma_{31} & \sigma_{33} \end{vmatrix}, \tag{6.8.14}$$

$$I_3 = \begin{vmatrix} \sigma_{11} & \sigma_{12} & \sigma_{13} \\ \sigma_{21} & \sigma_{22} & \sigma_{23} \\ \sigma_{31} & \sigma_{32} & \sigma_{33} \end{vmatrix}. \tag{6.8.15}$$

The three roots of Eq. (6.8.12) are the principal stresses, and will be designated by σ_I, σ_{II}, σ_{III} in an ordered fashion,

$$\sigma_I > \sigma_{II} > \sigma_{III}. \tag{6.8.16}$$

The cubic equation can also be exhibited in factored form:

$$(\tau - \sigma_I)(\tau - \sigma_{II})(\tau - \sigma_{III}) = 0. \tag{6.8.17}$$

We find, therefore [compare the expanded forms of Eqs. (6.8.17) with (6.8.12)], that the stress invariants, if expressed in terms of the principal stresses, have the forms

$$I_I = \sigma_I + \sigma_{II} + \sigma_{III}, \tag{6.8.18}$$

$$I_2 = \sigma_I\sigma_{II} + \sigma_{II}\sigma_{III} + \sigma_{III}\sigma_I, \tag{6.8.19}$$

$$I_3 = \sigma_I\sigma_{II}\sigma_{III}. \tag{6.8.20}$$

Since the principal stresses must be independent of the choice of the coordinate system of point p, the coefficients are invariant quantities, and hence are rightly called stress invariants.

In order to calculate the principal directions which correspond to the principal stresses, σ_I, σ_{II}, and σ_{III}, we must adopt a procedure similar to that used in determining the principal strain directions. The principal direction corresponding to σ_I, for example, is denoted by direction cosines labeled as $n_1^{(I)}$, $n_2^{(I)}$, and $n_3^{(I)}$. These three direction cosines must satisfy the set of homogeneous equations (6.8.9) in which τ is set equal to σ_I, and also the constraint equation (6.8.6). The three simultaneous equations to be solved for $n_m^{(I)}$ are

$$(\sigma_{11} - \sigma_I)n_1^{(I)} + \sigma_{12}n_2^{(I)} + \sigma_{13}n_3^{(I)} = 0, \tag{6.8.21}$$

$$\sigma_{21}n_1^{(I)} + (\sigma_{22} - \sigma_I)n_2^{(I)} + \sigma_{23}n_3^{(I)} = 0, \tag{6.8.22}$$

$$(n_1^{(I)})^2 + (n_2^{(I)})^2 + (n_3^{(I)})^2 = 1. \tag{6.8.23}$$

Actually, Eq. (6.8.9) represents three equations, but only two of them are independent, and the two exhibited in Eqs. (6.8.21) and (6.8.22) have been arbitrarily chosen.

We can also derive two other sets of equations of the form (6.8.21), (6.8.22), and (6.8.23) for σ_{II}, $n_m^{(II)}$, and σ_{III}, $n_m^{(III)}$.

6.9 THE EXTREME SHEAR STRESSES

Clearly, it is also pertinent to raise a question about the maximum and/or minimum values of the shear stress. This is most conveniently discussed by choosing the principal directions as the frame of reference and labeling the axes y_I, y_{II}, and y_{III}. (See Fig. 6.17.)

The stress transformation law with y_I, y_{II}, y_{III} as the axes assumes a particularly simple form because there are no shear stresses along the principal directions:

$$\tilde{\sigma}_{mn} = \sigma_I l_{\tilde{m}I} l_{\tilde{n}I} + \sigma_{II} l_{\tilde{m}II} l_{\tilde{n}II}$$
$$+ \sigma_{III} l_{\tilde{m}III} l_{\tilde{n}III}. \quad (6.9.1)$$

Let us choose as the object of our interest the shear stress $\tilde{\sigma}_{12}$. This does not entail any loss in generality since the orientation of $\tilde{y}_1$ and $\tilde{y}_2$ is arbitrary. For ease in writing we set

$$l_{\tilde{1}R} = n_R, \quad R = I, II, III, \quad (6.9.2)$$

$$l_{\tilde{2}R} = \lambda_R, \quad R = I, II, III, \quad (6.9.3)$$

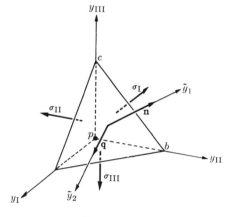

FIGURE 6.17

where n_R are the three direction cosines of the $\tilde{y}_1$-axis, and λ_R are the direction cosines of the $\tilde{y}_2$-axis (see Fig. 6.17). The transformation law for $\tilde{\sigma}_{12}$ is therefore written in the form

$$\tilde{\sigma}_{12} = \sigma_I n_I \lambda_I + \sigma_{II} n_{II} \lambda_{II} + \sigma_{III} n_{III} \lambda_{III}. \quad (6.9.4)$$

Our problem is now to find the extremum values of $\tilde{\sigma}_{12}$ subject to the constraints on the direction cosines:

$$C_1 = n_I \lambda_I + n_{II} \lambda_{II} + n_{III} \lambda_{III} = 0, \quad (6.9.5)$$

$$C_2 = (n_I)^2 + (n_{II})^2 + (n_{III})^2 - 1 = 0, \quad (6.9.6)$$

$$C_3 = (\lambda_I)^2 + (\lambda_{II})^2 + (\lambda_{III})^2 - 1 = 0. \quad (6.9.7)$$

Equation (6.9.5) is recognizable as the orthogonality condition between the $\tilde{y}_1$- and $\tilde{y}_2$-axes, whereas Eqs. (6.9.6) and (6.9.7) are the usual conditions on the direction cosines (see Eq. 6.8.6). We again use the Lagrange multiplier method and form the function

$$F = \tilde{\sigma}_{12} - \alpha C_1 - \beta C_2 - \gamma C_3, \tag{6.9.8}$$

where α, β, and γ are the Lagrange multipliers. The extremum of F is found in the standard fashion:

$$\frac{\partial F}{\partial n_I} = \sigma_I \lambda_I - \alpha \lambda_I - 2\beta n_I = 0, \tag{6.9.9}$$

$$\frac{\partial F}{\partial n_{II}} = \sigma_{II} \lambda_{II} - \alpha \lambda_{II} - 2\beta n_{II} = 0, \tag{6.9.10}$$

$$\frac{\partial F}{\partial n_{III}} = \sigma_{III} \lambda_{III} - \alpha \lambda_{III} - 2\beta n_{III} = 0, \tag{6.9.11}$$

$$\frac{\partial F}{\partial \lambda_I} = \sigma_I n_I - \alpha n_I - 2\gamma \lambda_I = 0, \tag{6.9.12}$$

$$\frac{\partial F}{\partial \lambda_{II}} = \sigma_{II} n_{II} - \alpha n_{II} - 2\gamma \lambda_{II} = 0, \tag{6.9.13}$$

$$\frac{\partial F}{\partial \lambda_{III}} = \sigma_{III} n_{III} - \alpha n_{III} - 2\gamma \lambda_{III} = 0, \tag{6.9.14}$$

$$\frac{\partial F}{\partial \alpha} = -C_1 = 0, \tag{6.9.15}$$

$$\frac{\partial F}{\partial \beta} = -C_2 = 0, \tag{6.9.16}$$

$$\frac{\partial F}{\partial \gamma} = -C_3 = 0. \tag{6.9.17}$$

These nine equations must be solved. We proceed by first determining the meaning of the Lagrange multipliers. If we multiply Eq. (6.9.9) by n_I, Eq. (6.9.10) by n_{II}, Eq. (6.9.11) by n_{III}, and then add, we get

$$2\beta = \tilde{\sigma}_{12}, \tag{6.9.18}$$

where Eqs. (6.9.4), (6.9.5), and (6.9.6) have been used. In a similar fashion it can be shown that

$$2\gamma = \tilde{\sigma}_{12}, \tag{6.9.19}$$

and

$$\alpha = \sigma_I (n_I)^2 + \sigma_{II} (n_{II})^2 + \sigma_{III} (n_{III})^2. \tag{6.9.20}$$

Next we note that Eqs. (6.9.9) and (6.9.12) are homogeneous equations for λ_I and n_I. Similarly, Eqs. (6.9.10) and (6.9.13) are homogeneous equations for λ_{II} and n_{II}, and Eqs. (6.9.11) and (6.9.14) are homogeneous equations for λ_{III} and n_{III}. There are two groups of solutions to these six equations. One group, representing what would generally be considered a trivial solution, leads to the following sets of direction cosines:

$$\text{(a)} \quad n_I = n_{II} = 0, \quad n_{III} = \pm 1,$$
$$\text{(b)} \quad n_I = n_{III} = 0, \quad n_{II} = \pm 1, \qquad (6.9.21)$$
$$\text{(c)} \quad n_{II} = n_{III} = 0, \quad n_I = \pm 1.$$

This group does not provide any new information, since it merely locates the principal planes on which the shear is zero. The other group, representing the necessary condition for the existence of a nontrivial solution to Eqs. (6.9.9) to (6.9.14), is given by

$$\tilde{\sigma}_{12} = \pm[\sigma_I - \sigma_I(n_I)^2 - \sigma_{II}(n_{II})^2 - \sigma_{III}(n_{III})^2], \qquad (6.9.22)$$
$$\tilde{\sigma}_{12} = \pm[\sigma_{II} - \sigma_I(n_I)^2 - \sigma_{II}(n_{II})^2 - \sigma_{III}(n_{III})^2], \qquad (6.9.23)$$
$$\tilde{\sigma}_{12} = \pm[\sigma_{III} - \sigma_I(n_I)^2 - \sigma_{II}(n_{II})^2 - \sigma_{III}(n_{III})^2]. \qquad (6.9.24)$$

The only consistent sets of direction cosines which will satisfy Eqs. (6.9.22), (6.9.23), and (6.9.24) as well as Eqs. (6.9.5), (6.9.6), and (6.9.7) are the following:

$$\text{(a)} \quad n_{II}^{(a)} = 0, \quad n_I^{(a)} = \pm\frac{1}{\sqrt{2}}, \quad n_{III}^{(a)} = \pm\frac{1}{\sqrt{2}},$$

$$\text{(b)} \quad n_I^{(b)} = 0, \quad n_{II}^{(b)} = \pm\frac{1}{\sqrt{2}}, \quad n_{III}^{(b)} = \pm\frac{1}{\sqrt{2}}, \qquad (6.9.25)$$

$$\text{(c)} \quad n_{III}^{(c)} = 0, \quad n_I^{(c)} = \pm\frac{1}{\sqrt{2}}, \quad n_{II}^{(c)} = \pm\frac{1}{\sqrt{2}}.$$

Case (a) locates a plane parallel to the y_{II}-axis (see Fig. 6.18) and intersecting the $y_I y_{II}$-plane at 45°. Similarly, case (b) locates a plane parallel to the y_I-axis and intersecting the $y_I y_{II}$-plane at 45°. Case (c) is a plane parallel to the y_{III}-axis and intersecting the $y_{III} y_I$-plane at 45°. By substituting the direction cosines for each case into any of the three Eqs. (6.9.22), (6.9.23), or (6.9.24), the extreme values of the shear stresses can be obtained:

$$\tilde{\sigma}_{12} \equiv S_a = \pm\tfrac{1}{2}(\sigma_I - \sigma_{III}), \quad (6.9.26)$$

$$\tilde{\sigma}_{12} \equiv S_b = \pm\tfrac{1}{2}(\sigma_{II} - \sigma_{III}), \quad (6.9.27)$$

$$\tilde{\sigma}_{12} \equiv S_c = \pm\tfrac{1}{2}(\sigma_I - \sigma_{II}). \quad (6.9.28)$$

Thus the maximum shear stress acts on the plane whose normal bisects the angle between the principal planes of the maximum and minimum principal stress, and has a magnitude equal to one-half the difference between these two principal stresses.

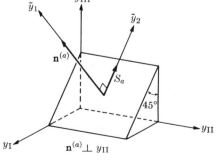

FIGURE 6.18

6.10 THE DEVIATOR AND SPHERICAL STRESS TENSORS

It will often be found convenient to express the stress tensor, σ_{mn}, as the sum of two other derived stress tensors called the *deviator* and *spherical stress tensors* (See Section 5.12). This is accomplished by writing

$$\sigma_{mn} = \hat{\sigma}_{mn} + \hat{\sigma}_{mn}, \tag{6.10.1}$$

where

$$\hat{\sigma}_{mn} = \sigma_{mn} - \tfrac{1}{3}\delta_{mn}\sigma_{rr}, \tag{6.10.2}$$

$$\hat{\sigma}_{mn} = \tfrac{1}{3}\delta_{mn}\sigma_{rr}. \tag{6.10.3}$$

The tensor, $\hat{\sigma}_{mn}$, is called the deviator stress tensor, and the other, $\hat{\sigma}_{mn}$, is called the spherical stress tensor. If Eqs. (6.10.2) and (6.10.3) are substituted in Eq. (6.10.1), an identity will result. It should be observed that σ_{rr} is the first stress invariant (cf., Eq. 6.8.13). In detail, the components of these two derived stress tensors are related to σ_{mn} as follows:

$$
\begin{aligned}
\hat{\sigma}_{11} &= \sigma_{11} - \tfrac{1}{3}(\sigma_{11} + \sigma_{22} + \sigma_{33}), & \hat{\sigma}_{12} &= \sigma_{12}, \\
\hat{\sigma}_{22} &= \sigma_{22} - \tfrac{1}{3}(\sigma_{11} + \sigma_{22} + \sigma_{33}), & \hat{\sigma}_{23} &= \sigma_{23}, \\
\hat{\sigma}_{33} &= \sigma_{33} - \tfrac{1}{3}(\sigma_{11} + \sigma_{22} + \sigma_{33}), & \hat{\sigma}_{31} &= \sigma_{31},
\end{aligned} \tag{6.10.4}
$$

and

$$
\begin{aligned}
\hat{\sigma}_{11} &= \hat{\sigma}_{22} = \hat{\sigma}_{33} = \tfrac{1}{3}(\sigma_{11} + \sigma_{22} + \sigma_{33}), \\
\hat{\sigma}_{12} &= \hat{\sigma}_{23} = \hat{\sigma}_{31} = 0.
\end{aligned} \tag{6.10.5}
$$

The concept of the stress invariants can also be applied to the deviator and the spherical stress tensors. There results (see Section 5.12)

$$\hat{I}_1 = 0, \quad \hat{I}_2 = I_2 - \tfrac{1}{3}(I_1)^2, \quad \hat{I}_3 = I_3 - \tfrac{1}{3}(I_1 I_2) + \tfrac{2}{27}(I_1)^3, \tag{6.10.6}$$

and

$$\hat{I}_1 = I_1, \quad \hat{I}_2 = \tfrac{1}{3}(I_1)^2, \quad \hat{I}_3 = \tfrac{1}{27}(I_1)^3. \tag{6.10.7}$$

6.11 SUMMARY

The stress equations derived in this chapter are summarized below in Cartesian coordinate form for convenient reference. The equations of equilibrium (6.11.3) and (6.11.4) are applicable only to linear strains. The other equations are applicable to general small strain conditions.

Stress transformation law

$$\bar{\sigma}_{rs}\,(\bar{y}_1, \bar{y}_2, \bar{y}_3) = \sigma_{mn}(y_1, y_2, y_3)l_{\bar{r}m}l_{\bar{s}n} \tag{6.11.1}$$

or

$$\tilde{\sigma}_{11} = \sigma_{11}(l_{\bar{1}1})^2 + \sigma_{22}(l_{\bar{1}2})^2 + \sigma_{33}(l_{\bar{1}3})^2$$
$$+ 2(\sigma_{12}l_{\bar{1}1}l_{\bar{1}2} + \sigma_{23}l_{\bar{1}2}l_{\bar{1}3} + \sigma_{31}l_{\bar{1}3}l_{\bar{1}1}),$$

$$\tilde{\sigma}_{22} = \sigma_{11}(l_{\bar{2}1})^2 + \sigma_{22}(l_{\bar{2}2})^2 + \sigma_{33}(l_{\bar{2}3})^2$$
$$+ 2(\sigma_{12}l_{\bar{2}1}l_{\bar{2}2} + \sigma_{23}l_{\bar{2}2}l_{\bar{2}3} + \sigma_{31}l_{\bar{2}3}l_{\bar{2}1}),$$

$$\tilde{\sigma}_{33} = \sigma_{11}(l_{\bar{3}1})^2 + \sigma_{22}(l_{\bar{3}2})^2 + \sigma_{33}(l_{\bar{3}3})^2$$
$$+ 2(\sigma_{12}l_{\bar{3}1}l_{\bar{3}2} + \sigma_{23}l_{\bar{3}2}l_{\bar{3}3} + \sigma_{31}l_{\bar{3}3}l_{\bar{3}1}),$$

$$\tilde{\sigma}_{12} = \sigma_{11}l_{\bar{1}1}l_{\bar{2}1} + \sigma_{22}l_{\bar{1}2}l_{\bar{2}2} + \sigma_{33}l_{\bar{1}3}l_{\bar{2}3} \tag{6.11.2}$$
$$+ \sigma_{12}(l_{\bar{1}1}l_{\bar{2}2} + l_{\bar{1}2}l_{\bar{2}1}) + \sigma_{23}(l_{\bar{1}2}l_{\bar{2}3} + l_{\bar{1}3}l_{\bar{2}2})$$
$$+ \sigma_{31}(l_{\bar{1}3}l_{\bar{2}1} + l_{\bar{1}1}l_{\bar{2}3}),$$

$$\tilde{\sigma}_{23} = \sigma_{11}l_{\bar{2}1}l_{\bar{3}1} + \sigma_{22}l_{\bar{2}2}l_{\bar{3}2} + \sigma_{33}l_{\bar{2}3}l_{\bar{3}3} + \sigma_{12}(l_{\bar{2}1}l_{\bar{3}2} + l_{\bar{2}2}l_{\bar{3}1})$$
$$+ \sigma_{23}(l_{\bar{2}2}l_{\bar{3}3} + l_{\bar{2}3}l_{\bar{3}2}) + \sigma_{31}(l_{\bar{2}3}l_{\bar{3}1} + l_{\bar{2}1}l_{\bar{3}3}),$$

$$\tilde{\sigma}_{31} = \sigma_{11}l_{\bar{3}1}l_{\bar{1}1} + \sigma_{22}l_{\bar{3}2}l_{\bar{1}2} + \sigma_{33}l_{\bar{3}3}l_{\bar{1}3} + \sigma_{12}(l_{\bar{3}1}l_{\bar{1}2} + l_{\bar{3}2}l_{\bar{1}1})$$
$$+ \sigma_{23}(l_{\bar{3}2}l_{\bar{1}3} + l_{\bar{3}3}l_{\bar{1}2}) + \sigma_{31}(l_{\bar{3}3}l_{\bar{1}1} + l_{\bar{3}1}l_{\bar{1}3}).$$

Equations of equilibrium

$$\frac{\partial \sigma_{mn}}{\partial y_m} + F_n = 0 \tag{6.11.3}$$

or

$$\frac{\partial \sigma_{11}}{\partial y_1} + \frac{\partial \sigma_{21}}{\partial y_2} + \frac{\partial \sigma_{31}}{\partial y_3} + F_1 = 0,$$

$$\frac{\partial \sigma_{12}}{\partial y_1} + \frac{\partial \sigma_{22}}{\partial y_2} + \frac{\partial \sigma_{32}}{\partial y_3} + F_2 = 0, \tag{6.11.4}$$

$$\frac{\partial \sigma_{13}}{\partial y_1} + \frac{\partial \sigma_{23}}{\partial y_2} + \frac{\partial \sigma_{33}}{\partial y_3} + F_3 = 0.$$

Equations for determination of principle planes

$$(\sigma_{mr} - \sigma_{\text{I}}\delta_{mr})n_r^{(\text{I})} = 0,$$
$$(\sigma_{mr} - \sigma_{\text{II}}\delta_{mr})n_r^{(\text{II})} = 0, \tag{6.11.5}$$
$$(\sigma_{mr} - \sigma_{\text{III}}\delta_{mr})n_r^{(\text{III})} = 0.$$

The characteristic equation of the system (6.11.5) is

$$-(\tau)^3 + I_1(\tau)^2 - I_2\tau + I_3 = 0, \tag{6.11.6}$$

where the roots of τ are $\tau = \sigma_{\text{I}}, \sigma_{\text{II}}, \sigma_{\text{III}}$.

Expressions for the stress invariants

$$I_1 = \sigma_{11} + \sigma_{22} + \sigma_{33} = \sigma_{\mathrm{I}} + \sigma_{\mathrm{II}} + \sigma_{\mathrm{III}}, \qquad (6.11.7)$$

$$I_2 = \begin{vmatrix} \sigma_{11} & \sigma_{12} \\ \sigma_{21} & \sigma_{22} \end{vmatrix} + \begin{vmatrix} \sigma_{22} & \sigma_{23} \\ \sigma_{32} & \sigma_{33} \end{vmatrix} + \begin{vmatrix} \sigma_{11} & \sigma_{13} \\ \sigma_{31} & \sigma_{33} \end{vmatrix} = \sigma_{\mathrm{I}}\sigma_{\mathrm{II}} + \sigma_{\mathrm{II}}\sigma_{\mathrm{III}} + \sigma_{\mathrm{III}}\sigma_{\mathrm{I}},$$

$$(6.11.8)$$

$$I_3 = \begin{vmatrix} \sigma_{11} & \sigma_{12} & \sigma_{13} \\ \sigma_{21} & \sigma_{22} & \sigma_{23} \\ \sigma_{31} & \sigma_{32} & \sigma_{33} \end{vmatrix} = \sigma_{\mathrm{I}}\sigma_{\mathrm{II}}\sigma_{\mathrm{III}}. \qquad (6.11.9)$$

Directions of normals to planes of extreme shear stresses

$$\begin{pmatrix} n_{\mathrm{I}}^{(a)} & n_{\mathrm{II}}^{(a)} & n_{\mathrm{III}}^{(a)} \\ n_{\mathrm{I}}^{(b)} & n_{\mathrm{II}}^{(b)} & n_{\mathrm{III}}^{(b)} \\ n_{\mathrm{I}}^{(c)} & n_{\mathrm{II}}^{(c)} & n_{\mathrm{III}}^{(c)} \end{pmatrix} = \begin{pmatrix} 0 & \pm 1/\sqrt{2} & \pm 1/\sqrt{2} \\ \pm 1/\sqrt{2} & 0 & \pm 1/\sqrt{2} \\ \pm 1/\sqrt{2} & \pm 1/\sqrt{2} & 0 \end{pmatrix}. \qquad (6.11.10)$$

(The above direction cosines refer to the principal axes.)

Values of extreme shear stresses

$$S_a = \pm\tfrac{1}{2}|\sigma_{\mathrm{II}} - \sigma_{\mathrm{III}}|, \qquad S_b = \pm\tfrac{1}{2}|\sigma_{\mathrm{III}} - \sigma_{\mathrm{I}}|, \qquad S_c = \pm\tfrac{1}{2}|\sigma_{\mathrm{I}} - \sigma_{\mathrm{II}}|.$$

$$(6.11.11)$$

Deviator components of stress

$$\hat{\sigma}_{mn} = \sigma_{mn} - \tfrac{1}{3}\delta_{mn}\sigma_{rr}, \qquad (6.11.12)$$

or

$$\begin{aligned} \hat{\sigma}_{11} &= \tfrac{2}{3}\sigma_{11} - \tfrac{1}{3}(\sigma_{22} + \sigma_{33}), & \hat{\sigma}_{12} &= \sigma_{12}, \\ \hat{\sigma}_{22} &= \tfrac{2}{3}\sigma_{22} - \tfrac{1}{3}(\sigma_{11} + \sigma_{33}), & \hat{\sigma}_{23} &= \sigma_{23}, \\ \hat{\sigma}_{33} &= \tfrac{2}{3}\sigma_{33} - \tfrac{1}{3}(\sigma_{11} + \sigma_{22}), & \hat{\sigma}_{31} &= \sigma_{31}. \end{aligned} \qquad (6.11.13)$$

Spherical components of stress

$$\hat{\hat{\sigma}}_{mn} = \tfrac{1}{3}\delta_{mn}\sigma_{rr}, \qquad (6.11.14)$$

or

$$\hat{\hat{\sigma}}_{11} = \hat{\hat{\sigma}}_{22} = \hat{\hat{\sigma}}_{33} = \tfrac{1}{3}(\sigma_{11} + \sigma_{22} + \sigma_{33}), \qquad (6.11.15)$$

$$\hat{\hat{\sigma}}_{12} = \hat{\hat{\sigma}}_{23} = \hat{\hat{\sigma}}_{31} = 0. \qquad (6.11.16)$$

Deviator stress invariants

$$\hat{I}_1 = 0, \qquad \hat{I}_2 = I_2 - \tfrac{1}{3}(I_1)^2, \qquad \hat{I}_3 = I_3 - \tfrac{1}{3}(I_1 I_2) + \tfrac{2}{27}(I_1)^3. \qquad (6.11.17)$$

Spherical stress invariants

$$\hat{\hat{I}}_1 = I_1, \qquad \hat{\hat{I}}_2 = \tfrac{1}{3}(I_1)^2, \qquad \hat{\hat{I}}_3 = \tfrac{1}{27}(I_1)^3. \qquad (6.11.18)$$

PROBLEMS

6.1 Expand the stress transformation law, Eq. (6.4.9)

$$\tilde{\sigma}_{rs}(\tilde{y}_1, \tilde{y}_2, \tilde{y}_3) = \sigma_{mn}(y_1, y_2, y_3)l_{\tilde{r}m}l_{\tilde{s}n}$$

for $\tilde{\sigma}_{11}$ and $\tilde{\sigma}_{23}$.

6.2 At a given point in a body, $\sigma_{11} = \sigma_{22} = \sigma_{33} = \sigma$, and $\sigma_{12} = \sigma_{13} = \sigma_{23} = 0$. (Note that the y_1-, y_2-, and y_3-axes are principal axes.) By means of the stress transformation law show that in any coordinate system $\tilde{y}_1, \tilde{y}_2$, and $\tilde{y}_3$, the normal stresses are equal to σ and the shear stresses are zero.

Such a state of stress, when σ is negative, is called *hydrostatic compression* (or pressure), and when σ is positive, *hydrostatic tension*.

6.3 At a given point in a body $\sigma_{11} = \sigma_{22} = \sigma$, and $\sigma_{12} = \sigma_{23} = \sigma_{13} = 0$. (Note that the y_1-, y_2-, and y_3-axes, are principal axes.) By means of the stress transformation law show that when the $\tilde{y}_1$- and $\tilde{y}_2$-axes are in the y_1y_2-plane, $\tilde{\sigma}_{11} = \tilde{\sigma}_{22} = \sigma$ and $\tilde{\sigma}_{12} = 0$. What is the value of the maximum shear stress (in terms of σ_{33}) when $\sigma = k\sigma_{33}$?

6.4 A state of plane stress is defined by the condition that three of the stress components are zero, for example, σ_{33}, σ_{13}, and σ_{23} are zero. Assume that the nonzero components are

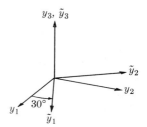

$$\sigma_{11} = 10{,}000 \text{ psi,}$$
$$\sigma_{22} = -5{,}000 \text{ psi,}$$
$$\sigma_{12} = 5{,}000 \text{ psi.}$$

Determine $\tilde{\sigma}_{11}, \tilde{\sigma}_{22}$, and $\tilde{\sigma}_{12}$ given that $\tilde{y}_1$ and $\tilde{y}_2$ are in the y_1y_2-plane and the angle between y_1 and $\tilde{y}_1$ is 30°. Determine the two principal stresses, σ_I and σ_{II}. *Note* that the idea of Mohr's circle can be applied to the transformation of stress (see Problem 5.6).

FIGURE P.6.4

6.5 Show that the state of stress at a stress-free surface is a plane stress state if the y_3-axis is normal to the surface, and y_1 and y_2 are tangent to the surface (Fig. P.6.5).

6.6 Figure P.6.6 illustrates the principal stresses acting on the principal planes of an infinitesimal cube surrounding a point p in a body. What is the maximum shear stress?

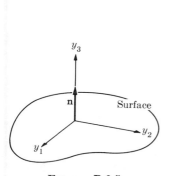

FIGURE P.6.5

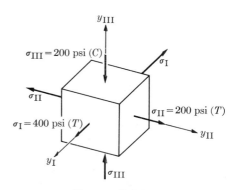

FIGURE P.6.6

6.7   A round bar is loaded as a tension specimen with an axial stress $\sigma_{11} = \sigma$. Then a radial pressure p is applied (by a fluid) so that $p = \sigma/4$. By what factor is the maximum shear stress increased with the application of this pressure?

6.8   An *octahedral plane* is a plane which intersects the principal axes at equal angles (Fig. P.6.8). (a) Show that the shear stress on this plane (called the *octahedral shear stress*) is given by

$$\sigma_{oct} = \tfrac{1}{3}\sqrt{(\sigma_I - \sigma_{II})^2 + (\sigma_{II} - \sigma_{III})^2 + (\sigma_{III} - \sigma_I)^2}.$$

(b) Show that the octahedral shear stress is related to the second deviator stress invariant by

$$\sigma_{oct} = \sqrt{-\tfrac{2}{3}\hat{I}_2}.$$

6.9   A pressure vessel is shown in Fig. P.6.9. If the circumferential normal stress (hoop stress) σ_{33} and the longitudinal normal stress σ_{11} are assumed uniform, while the normal stress σ_{22} along the radial direction varies linearly, we have

$$\sigma_{11} = pr/2t, \qquad \sigma_{22} = -p(1 - y_2/t), \qquad \sigma_{33} = pr/t,$$

where r is the radius of the vessel. In this case, the shear stresses corresponding to these axes are zero. Determine the location where the shear stress is maximum. Calculate the magnitude of this shear stress and indicate the plane where this maximum shear stress exists.

6.10   The components of a stress tensor at a point p are shown in the following matrix:

$$\sigma_{mn} = \begin{bmatrix} 12{,}000 & 4000 & 0 \\ 4000 & -8000 & 2000 \\ 0 & 2000 & -4000 \end{bmatrix}.$$

Determine the principal stresses, their directions, and the extremum value of each of the shear stresses.

6.11   (a) Verify that the equations of equilibrium (6.6.6) through (6.6.8) reduce to the form:

$$\frac{\partial \sigma_{11}}{\partial y_1} + \frac{\partial \sigma_{21}}{\partial y_2} + F_1 = 0, \qquad \frac{\partial \sigma_{12}}{\partial y_1} + \frac{\partial \sigma_{22}}{\partial y_2} + F_2 = 0,$$

for plane stress in the y_1y_2-plane. (b) Verify that the function $\phi(y_1, y_2)$, defined by

$$\sigma_{11} = \frac{\partial^2 \phi}{\partial y_2^2}, \qquad \sigma_{22} = \frac{\partial^2 \phi}{\partial y_1^2}, \qquad \sigma_{12} = -\frac{\partial^2 \phi}{\partial y_1 \, \partial y_2},$$

satisfies the equations of equilibrium identically for plane stress with no body forces. The function $\phi(y_1, y_2)$ is called the *Airy stress function*.

6.12   The triangular plate shown in Fig. P.6.12 is in a state of plane stress under a certain set of boundary conditions. The Airy stress function which describes the stress distribution is given by $\phi(y_1, y_2) = A(\tfrac{1}{12}y_1^4 + \tfrac{1}{4}y_1^2 y_2^2 - \tfrac{1}{6}y_2^4)$, where A is a constant. (a) Show that this stress function is a possible solution of the problem, given that no

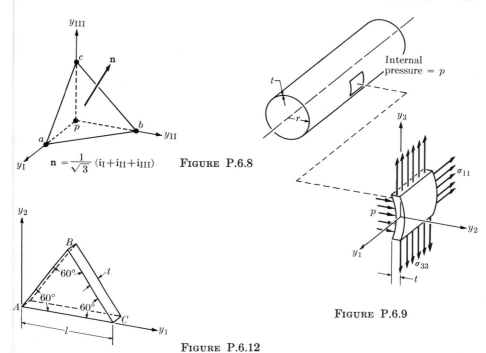

$$\mathbf{n} = \frac{1}{\sqrt{3}} (\mathbf{i}_I + \mathbf{i}_{II} + \mathbf{i}_{III}) \qquad \text{FIGURE P.6.8}$$

FIGURE P.6.9

FIGURE P.6.12

body forces act on the plate. (b) Determine the stress distribution in the plate, and the distribution of the normal component of the boundary force along edge AB ("normal" component means normal to edge AB).

6.13 A thin slab hangs under the influence of gravity g as shown in Fig. P.6.13. It can be assumed here that the body is in a state of 2-dimensional *plane stress*, that is,

$$\sigma_{31} = \sigma_{32} = \sigma_{33} = 0, \qquad \text{and} \qquad \partial/\partial y_3 = 0.$$

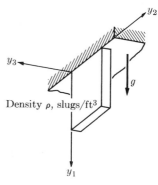

Write out the appropriate equations of equilibrium for this stress state. Show that a stress function $\Phi(y_1, y_2)$ defined as

$$\sigma_{11} = \frac{\partial^2 \Phi}{\partial y_2^2} - \rho g y_1,$$

$$\sigma_{22} = \frac{\partial^2 \Phi}{\partial y_1^2},$$

Density ρ, slugs/ft³

$$\sigma_{12} = - \frac{\partial^2 \Phi}{\partial y_1 \, \partial y_2}$$

FIGURE P.6.13

satisfies these equilibrium equations identically.

6.14 Show that the conventional beam relations $dS/dy_1 = p$, $dM/dy_1 = S$, can be obtained by integrating the appropriate differential equations of force equilibrium, $\partial \sigma_{mn}/\partial y_m + F_n = 0$, over the thickness h and width b of a beam of rectangular cross-

section. See Fig. P.6.14. The bending moment
and shear are the equipollent of the normal
stress σ_{11} and shear stress σ_{13} over the cross section of the beam:

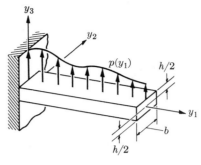

Bending moment:

$$M = -b \int_{-h/2}^{h/2} y_3\sigma_{11}\, dy_3,$$

Shear force:

$$S = -b \int_{-h/2}^{h/2} \sigma_{13}\, dy_3.$$

FIGURE P.6.14

The lateral loading may be considered as a distributed traction σ_{33} acting on the
upper face of the beam:

$$p = b\sigma_{33} \qquad \text{at} \quad y_3 = h/2.$$

6.15 The nine components of stress for cylindrical polar coordinates are shown
(Fig. P.6.15) acting on the appropriate element of volume which has edges of length
$r\,d\theta$, dr, and dz.

(a) Show that the unit vectors $\mathbf{t}_r$, $\mathbf{t}_\theta$, and $\mathbf{t}_z$ in cylindrical coordinates are related
to the unit vectors $\mathbf{i}_1$, $\mathbf{i}_2$, and $\mathbf{i}_3$ by

$$\mathbf{t}_r = \cos\theta\,\mathbf{i}_1 + \sin\theta\,\mathbf{i}_2, \qquad \mathbf{t}_\theta = -\sin\theta\,\mathbf{i}_1 + \cos\theta\,\mathbf{i}_2, \qquad \mathbf{t}_z = \mathbf{i}_3.$$

(b) Show that the force vectors acting on three of the faces of the element of volume
are
$$\mathbf{S}_r = (\sigma_{rr}\mathbf{t}_r + \sigma_{r\theta}\mathbf{t}_\theta + \sigma_{rz}\mathbf{t}_z)r\,d\theta\,dz,$$
$$\mathbf{S}_\theta = (\sigma_{\theta r}\mathbf{t}_r + \sigma_{\theta\theta}\mathbf{t}_\theta + \sigma_{\theta z}\mathbf{t}_z)\,dr\,dz,$$
$$\mathbf{S}_z = (\sigma_{zr}\mathbf{t}_r + \sigma_{z\theta}\mathbf{t}_\theta + \sigma_{zz}\mathbf{t}_z)r\,d\theta\,dr.$$

(c) Let the body force vector per unit of volume acting on the element of volume
be defined as $\mathbf{F} = F_r\mathbf{t}_r + F_\theta\mathbf{t}_\theta + F_z\mathbf{t}_z$. Show that the vector equation of force
equilibrium is

$$\frac{\partial \mathbf{S}_r}{\partial r}\,dr + \frac{\partial \mathbf{S}_\theta}{\partial \theta}\,d\theta + \frac{\partial \mathbf{S}_z}{\partial z}\,dz + \mathbf{F}r\,dr\,d\theta\,dz = 0.$$

(d) Show that the components of stress are symmetric, i.e., show that $\sigma_{r\theta} = \sigma_{\theta r}$,
$\sigma_{rz} = \sigma_{zr}$, $\sigma_{z\theta} = \sigma_{\theta z}$.

(e) From the results of part (c), show that the three scalar equations of equilibrium
in cylindrical polar coordinates are

$$\frac{\partial \sigma_{rr}}{\partial r} + \frac{1}{r}\frac{\partial \sigma_{r\theta}}{\partial \theta} + \frac{\partial \sigma_{rz}}{\partial z} + \frac{1}{r}(\sigma_{rr} - \sigma_{\theta\theta}) + F_r = 0,$$

$$\frac{\partial \sigma_{\theta r}}{\partial r} + \frac{1}{r}\frac{\partial \sigma_{\theta\theta}}{\partial \theta} + \frac{\partial \sigma_{\theta z}}{\partial z} + \frac{2}{r}\sigma_{\theta r} + F_\theta = 0,$$

$$\frac{\partial \sigma_{zr}}{\partial r} + \frac{1}{r}\frac{\partial \sigma_{z\theta}}{\partial \theta} + \frac{\partial \sigma_{zz}}{\partial z} + \frac{1}{r}\sigma_{zr} + F_z = 0.$$

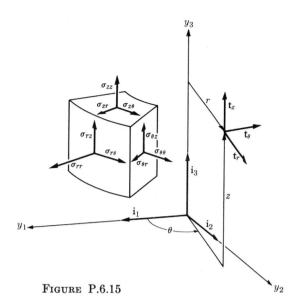

FIGURE P.6.15

Note: The components $\sigma_{\theta\theta}$, $\sigma_{\theta r}$, etc., used in this problem are the physical components of the stress tensors, and as such they all have dimensions of force per unit area. There are more elegant tensor methods for obtaining these equations of equilibrium. The interested reader is referred to Green and Zerna, Ref. 6.3 in the Bibliography.

6.16   Follow the steps outlined in Problem 6.15 and verify that the scalar equations of equilibrium in spherical polar coordinates ρ, ϕ, θ are given by (see Problem 5.17 for unit tangent vectors)

$$\frac{\partial \sigma_{\rho\rho}}{\partial \rho} + \frac{1}{\rho}\frac{\partial \sigma_{\rho\phi}}{\partial \phi} + \frac{1}{\rho \sin \phi}\frac{\partial \sigma_{\rho\theta}}{\partial \theta} + \frac{\cos \phi}{\rho \sin \phi}\sigma_{\rho\phi} + \frac{1}{\rho}[2\sigma_{\rho\rho} - \sigma_{\phi\phi} - \sigma_{\theta\theta}] + F_{\rho} = 0,$$

$$\frac{\partial \sigma_{\phi\rho}}{\partial \rho} + \frac{1}{\rho}\frac{\partial \sigma_{\phi\phi}}{\partial \phi} + \frac{1}{\rho \sin \phi}\frac{\partial \sigma_{\phi\theta}}{\partial \theta} + \frac{3}{\rho}\sigma_{\phi\phi} + \frac{\cos \phi}{\rho \sin \phi}(\sigma_{\phi\phi} - \sigma_{\theta\theta}) + F_{\phi} = 0,$$

$$\frac{\partial \sigma_{\theta\rho}}{\partial \rho} + \frac{1}{\rho}\frac{\partial \sigma_{\theta\phi}}{\partial \phi} + \frac{1}{\rho \sin \phi}\frac{\partial \sigma_{\theta\theta}}{\partial \theta} + \frac{3}{\rho}\sigma_{\theta\rho} + \frac{2 \cos \phi}{\rho \sin \phi}\sigma_{\theta\phi} + F_{\theta} = 0.$$

See *note* at the end of Problem 6.15.

REFERENCES

(1) E. et F. COSSERAT, *Théorie des Corps Déformables*, A. Hermann et Fils, Paris, 1909.
(2) R. D. MINDLIN and H. F. TIERSTEN, "Effects of Couple Stresses in Linear Elasticity, *Archives for Rational Mechanics and Analysis*, **11**, 415–448 (1962).

Elasticity

7.1 INTRODUCTION

In Chapter 4 some simple examples were used to illustrate problems which require the simultaneous application of force equations and displacement equations. This necessitated the formulation of simple constitutive relations which connected force and displacement. Thus far, the study of strain in Chapter 5 and the study of stress in Chapter 6 have been pursued independently. Certain engineering problems can be solved without the need for relating stress to strain, problems analogous to those discussed in Chapters 2 and 3. However, many problems do require the simultaneous consideration of stress and strain, and hence the connection between stress and strain needs to be formalized. This chapter will be devoted to an exposition of the basic ideas governing an elementary class of constitutive relations known as the *elastic stress-strain relations*. With these relations, which will contain experimentally determined constants describing the material, the constants k used in Chapter 4 to relate force to displacement can be determined. Additionally, the concept of strain energy will be introduced. Finally, the complete set of the equations of elasticity will be summarized.

7.2 THE GENERALIZED HOOKE'S LAW—ANISOTROPY

It is an experimental fact that the relative elongations, E_n, and angular changes, ϕ_{nm}, are linearly proportional to applied stresses for most materials used for structural purposes providing the stresses are kept to reasonable magnitudes. Furthermore, if the applied stresses are removed, the relative elongations and angular changes vanish, and the body reverts to an unstrained state, i.e., there is no macroscopic evidence that the body has been subjected to loads. A body which exhibits this type of behavior is called a *linearly elastic body*.

The Cartesian strain tensor, γ_{mn}, and the Cartesian stress tensor, σ_{mn}, have been defined in previous chapters. We discovered in Chapter 5 that a rather complicated relationship exists between relative elongation and the extensional component of the strain tensor (see Section 5.4). The relationship between the angular changes and the shearing components of the strain tensor is equally complex. It will not be possible for us to express the connection between the general strain tensor and the stress tensor in a simple or meaningful manner. Fortunately, the operating range of strain for the vast majority of engineering materials is covered by the case which we have called "small" strain. We have seen in Chapter 5 that if the strains are small then the extensional components of strain are physically the relative elongations of an element, and the shearing components are the angular changes. The developments in this chapter concerning the constitutive relations will, therefore, be restricted to small strains which then will automatically also apply to linear strains.

The most general *linear* relationship which connects stress to strain is known as the *generalized Hooke's Law* and can be expressed mathematically as

$$\sigma_{mn} = E_{mnpr}\gamma_{pr}, \tag{7.2.1}$$

where the components of the fourth-order tensor E_{mnpr} are known as the *elastic constants*. It is seen that each stress component is related to all nine components of the strain tensor. For example, one of the nine equations collectively represented by Eq. (7.2.1) has the form

$$\begin{aligned}
\sigma_{13} = E_{1311}\gamma_{11} &+ E_{1312}\gamma_{12} + E_{1313}\gamma_{13} \\
&+ E_{1321}\gamma_{21} + E_{1322}\gamma_{22} + E_{1323}\gamma_{23} \\
&+ E_{1331}\gamma_{31} + E_{1332}\gamma_{32} + E_{1333}\gamma_{33}.
\end{aligned} \tag{7.2.2}$$

There are 81 components of the tensor E_{mnpr}, but fortunately this tensor exhibits certain symmetry properties which reduce the total number of independent components to 21 for a material which does not have any axes of symmetry. Such a material is called *aeolotropic* or *anisotropic*.

If a body is in a state of strain such that the only strain components different from zero are γ_{12} and γ_{21}, then the stress-strain relations for this special situation would be written as

$$\sigma_{mn} = E^*_{mn12}\gamma_{12} + E^*_{mn21}\gamma_{21}. \tag{7.2.3}$$

However, the strain tensor is symmetric (see Section 5.3) and hence the above equation can be put into the form

$$\sigma_{mn} = (E^*_{mn12} + E^*_{mn21})\gamma_{12}. \tag{7.2.4}$$

We can now introduce a new fourth-order tensor,

$$E_{mn12} = \frac{E^*_{mn12} + E^*_{mn21}}{2}, \tag{7.2.5}$$

which is symmetric with respect to 1 and 2. Thus the stress-strain relation can be written as

$$\sigma_{mn} = E_{mn12}\gamma_{12} + E_{mn21}\gamma_{21}, \tag{7.2.6}$$

with

$$E_{mn12} = E_{mn21}. \tag{7.2.7}$$

In more general terms, we have

$$E_{mnpr} = E_{mnrp}. \tag{7.2.8}$$

Thus the tensor E_{mnpr} is unchanged when the last two indices are interchanged. This reduces the number of independent elastic constants by 27 to a total of 54.

If the state of strain in a body is such that the only nonzero component is γ_{11}, then the stress-strain relation becomes

$$\sigma_{mn} = E_{mn11}\gamma_{11}. \tag{7.2.9}$$

Two of the scalar equations contained in Eq. (7.2.9) are

$$\sigma_{12} = E_{1211}\gamma_{11}, \tag{7.2.10}$$

$$\sigma_{21} = E_{2111}\gamma_{11}. \tag{7.2.11}$$

In this case the symmetry of the stress tensor requires that

$$E_{1211} = E_{2111}, \tag{7.2.12}$$

or, in more general terms,

$$E_{mnpr} = E_{nmpr}. \tag{7.2.13}$$

Thus the tensor E_{mnpr} is unchanged if the first two indices are interchanged. This symmetry property reduces the number of independent elastic constants by 18 to a total of 36.

The further reduction of the number of independent constants to the final total of 21 can be accomplished only by thermodynamical considerations. Let us consider a unit cube of material which is in a state of stress and strain characterized by σ_{mn} and γ_{mn}. If we let the strain components increase by an amount $d\gamma_{mn}$, the work done by the stress components is given by

$$dU^* = \sigma_{mn}\,d\gamma_{mn}, \tag{7.2.14}$$

where U^* is called the *strain-energy density*. The validity of Eq. (7.2.14) can be ascertained by visualizing the deformation of a unit cube for each of the increases $d\gamma_{11}$, $d\gamma_{22}$, etc. Actually, Eq. (7.2.14) will not be fully appreciated until energy theorems are discussed in Chapter 9. However, its use is needed at this stage.

If the deformation process is isothermal and reversible, then according to the First Law of Thermodynamics, the work done is equal to the increase in the Helmholtz free energy, $d\psi$. This is stated as

$$d\psi = dU^* = \sigma_{mn}\, d\gamma_{mn}. \tag{7.2.15}$$

With the substitution of the generalized Hooke's law, Eq. (7.2.1), we have

$$d\psi = E_{mnpr}\gamma_{pr}\, d\gamma_{mn}. \tag{7.2.16}$$

Since the free energy is a function of γ_{mn}, the total differential $d\psi$ can also be written as

$$d\psi = \frac{\partial \psi}{\partial \gamma_{mn}}\, d\gamma_{mn}. \tag{7.2.17}$$

Comparing Eqs. (7.2.16) and (7.2.17), we find that

$$\frac{\partial \psi}{\partial \gamma_{mn}} = E_{mnpr}\gamma_{pr}. \tag{7.2.18}$$

The partial differentiation of Eq. (7.2.18) with respect to γ_{pr} yields

$$\frac{\partial}{\partial \gamma_{pr}}\left(\frac{\partial \psi}{\partial \gamma_{mn}}\right) = E_{mnpr}. \tag{7.2.19}$$

Interchanging the indices in Eq. (7.2.19), we obtain

$$\frac{\partial}{\partial \gamma_{mn}}\left(\frac{\partial \psi}{\partial \gamma_{pr}}\right) = E_{prmn}. \tag{7.2.20}$$

Since the order of partial differential is unimportant, that is,

$$\frac{\partial}{\partial \gamma_{pr}}\left(\frac{\partial \psi}{\partial \gamma_{mn}}\right) = \frac{\partial}{\partial \gamma_{mn}}\left(\frac{\partial \psi}{\partial \gamma_{pr}}\right), \tag{7.2.21}$$

it is clear that†

$$E_{mnpr} = E_{prmn}. \tag{7.2.22}$$

Thus the first pair of indices in the elasticity tensor can be interchanged with

---

† It can also be demonstrated that if the deformation process is adiabatic and reversible, the same conclusion is valid. Thus the assumptions of reversible isothermal or adiabatic processes are implicit in the use of a strain-energy function. Loads applied very slowly represent nearly isothermal conditions, and loads applied very rapidly, adiabatic conditions. The elastic constants $E_{mnpr} = \partial^2 U^*/\partial \gamma_{mn}\partial \gamma_{pr}$ are only slightly different for the two cases. For example, for steel in tension, the modulus of elasticity for adiabatic stretching is only 1 percent larger than for isothermal stretching.

the second pair without changing the values. This operation further reduces the number of independent elastic constants to a total of 21.†

Quite clearly, since the components both of the stress tensor and the strain tensor are functions of the orientation of the axis system, the elastic constants will also be functions of axis orientation. The transformation law for the elasticity tensor can be derived in a relatively simple fashion. If another co-ordinate system, $\tilde{y}_n$, is used then Eq. (7.2.1) is written as

$$\tilde{\sigma}_{kl}(\tilde{y}_n) = \widetilde{E}_{klst}(\tilde{y}_n)\tilde{\gamma}_{st}(\tilde{y}_n). \tag{7.2.23}$$

The tensor nature for both γ_{pr} and σ_{mn} has been established. Their transformation laws (see Eqs. 5.14.5a and 6.11.1) have been established:

$$\tilde{\sigma}_{kl} = \sigma_{mn}l_{\tilde{k}m}l_{\tilde{l}n}, \tag{7.2.24}$$

$$\gamma_{pr} = \tilde{\gamma}_{st}l_{p\tilde{s}}l_{r\tilde{t}}. \tag{7.2.25}$$

We can substitute Eq. (7.2.25) in Eq. (7.2.1), and the result into Eq. (7.2.24) to yield

$$\tilde{\sigma}_{kl} = E_{mnpr}\tilde{\gamma}_{st}l_{\tilde{k}m}l_{\tilde{l}n}l_{\tilde{s}p}l_{\tilde{t}r}, \tag{7.2.26}$$

where we have reversed the order of some of the subscripts in Eq. (7.2.25). This is permissible because $l_{p\tilde{s}}$ merely represents the cosine of the angle between y_p and $\tilde{y}_s$. Now, a comparison of Eqs. (7.2.23) and (7.2.26) reveals the transformation law for the elasticity tensor:

$$\widetilde{E}_{klst}(\tilde{y}_n) = E_{mnpr}(y_n)l_{\tilde{k}m}l_{\tilde{l}n}l_{\tilde{s}p}l_{\tilde{t}r}. \tag{7.2.27}$$

The preceding development would be considered superfluous in tensor analysis. Once the tensor natures of σ_{mn} and γ_{pr} have been established, the only permissible tensor connection between them is an equation of the form shown in Eq. (7.2.1); furthermore, the connecting function must obey the transformation law given by Eq. (7.2.27). However, since the complete tensor calculus has not been developed here, it is felt that the few additional equations would be informative and educational.

The generalized Hooke's law, in view of the fundamental symmetries which have been shown to exist, is most conveniently written as

$$\sigma_{11} = E_{1111}\gamma_{11} + E_{1122}\gamma_{22} + E_{1133}\gamma_{33} + 2(E_{1112}\gamma_{12} + E_{1113}\gamma_{13} + E_{1123}\gamma_{23}), \tag{7.2.28}$$

$$\sigma_{22} = E_{2211}\gamma_{11} + E_{2222}\gamma_{22} + E_{2233}\gamma_{33} + 2(E_{2212}\gamma_{12} + E_{2213}\gamma_{13} + E_{2233}\gamma_{23}), \tag{7.2.29}$$

---

† Historically, Cauchy thought that there were only 15 constants; Green, through energy considerations, arrived at the correct number of 21.

$$\sigma_{33} = E_{3311}\gamma_{11} + E_{3322}\gamma_{22} + E_{3333}\gamma_{33} + 2(E_{3312}\gamma_{12} + E_{3313}\gamma_{13} + E_{3323}\gamma_{23}),$$
$$(7.2.30)$$

$$\sigma_{12} = E_{1211}\gamma_{11} + E_{1222}\gamma_{22} + E_{1233}\gamma_{33} + 2(E_{1212}\gamma_{12} + E_{1213}\gamma_{13} + E_{1223}\gamma_{23}),$$
$$(7.2.31)$$

$$\sigma_{13} = E_{1311}\gamma_{11} + E_{1322}\gamma_{22} + E_{1333}\gamma_{33} + 2(E_{1312}\gamma_{12} + E_{1313}\gamma_{13} + E_{1323}\gamma_{23}),$$
$$(7.2.32)$$

$$\sigma_{23} = E_{2311}\gamma_{11} + E_{2322}\gamma_{22} + E_{2333}\gamma_{33} + 2(E_{2312}\gamma_{12} + E_{2313}\gamma_{13} + E_{2323}\gamma_{23}).$$
$$(7.2.33)$$

Note that there are three other relations which have not been displayed, since these are identical with Eqs. (7.2.31), (7.2.32), and (7.2.33). In addition, one should note the presence of the factor 2, which is due to the symmetries of the tensors.

For purposes of further discussion, it will be found convenient to exhibit the constants E_{mnpr} in a square array, or matrix form:

$$(E_{mnpr}) = \begin{pmatrix} E_{1111} & E_{1122} & E_{1133} & E_{1112} & E_{1113} & E_{1123} \\ E_{2211} & E_{2222} & E_{2233} & E_{2212} & E_{2213} & E_{2223} \\ E_{3311} & E_{3322} & E_{3333} & E_{3312} & E_{3313} & E_{3323} \\ E_{1211} & E_{1222} & E_{1233} & E_{1212} & E_{1213} & E_{1223} \\ E_{1311} & E_{1322} & E_{1333} & E_{1312} & E_{1313} & E_{1323} \\ E_{2311} & E_{2322} & E_{2333} & E_{2312} & E_{2313} & E_{2323} \end{pmatrix}, \qquad (7.2.34)\dagger$$

where $E_{mnpr} = E_{prmn}$.

The stress-strain relations given by Eq. (7.2.1) can be expressed in the inverted form

$$\gamma_{mn} = S_{mnpr}\sigma_{pr}, \qquad (7.2.35)$$

where S_{mnpr} is known as the *compliance tensor*. It should be evident, after just a few moments of thought, that S_{mnpr} has the same symmetry properties as E_{mnpr}, and the same type of transformation law. Furthermore, a set of equations analogous to Eqs. (7.2.28) through (7.2.33) can be written with the roles of σ_{mn} and γ_{mn} reversed, and the components of E_{mnpr} replaced by those of S_{mnpr} with the same arrangement of indices. In other words, an equation such as Eq. (7.2.34) can be exhibited for S_{mnpr} merely by replacing E by S.

---

† The reader who wants to write the stress-strain laws in matrix form is cautioned against using the 6 × 6 array shown in this equation. Read the sentence underneath Eq. (7.2.33).

The macroscopic structure of the material may introduce additional symmetries into the elasticity and compliance tensors. A tree, which grows by the addition of rings, yields a material with elastic properties preferentially oriented with and against the grain. A single crystal of silver, which is a face-centered cubic material, has preferred axes along the crystal lattice. Most of the common metallic engineering materials, however, consist of the random orientation of many single crystals. This randomness results in a material with properties that are macroscopically independent of the orientation of the rectangular Cartesian axes. It is an interesting exercise in the use of the transformation laws to derive the number of independent elastic constants for the various classes of materials.

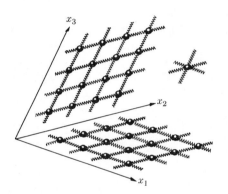

FIG. 7.1 Spring-ball model—three oblique axes.

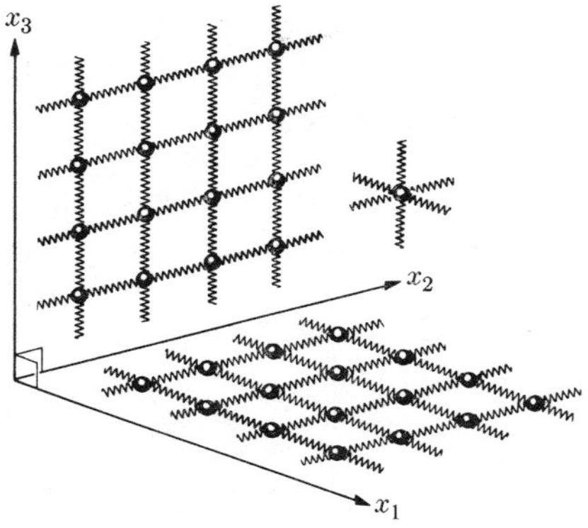

FIG. 7.2 Spring-ball model for monoclinic materials.

Before we proceed with the analysis of the symmetry relations, a few additional remarks concerning the aeolotropic or anisotropic material may be helpful. Such a material may be visualized on the atomic scale as one in which the atoms are arranged along three oblique axes (see Fig. 7.1). In place of atoms, we may consider rigid balls connected by springs with spring constants which are different along the three directions. The response of the material to impressed forces is different along each of the axes. If one pushes along the x_1-direction, changes in length as well as in angle will occur along and between all the axes. There are no planes of symmetry, and 21 constants are required to express the connection between stress and strain.

The following discussion will focus on the components of E_{mnpr}, but whatever results are found to be true for the elasticity tensor will also be true for the corresponding component of S_{mnpr}. For example, we will discover in the next section that E_{1123} must be zero for a monoclinic material. Since this result will be shown to depend upon the transformation law for E_{mnpr}, and since S_{mnpr} has the same transformation law, then S_{1123} also is zero.

7.3 MONOCLINIC MATERIAL: THIRTEEN CONSTANTS

A *monoclinic* material exhibits symmetry of its elastic properties with respect to one plane which in this section will be taken as the $y_1 y_2$-plane. If we apply the simple spring-ball model of Fig. 7.2, to a monoclinic material, its x_3-axis will be perpendicular to the other two which will intersect each other at an oblique angle. The spring constants will be different for each of the three directions, x_1, x_2, x_3. This symmetry is expressed by the requirement that the elastic constants do not change under the following coordinate transformation:

$$\tilde{y}_1 = y_1, \qquad \tilde{y}_2 = y_2, \qquad \tilde{y}_3 = -y_3. \qquad (7.3.1)$$

The direction cosines which correspond to this coordinate transformation can be expressed as follows:

	y_1	y_2	y_3
$\tilde{y}_1$	1	0	0
$\tilde{y}_2$	0	1	0
$\tilde{y}_3$	0	0	-1

$$(7.3.2)$$

Under the coordinate transformation given by Eq. (7.3.1), it is required that

$$\widetilde{E}_{mnpr}(\tilde{y}_1, \tilde{y}_2, \tilde{y}_3) = E_{mnpr}(y_1, y_2, y_3). \qquad (7.3.3)$$

Clearly, the additional symmetry must further reduce the number of independent constants from the 21 required to describe an aeolotropic material. The application of the transformation law for the elasticity tensor (Eq. 7.2.27) will determine the conditions necessary to satisfy Eq. (7.3.3). For example, let us determine the dependence of $\widetilde{E}_{1111}$ on E_{klst}:

$$\widetilde{E}_{1111} = E_{klst} l_{\tilde{1}k} l_{\tilde{1}l} l_{\tilde{1}s} l_{\tilde{1}t} = E_{1111} l_{\tilde{1}1} l_{\tilde{1}1} l_{\tilde{1}1} l_{\tilde{1}1} = E_{1111}. \qquad (7.3.4)$$

Since there are only three nonzero direction cosines (Eq. 7.3.2), the expansion of the transformation law is simplified. The result given by Eq. (7.3.4) states that Eq. (7.3.3) is satisfied for this one component. Similarly,

$$\widetilde{E}_{1123} = E_{1123} l_{\tilde{1}1} l_{\tilde{1}1} l_{\tilde{2}2} l_{\tilde{3}3} = -E_{1123}. \qquad (7.3.5)$$

The result stated in Eq. (7.3.5) is at odds with the invariance condition (Eq. 7.3.3); in order to meet the condition, the component E_{1123} is set equal to zero. It can be verified that of the 21 components of E_{mnpr} which must be tested against the transformation law, 7 others will yield a result similar to that given by Eq. (7.3.5). Thus a monoclinic material possesses only 13 nonzero independent elastic constants. Where the plane of symmetry is the $y_1 y_2$-plane, the components which must be set equal to zero in order to satisfy Eq. (7.3.3) are:

$$E_{1123}, \quad E_{2223}, \quad E_{3323}, \quad E_{1131}, \quad E_{2231}, \quad E_{3331}, \quad E_{1223}, \quad E_{1231}. \qquad (7.3.6)$$

The array of elastic constants for a monoclinic material with symmetry about the y_1y_2-plane becomes

$$
(E_{mnpr}) = \begin{pmatrix}
E_{1111} & E_{1122} & E_{1133} & E_{1112} & 0 & 0 \\
E_{1122} & E_{2222} & E_{2233} & E_{2212} & 0 & 0 \\
E_{1133} & E_{2233} & E_{3333} & E_{3312} & 0 & 0 \\
E_{1112} & E_{2212} & E_{3312} & E_{1212} & 0 & 0 \\
0 & 0 & 0 & 0 & E_{1313} & E_{1323} \\
0 & 0 & 0 & 0 & E_{1323} & E_{2323}
\end{pmatrix}. \tag{7.3.7}
$$

The physical meaning of the array shown in Eq. (7.3.7) is best appreciated by using the compliance tensor which, as we previously indicated, will have the same zero components as does E_{mnpr}. Thus, according to Eq. (7.3.7), the absence of S_{3313} means that a stress component σ_{33} will not cause any shearing strain γ_{13}. This can be seen by looking at the x_1x_3-plane of Fig. 7.2. Since x_1 is perpendicular to x_3, these axes also can be labeled as y_1, y_3. From Fig. 7.3 it can be seen that a negative stress σ_{33} will not cause the lines of action of the springs to alter their orthogonal relationship; this means that γ_{13} is zero.

Conversely, the nonzero component S_{1211} means that a stress σ_{11} will cause a shear. This is evident in Fig. 7.4, where we can see that σ_{11} will cause changes in the angles between the lines of action of the springs.

The compounds di-potassium tartrate, $K_2(C_4H_4O_6) \cdot \frac{1}{2}H_2O$, and sodium thiosulfate, $Na_2S_2O_3$, are monoclinic materials.

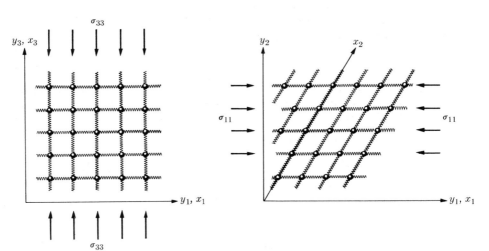

FIG. 7.3 Spring-ball model: x_1- and x_2-axes are perpendicular.

FIG. 7.4 Spring-ball model: x_1- and x_2-axes are not perpendicular.

7.4 ORTHOTROPIC MATERIAL: NINE CONSTANTS

An *orthotropic* material exhibits symmetry of its elastic properties with respect to two orthogonal planes, the y_1y_2- and y_2y_3-planes. (The spring-ball model shown in Fig. 7.5 now consists of axes which are mutually orthogonal, but which have different spring constants.) This symmetry is best expressed as an additional requirement, on a monoclinic material, that the constants of the previous section do not change under the following coordinate transformation:

$$\tilde{y}_1 = -y_1,$$
$$\tilde{y}_2 = y_2, \qquad (7.4.1)$$
$$\tilde{y}_3 = y_3.$$

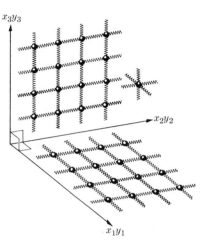

The direction cosines which correspond to Eq. (7.4.1) can be expressed as follows:

	y_1	y_2	y_3
$\tilde{y}_1$	-1	0	0
$\tilde{y}_2$	0	1	0
$\tilde{y}_3$	0	0	1

(7.4.2)

FIG. 7.5 Spring-ball model: x_1-, x_2-, and x_3-axes are mutually perpendicular.

As in Section 7.3, the application of the transformation law (Eq. 7.2.27) subject to the direction cosines given by Eq. (7.4.2) will lead to contradictions of the type shown in Eq. (7.3.5). These are resolved by setting the elastic constant equal to zero. For the orthotropic material, it is easily verified that the following constants, in addition to the ones shown in Eq. (7.3.6), must be zero:

$$E_{1112}, \quad E_{2212}, \quad E_{3312}, \quad E_{1323}. \qquad (7.4.3)$$

Thus an orthotropic material whose planes of symmetry are the y_1y_2- and y_2y_3-planes can be described by the following array of nine constants:

$$(E_{mnpr}) = \begin{pmatrix} E_{1111} & E_{1122} & E_{1133} & 0 & 0 & 0 \\ E_{1122} & E_{2222} & E_{2233} & 0 & 0 & 0 \\ E_{1133} & E_{2233} & E_{3333} & 0 & 0 & 0 \\ 0 & 0 & 0 & E_{1212} & 0 & 0 \\ 0 & 0 & 0 & 0 & E_{1313} & 0 \\ 0 & 0 & 0 & 0 & 0 & E_{2323} \end{pmatrix}. \qquad (7.4.4)$$

Note that since the y_n-axes coincide with the x_n-axes which are now orthogonal, normal stresses do not cause shearing strains, and shearing stresses do not cause extensional strains.

In the study of crystals, an orthotropic material is also called an ortho-rhombic material. An example of such a material is aragonite, $CaCO_3$, which has the following values for its elastic constants (in millions of pounds per square inch):

$$E_{1111} = 23.2, \quad E_{1212} = 6.10, \quad E_{1122} = 5.41,$$
$$E_{2222} = 12.6, \quad E_{1313} = 3.71, \quad E_{1133} = 0.25,$$
$$E_{3333} = 12.3, \quad E_{2323} = 6.19, \quad E_{2233} = 2.28.$$

Another example is topaz, $Al_2(F, OH)_2,SiO_4$:

$$E_{1111} = 40.1, \quad E_{1212} = 15.7, \quad E_{1122} = 18.3,$$
$$E_{2222} = 50.6, \quad E_{1313} = 19.3, \quad E_{1133} = 12.3,$$
$$E_{3333} = 42.8, \quad E_{2323} = 19.0, \quad E_{2233} = 12.8.$$

7.5 TETRAGONAL MATERIAL: SIX CONSTANTS

A *tetragonal* material is an orthotropic material which has the same properties along two of the axes but different properties along the third axis. For our purposes, the y_2- and y_3-axes will have the same properties. Then the additional requirement (beyond those needed for an orthotropic material) is that the elastic constants remain unchanged under a rotation of the y_2- and y_3-axes about the y_1-axis. This transformation is given by

$$\tilde{y}_1 = y_1, \quad \tilde{y}_2 = y_3, \quad \tilde{y}_3 = -y_2. \tag{7.5.1}$$

The direction cosines may be expressed as

	y_1	y_2	y_3
$\tilde{y}_1$	1	0	0
$\tilde{y}_2$	0	0	1
$\tilde{y}_3$	0	-1	0

$$(7.5.2)$$

To preserve invariance under the transformation given by Eqs. (7.5.1), it is necessary to require that

$$E_{2222} = E_{3333}, \tag{7.5.3}$$

$$E_{1122} = E_{1133}, \tag{7.5.4}$$

$$E_{1212} = E_{1313}. \tag{7.5.5}$$

Now there are only six independent constants:

$$(E_{mnpr}) = \begin{pmatrix} E_{1111} & E_{1122} & E_{1122} & 0 & 0 & 0 \\ E_{1122} & E_{2222} & E_{2233} & 0 & 0 & 0 \\ E_{1122} & E_{2233} & E_{2222} & 0 & 0 & 0 \\ 0 & 0 & 0 & E_{1212} & 0 & 0 \\ 0 & 0 & 0 & 0 & E_{1212} & 0 \\ 0 & 0 & 0 & 0 & 0 & E_{2323} \end{pmatrix}. \tag{7.5.6}$$

7.6   CUBIC MATERIAL: THREE CONSTANTS

The next step is to consider a material in which the properties are the same along three orthogonal directions. Such a material is known as a *cubic* material, and it can be considered as a tetragonal material with the additional constraint that invariance is preserved if the y_1- and y_2-axes are rotated 90° about the y_3-axis:

$$\tilde{y}_1 = y_2, \qquad \tilde{y}_2 = -y_1, \qquad \tilde{y}_3 = y_3. \tag{7.6.1}$$

The direction cosines are

	y_1	y_2	y_3
$\tilde{y}_1$	0	1	0
$\tilde{y}_2$	-1	0	0
$\tilde{y}_3$	0	0	1

$$\tag{7.6.2}$$

In order to preserve invariance under Eq. (7.6.1), it is necessary to require that

$$E_{1111} = E_{2222}, \tag{7.6.3}$$

$$E_{2233} = E_{1122}, \tag{7.6.4}$$

$$E_{2323} = E_{1212}. \tag{7.6.5}$$

Now there are only three independent constants:

$$(E_{mnpr}) = \begin{pmatrix} E_{1111} & E_{1122} & E_{1122} & 0 & 0 & 0 \\ E_{1122} & E_{1111} & E_{1122} & 0 & 0 & 0 \\ E_{1122} & E_{1122} & E_{1111} & 0 & 0 & 0 \\ 0 & 0 & 0 & E_{1212} & 0 & 0 \\ 0 & 0 & 0 & 0 & E_{1212} & 0 \\ 0 & 0 & 0 & 0 & 0 & E_{1212} \end{pmatrix}. \tag{7.6.6}$$

7.7 ISOTROPIC MATERIAL: TWO CONSTANTS

The simplest type of material is called an *isotropic* material; it possesses elastic properties independent of the orientation of the axes. If the cubic material is used as a starting point, then the additional constraint to show invariance is a rotation of the y_1- and y_2-axes through an angle of 45° about the y_3-axis. We arrive at the following equations:

$$\tilde{y}_1 = y_1 \cos 45° + y_2 \cos 45°,$$
$$\tilde{y}_2 = -y_1 \cos 45° + y_2 \cos 45°, \qquad (7.7.1)$$
$$\tilde{y}_3 = y_3.$$

The direction cosines are

	y_1	y_2	y_3
$\tilde{y}_1$	$\sqrt{2}/2$	$\sqrt{2}/2$	0
$\tilde{y}_2$	$-\sqrt{2}/2$	$\sqrt{2}/2$	0
$\tilde{y}_3$	0	0	1

$$(7.7.2)$$

The final requirement is found to be

$$E_{1212} = \tfrac{1}{2}(E_{1111} - E_{1122}). \qquad (7.7.3)$$

It can now be verified that further coordinate transformations of any kind will not change the form of the stress-strain laws. Thus two constants are required to define the elasticity of an isotropic material.

The elastic constants for an isotropic material are usually written in the notation

$$E_{1122} = \lambda, \qquad (7.7.4)$$
$$E_{1111} = \lambda + 2\mu, \qquad (7.7.5)$$
$$E_{1212} = \mu. \qquad (7.7.6)$$

The pair of constants λ and μ are called *Lamé's constants*, and μ is often referred to as the *shear modulus*.

The array of constants for an isotropic material is thus reduced to

$$(E_{mnpr}) = \begin{pmatrix} \lambda + 2\mu & \lambda & \lambda & 0 & 0 & 0 \\ \lambda & \lambda + 2\mu & \lambda & 0 & 0 & 0 \\ \lambda & \lambda & \lambda + 2\mu & 0 & 0 & 0 \\ 0 & 0 & 0 & \mu & 0 & 0 \\ 0 & 0 & 0 & 0 & \mu & 0 \\ 0 & 0 & 0 & 0 & 0 & \mu \end{pmatrix}. \qquad (7.7.7)$$

The stress-strain relations for an isotropic material, when written out explicitly in terms of the elasticity tensor, have the following forms:

$$\sigma_{11} = 2\mu\gamma_{11} + \lambda(\gamma_{11} + \gamma_{22} + \gamma_{33}), \tag{7.7.8}$$

$$\sigma_{22} = 2\mu\gamma_{22} + \lambda(\gamma_{11} + \gamma_{22} + \gamma_{33}), \tag{7.7.9}$$

$$\sigma_{33} = 2\mu\gamma_{33} + \lambda(\gamma_{11} + \gamma_{22} + \gamma_{33}), \tag{7.7.10}$$

$$\sigma_{12} = 2\mu\gamma_{12}, \tag{7.7.11}$$

$$\sigma_{13} = 2\mu\gamma_{13}, \tag{7.7.12}$$

$$\sigma_{23} = 2\mu\gamma_{23}. \tag{7.7.13}$$

Note that these can be put into the compact tensor equation

$$\sigma_{mn} = 2\mu\gamma_{mn} + \lambda\delta_{mn}\gamma_{rr}. \tag{7.7.14}$$

The two independent elastic constants in the engineering terminology are E, *Young's modulus* or the *modulus of elasticity*, and ν, *Poisson's ratio*. Under uniaxial stress, say σ_{11}, the normal strain component γ_{11} is given by $\gamma_{11} = \sigma_{11}/E$, while the other two strain components, γ_{22} and γ_{33}, are given by

$$\gamma_{22} = -\nu\gamma_{11} = -\frac{\nu}{E}\sigma_{11}, \tag{7.7.15}$$

$$\gamma_{33} = -\nu\gamma_{11} = -\frac{\nu}{E}\sigma_{11}. \tag{7.7.16}$$

The Poisson ratio thus is the ratio between the lateral contraction and the axial elongation under a uniaxial stress condition.

Under combined stresses σ_{11}, σ_{22}, and σ_{33} the three components of strain are given by superposition,

$$\gamma_{11} = \frac{1}{E}(\sigma_{11} - \nu\sigma_{22} - \nu\sigma_{33}), \tag{7.7.17}$$

$$\gamma_{22} = \frac{1}{E}(-\nu\sigma_{11} + \sigma_{22} - \nu\sigma_{33}), \tag{7.7.18}$$

$$\gamma_{33} = \frac{1}{E}(-\nu\sigma_{11} - \nu\sigma_{22} + \sigma_{33}). \tag{7.7.19}$$

These equations can be solved to yield expressions for stresses in terms of strains as follows:

$$\sigma_{11} = \frac{E}{(1+\nu)(1-2\nu)}[(1-\nu)\gamma_{11} + \nu\gamma_{22} + \nu\gamma_{33}], \tag{7.7.20}$$

$$\sigma_{22} = \frac{E}{(1+\nu)(1-2\nu)}[\nu\gamma_{11} + (1-\nu)\gamma_{22} + \nu\gamma_{33}], \tag{7.7.21}$$

$$\sigma_{33} = \frac{E}{(1+\nu)(1-2\nu)}[\nu\gamma_{11} + \nu\gamma_{22} + (1-\nu)\gamma_{33}]. \tag{7.7.22}$$

By comparing the above expressions with Eqs. (7.7.8), (7.7.9), and (7.7.10) one can conclude that

$$\lambda = \frac{\nu E}{(1 + \nu)(1 - 2\nu)}, \tag{7.7.23}$$

and

$$\mu = \frac{E}{2(1 + \nu)}. \tag{7.7.24}$$

The relation between shear stress-strain relations can be written in terms of E and ν as follows:†

$$\gamma_{12} = \frac{1}{E}(1 + \nu)\sigma_{12}, \tag{7.7.25}$$

$$\gamma_{23} = \frac{1}{E}(1 + \nu)\sigma_{23}, \tag{7.7.26}$$

$$\gamma_{31} = \frac{1}{E}(1 + \nu)\sigma_{31}. \tag{7.7.27}$$

Note that Eqs. (7.7.20) through (7.7.22), and (7.7.25) through (7.7.27) can be put into the following compact form:

$$\gamma_{mn} = \frac{1}{E}[(1 + \nu)\sigma_{mn} - \nu\delta_{mn}\sigma_{rr}]. \tag{7.7.28}$$

Some forms of the stress-strain relations are of particular interest. Substituting the deviator and spherical tensors in Eq. (7.7.14) yields

$$\hat{\sigma}_{mn} + \mathring{\sigma}_{mn} = \lambda\,\delta_{mn}(\hat{\gamma}_{rr} + \mathring{\gamma}_{rr}) + 2\mu(\hat{\gamma}_{mn} + \mathring{\gamma}_{mn}). \tag{7.7.29}$$

First (see Eq. 6.10.3), we find that the spherical stress $\mathring{\sigma}_{mn}$ can be written as

$$\mathring{\sigma}_{mn} = \tfrac{1}{3}\,\delta_{mn}\sigma_{rr}. \tag{7.7.30}$$

From Eqs. (7.7.8) through (7.7.10) we obtain

$$\sigma_{rr} = (2\mu + 3\lambda)\gamma_{rr}, \tag{7.7.31}$$

and from Eq. (5.12.3) we have

$$\tfrac{1}{3}\,\delta_{mn}\gamma_{rr} = \mathring{\gamma}_{mn} \tag{7.7.32}$$

and

$$\gamma_{rr} = \mathring{\gamma}_{rr}. \tag{7.7.33}$$

With these relations we can write Eq. (7.7.30) as

$$\mathring{\sigma}_{mn} = \lambda\,\delta_{mn}\mathring{\gamma}_{rr} + 2\mu\mathring{\gamma}_{mn}. \tag{7.7.34}$$

---

† In conventional engineering notation (see Section 5.5) the shear stress-strain relations are usually expressed as $\gamma_{xy} = \tau_{xy}/G$, $\gamma_{yz} = \tau_{yz}/G$, $\gamma_{xz} = \tau_{xz}/G$, where G is called the shear modulus. The symbol τ is used to denote the shear stresses. It is seen that $G = E/2(1 + \nu) = \mu$.

It is now clear that Eq. (7.7.29) can be split into two equations, the one above, and the corresponding one for the deviator component,

$$\hat{\sigma}_{mn} = \lambda \, \delta_{mn} \hat{\gamma}_{rr} + 2\mu \hat{\gamma}_{mn}. \tag{7.7.35}$$

In Eq. (7.7.35) it is observed that the dilatation $\hat{\gamma}_{rr}$, associated with the deviator strain, is zero, and hence the relations between deviator stresses and strains become

$$\hat{\sigma}_{mn} = 2\mu \hat{\gamma}_{mn}. \tag{7.7.36}$$

The stress-strain relations, Eq. (7.7.34), between the spherical stress and strain tensors are in actuality only one relation because there is only one non-zero value of the spherical components [see Eqs. (5.14.15b) and (6.10.5)]. Thus, Eq. (7.7.34) can be replaced by

$$\hat{\sigma}_{rr} = 3K\hat{\gamma}_{rr}, \tag{7.7.37}$$

where the constant K is called the *bulk modulus*, and is related to the Lamé constants by

$$K = \frac{3\lambda + 2\mu}{3}. \tag{7.7.38}$$

The bulk modulus can be shown to be the ratio of a hydrostatic pressure, Δp, and the resulting dilatation,

$$K = -\frac{\Delta p}{(dV - dV^{(0)})/dV^{(0)}} = -\frac{\Delta p}{J_1}. \tag{7.7.39}$$

A special form of the stress-strain relations can be obtained by substituting the definitions for the deviator components [see Eqs. (5.12.2) and (6.10.2)] into Eq. (7.7.28). We obtain

$$\gamma_{mn} = \frac{1}{2\mu} [\sigma_{mn} - \tfrac{1}{3} \delta_{mn}\sigma_{rr}] + \frac{1}{9K} \delta_{mn}\sigma_{rr}. \tag{7.7.40}$$

7.8   THERMOELASTIC STRESS-STRAIN RELATION

Adding the effect of temperature changes to the stress-strain relations is accomplished in a straightforward fashion for a material that is thermodynamically as well as elastically isotropic. A thermodynamically isotropic material is one in which the expansion caused by a change in temperature is the same in all directions. It is an experimentally observed fact that a body which is stress free will experience relative elongations E_n, but no angular changes ϕ_{nm}, due to an increase or decrease in temperature. Let α signify the coefficient of thermal expansion with dimensions of $(T)^{-1}$, where T is temperature measured in any scale. For most materials the relative elongations are small compared to unity, even for temperature changes approaching the melting point. Thus we can write

$$\gamma_{mn}^{(T)} = \alpha \, \Delta T \, \delta_{mn}, \tag{7.8.1}$$

where $\gamma_{mn}^{(T)}$ represents the thermally induced strains and ΔT represents the change in temperature from some reference state. This thermally induced strain can be superimposed on the stress-induced strains provided the reference temperature is the temperature at which the body is stress free. In addition, we assume that the application of stress by itself will not cause appreciable changes in temperature. For thermoelastic problems, then, the constitutive relation of Eq. (7.7.28) is replaced by

$$\gamma_{mn} = \frac{1}{E}\left[(1 + \nu)\sigma_{mn} - \nu\,\delta_{mn}\sigma_{rr}\right] + \alpha\,\delta_{mn}\,\Delta T. \tag{7.8.2}$$

The individual relations are

$$\gamma_{11} = \frac{1}{E}\left[\sigma_{11} - \nu(\sigma_{22} + \sigma_{33})\right] + \alpha\,\Delta T, \tag{7.8.3}$$

$$\gamma_{22} = \frac{1}{E}\left[\sigma_{22} - \nu(\sigma_{33} + \sigma_{11})\right] + \alpha\,\Delta T, \tag{7.8.4}$$

$$\gamma_{33} = \frac{1}{E}\left[\sigma_{33} - \nu(\sigma_{11} + \sigma_{22})\right] + \alpha\,\Delta T, \tag{7.8.5}$$

$$\gamma_{12} = \frac{1}{E}(1 + \nu)\sigma_{12}, \tag{7.8.6}$$

$$\gamma_{23} = \frac{1}{E}(1 + \nu)\sigma_{23}, \tag{7.8.7}$$

$$\gamma_{31} = \frac{1}{E}(1 + \nu)\sigma_{31}. \tag{7.8.8}$$

The inverse relationships are obtained by considering Eqs. (7.8.3) through (7.8.8) as a set of simultaneous equations for σ_{mn}. We obtain

$$\sigma_{mn} = 2\mu\gamma_{mn} + \lambda\,\delta_{mn}\gamma_{rr} - (3\lambda + 2\mu)\,\delta_{mn}\alpha\,\Delta T. \tag{7.8.9}$$

The individual relations are

$$\sigma_{11} = 2\mu\gamma_{11} + \lambda(\gamma_{11} + \gamma_{22} + \gamma_{33}) - \frac{E}{1 - 2\nu}\alpha\,\Delta T, \tag{7.8.10}$$

$$\sigma_{22} = 2\mu\gamma_{22} + \lambda(\gamma_{11} + \gamma_{22} + \gamma_{33}) - \frac{E}{1 - 2\nu}\alpha\,\Delta T, \tag{7.8.11}$$

$$\sigma_{33} = 2\mu\gamma_{33} + \lambda(\gamma_{11} + \gamma_{22} + \gamma_{33}) - \frac{E}{1 - 2\nu}\alpha\,\Delta T, \tag{7.8.12}$$

$$\sigma_{12} = 2\mu\gamma_{12}, \tag{7.8.13}$$

$$\sigma_{23} = 2\mu\gamma_{23}, \tag{7.8.14}$$

$$\sigma_{31} = 2\mu\gamma_{31}. \tag{7.8.15}$$

We observe from an examination of Eqs. (7.8.10) through (7.8.15) that if a small cube of material, so restrained that changes in length are not possible, is heated uniformly to a temperature of $T + \Delta T$, the following thermal stresses will be induced:

$$\sigma_{11}^{(T)} = - \frac{E}{1 - 2\nu} \alpha \, \Delta T, \tag{7.8.16}$$

$$\sigma_{22}^{(T)} = - \frac{E}{1 - 2\nu} \alpha \, \Delta T, \tag{7.8.17}$$

$$\sigma_{33}^{(T)} = - \frac{E}{1 - 2\nu} \alpha \, \Delta T. \tag{7.8.18}$$

7.9 ELASTIC CONSTANTS FOR SOME MATERIALS OF ENGINEERING INTEREST

There are shown in Tables 7.1 and 7.2 the elastic constants of a few materials in their elemental single crystalline state. Some of the metals most useful to our civilization are included. Note that these are all members of the hexagonal or cubic crystal systems.

(a) Hexagonal system. Some common materials belong to the hexagonal class. Their elastic properties can be described by the following array of five constants:

$$(E_{mnpr}) = \begin{pmatrix} E_{1111} & E_{1122} & E_{1133} & 0 & 0 & 0 \\ E_{1122} & E_{1111} & E_{1133} & 0 & 0 & 0 \\ E_{1133} & E_{1133} & E_{3333} & 0 & 0 & 0 \\ 0 & 0 & 0 & \frac{1}{2}(E_{1111}-E_{1122}) & 0 & 0 \\ 0 & 0 & 0 & 0 & E_{1313} & 0 \\ 0 & 0 & 0 & 0 & 0 & E_{1313} \end{pmatrix}.$$

Table 7.1 lists the five constants for four of these materials.

TABLE 7.1 *Elastic Constants of Some Hexagonal Crystals*

Material[a]	E_{1111}	E_{3333}	E_{1122}	E_{1133}	E_{1313}
Beryllium	40.7	43.8	−3.62	–	22.5
Magnesium	8.66	8.95	3.80	3.14	2.38
Zinc	23.4	8.85	4.96	7.26	5.55
Ice	2.01	2.17	1.03	0.844	0.463

[a] Multiply values by 10^6 to obtain results in lb/in^2.

TABLE 7.2   *Elastic Constants of Some Cubic Crystals*

Material[a]	E_{1111}	E_{1122}	E_{1212}
Aluminum	15.7	8.89	4.13
Copper	24.4	17.6	10.9
Iron	29.0	15.7	16.8
Molybdenum	66.7	25.3	16.0
Nickel	35.7	21.4	18.1
Tungsten	72.7	28.7	22.0

[a] Multiply values by 10^6 to obtain results in lb/in^2.

TABLE 7.3   *Elementary Properties of Some Materials in Polycrystalline Form*

MATERIAL	$\alpha \times 10^6$ per °F	$E \times 10^{-6}$ psi	ρ lb/in^3	$(E/\rho) \times 10^{-6}$ in.
Aluminum (2024–T3)	13.0	10.5	0.100	105
Beryllium	6.9	36.8	.066	558
Boron	8.3	55.0	.085	648
Brass[a]	10.6	13.1	.309	42.4
Bronze	11.1	14.5	.318	45.6
Chromium	3.8	36.0	.239	151
Concrete[b]	6.0	3.4	.090	37.8
Copper[a]	7.8	18.1	.321	56.4
Douglas fir[c]	3.0	1.9	.0208	91.4
Glass	4.0	9.4	.093	101
Gun metal	10.2	10.0	.317	31.6
Magnesium (AZ61A)	16.0	6.5	.065	100
Molybdenum	2.7	42.7	.326	131
Nickel	7.2	30.0	.316	95
Polyethylene	170	1.9	.033	57.6
Steel (SAE 4130)	6.5	29.0	.283	103
Titanium	6.0	17.0	.163	104
Tungsten	2.4	51.5	.680	75.8
Zinc[a]	14.7	13.1	.260	50.4

[a] Cold-rolled; [b] typical medium-strength concrete; [c] with the grain.

(b) Cubic system. Some other common materials belong to the cubic class. Their elastic properties are described by the array of constants in Section 7.6. Table 7.2 lists the three constants for six of these materials.

In order to obtain the values shown in Tables 7.1 and 7.2, single crystals of the material had to be grown. This is a laboratory process. In their useful form as found in structures, the metals are polycrystalline and on a macroscopic scale can be considered as isotropic materials. Shown in Table 7.3 are the isotropic elastic properties of a number of useful materials. Also shown are the coefficients of thermal expansion, the weight densities, and an elasticity efficiency factor E/ρ. This latter number is a measure of the weight of material required to attain a certain level of stiffness; the material with the larger number is deemed more efficient in this regard. Note that the common structural materials (aluminum, magnesium, steel and titanium) used in flight vehicles all have about the same value of E/ρ. The tantalizing promise of beryllium and boron are shown by their extremely large values. Of interest, also, is the fact that glass is the equal of most metals in this respect.

7.10 STRAIN ENERGY

The forces which are applied to a structure will perform work on the structure, since the points of application of the forces will undergo motion as the structure suffers distortions. In general, the change in temperature of most engineering materials due to deformation is negligibly small, and the assumption of isothermal behavior can be made. Then, the First Law of Thermodynamics states that the work done by the external forces results in an increase in the internal energy of the structure. In the theory of elasticity, the increase in internal energy is called the *strain energy*, and is denoted by U_i.

A simple example will suffice to reveal the mathematical form of the strain energy. Consider a member of length L which is in a uniaxial state of stress under the action of a total load P uniformly distributed over the free end (see Fig. 7.6). The state of stress in the bar is characterized by σ_{11} alone, and the state of deformation by γ_{11}, $\gamma_{22} = -\nu\gamma_{11}$ and $\gamma_{33} = -\nu\gamma_{11}$ (see Eqs. 7.7.15 and 7.7.16). As the load is gradually increased to its final magnitude of P, the end of the bar, which is the point of application of the load, will extend by the value of ΔL. A linear relationship (see Section 7.2) between stress and strain, and hence between load and deflection, is assumed. Thus the work done by the force P is

$$W = \tfrac{1}{2}P\,\Delta L. \tag{7.10.1}$$

An element of volume in the bar experiences an increase in the state of stress to a final value $\sigma_{11} = P/A$. On the planes which experience a stress, σ_{11}, there is a corresponding deformation characterized by γ_{11}. The relationship between σ_{11} and γ_{11} is linear, and is represented in Fig. 7.7. The strain energy due to

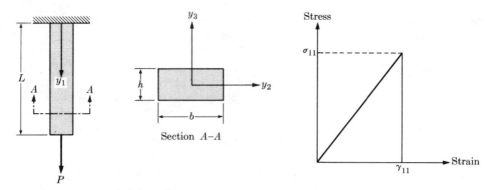

FIG. 7.6   Prismatic bar under uniaxial tension.   FIG. 7.7   Diagram of linear stress-strain relation.

this uniaxial state of stress is given by the following expression:

$$U = \int_0^L \int_{-b/2}^{b/2} \int_{-h/2}^{h/2} \tfrac{1}{2}\sigma_{11}\gamma_{11}\, dy_1\, dy_2\, dy_3. \tag{7.10.2}$$

Thus the strain energy per unit of volume is proportional to the area under the stress-strain curve of Fig. 7.7.

In a similar fashion, the strain energy stored per unit of volume in a structure which experiences a general state of stress and strain is given by

$$U^* = \tfrac{1}{2}\sigma_{mn}\gamma_{mn}, \tag{7.10.3}$$

where the * indicates that U^* is a strain-energy density. Note that there are nine terms in Eq. (7.10.3). The total strain energy in a body is

$$U = \iiint U^*\, dy_1\, dy_2\, dy_3 = \iiint \tfrac{1}{2}\sigma_{mn}\gamma_{mn}\, dy_1\, dy_2\, dy_3. \tag{7.10.4}$$

There are several alternative forms of Eq. (7.10.3). The strain tensor γ_{mn} can be eliminated by substitution of the pertinent stress relation (see Eq. 7.7.28). The result is

$$U^* = (1/2E)[(1 + \nu)\sigma_{mn}\sigma_{mn} - \nu(\sigma_{rr})^2]. \tag{7.10.5}$$

In explicit detail, Eq. (7.10.5) has the form

$$U^* = (1/2E)[(\sigma_{11}^2 + \sigma_{22}^2 + \sigma_{33}^2) - 2\nu(\sigma_{11}\sigma_{22} + \sigma_{22}\sigma_{33} + \sigma_{33}\sigma_{11}) \\ + 2(1 + \nu)(\sigma_{12}^2 + \sigma_{23}^2 + \sigma_{31}^2)]. \tag{7.10.6}$$

The strain energy can also be expressed solely in terms of the strains. If Eq. (7.7.14) is employed as the stress-strain relation, we find that

$$U^* = \tfrac{1}{2}\lambda(\gamma_{rr})^2 + \mu\gamma_{mn}\gamma_{mn}. \tag{7.10.7}$$

In explicit detail, Eq. (7.10.7) has the form

$$U^* = \tfrac{1}{2}\lambda(\gamma_{11} + \gamma_{22} + \gamma_{33})^2 + \mu(\gamma_{11}^2 + \gamma_{22}^2 + \gamma_{33}^2) + 2\mu(\gamma_{12}^2 + \gamma_{23}^2 + \gamma_{31}^2).$$
$$(7.10.8)$$

It is also of interest to present the strain-energy density in a form based on the stress-strain relation, Eq. (7.7.40). If we substitute this equation in Eq. (7.10.3), we get

$$U^* = \frac{1}{4\mu}[\sigma_{mn}\sigma_{mn} - \tfrac{1}{3}(\sigma_{rr})^2] + \frac{1}{18K}(\sigma_{rr})^2. \qquad (7.10.9)$$

We sometimes need to calculate the strain energies associated with the deviator and spherical components of the stress and strain tensors. The energy associated with the deviator components is called the *energy of distortion*, because there is no volume change associated with the deviator strains. The energy of distortion per unit volume is designated by $\hat{U}^*$, and is defined as

$$\hat{U}^* = \tfrac{1}{2}\hat{\sigma}_{mn}\hat{\gamma}_{mn}, \qquad (7.10.10)$$

and in terms of the stresses can be found by substituting Eqs. (5.12.2) and (6.10.2) into Eqs. (7.10.10):

$$\hat{U}^* = \frac{1}{4\mu}[\sigma_{mn}\sigma_{mn} - \tfrac{1}{3}(\sigma_{rr})^2]. \qquad (7.10.11)$$

Similarly, the energy associated with the spherical components is called the *energy of dilatation* because the entire volume change associated with a given state of infinitesimal strain is contained in the spherical component. The energy of dilatation per unit volume is designated by $\check{U}^*$, and is defined as

$$\check{U}^* = \tfrac{1}{2}\check{\sigma}_{mn}\check{\gamma}_{mn}. \qquad (7.10.12)$$

This equation can be put into the form

$$\check{U}^* = \frac{1}{18K}(\sigma_{rr})^2. \qquad (7.10.13)$$

It should be pointed out that the result,

$$U^* = \hat{U}^* + \check{U}^*, \qquad (7.10.14)$$

is a most convenient, albeit a totally unexpected one. In general, however,

$$U^*(\sigma) \neq U^*(\sigma_a) + U^*(\sigma_b), \qquad (7.10.15)$$

where

$$\dot{\sigma} = \sigma_a + \sigma_b. \qquad (7.10.16)$$

That is, the total strain energy associated with a stress σ is not the sum of the energies associated with an arbitrary partitioning of the stress.

7.11 SUMMARY: ISOTROPIC STRESS-STRAIN LAW AND ENERGY RELATIONS

The stress-strain laws derived in this chapter, as well as the energy relations, are restricted to the case of small strain. Nowhere has it been necessary to bring in the strain-displacement relations; hence the geometrical nonlinearities due to the presence of appreciable rotations have played no part in the formulation. Thus, in a certain sense, the stress-strain and energy relations are more general than the equilibrium equations derived in Chapter 6, since the latter are based on linear strain, i.e., on negligible rotations. It is, however, only a short step from the energy relations to equilibrium equations which do account for rotations, and that is the reason for not immediately going to the case of linear strain. At this stage of our study in solid mechanics, there is only a notational difference between the relations summarized in this section and those in the following section which will deal with linear elasticity.

We will not summarize all the various types of materials, but merely those pertaining to isotropic materials.

Stress-strain relations—elasticity tensor form

$$\sigma_{mn} = 2\mu\gamma_{mn} + \lambda\,\delta_{mn}\gamma_{rr} - (3\lambda + 2\mu)\,\delta_{mn}\alpha\,\Delta T, \tag{7.11.1}$$

$$\sigma_{11} = 2\mu\gamma_{11} + \lambda(\gamma_{11} + \gamma_{22} + \gamma_{33}) - \frac{E}{1 - 2\nu}\,\alpha\,\Delta T, \tag{7.11.2}$$

$$\sigma_{22} = 2\mu\gamma_{22} + \lambda(\gamma_{11} + \gamma_{22} + \gamma_{33}) - \frac{E}{1 - 2\nu}\,\alpha\,\Delta T, \tag{7.11.3}$$

$$\sigma_{33} = 2\mu\gamma_{33} + \lambda(\gamma_{11} + \gamma_{22} + \gamma_{33}) - \frac{E}{1 - 2\nu}\,\alpha\,\Delta T, \tag{7.11.4}$$

$$\sigma_{12} = 2\mu\gamma_{12}, \tag{7.11.5}$$

$$\sigma_{23} = 2\mu\gamma_{23}, \tag{7.11.6}$$

$$\sigma_{31} = 2\mu\gamma_{31}. \tag{7.11.7}$$

Stress-strain relations—compliance tensor form

$$\gamma_{mn} = (1/E)\,[(1 + \nu)\sigma_{mn} - \nu\,\delta_{mn}\sigma_{rr}] + \alpha\,\delta_{mn}\,\Delta T, \tag{7.11.8}$$

$$\gamma_{11} = (1/E)\,[\sigma_{11} - \nu\sigma_{22} - \nu\sigma_{33}] + \alpha\,\Delta T, \tag{7.11.9}$$

$$\gamma_{22} = (1/E)\,[-\nu\sigma_{11} + \sigma_{22} - \nu\sigma_{33}] + \alpha\,\Delta T, \tag{7.11.10}$$

$$\gamma_{33} = (1/E)\,[-\nu\sigma_{11} - \nu\sigma_{22} + \sigma_{33}] + \alpha\,\Delta T, \tag{7.11.11}$$

$$\gamma_{12} = \frac{1 + \nu}{E}\,\sigma_{12}, \tag{7.11.12}$$

$$\gamma_{23} = \frac{1 + \nu}{E}\,\sigma_{23}, \tag{7.11.13}$$

$$\gamma_{31} = \frac{1 + \nu}{E}\,\sigma_{31}. \tag{7.11.14}$$

Special forms for deviator and spherical components

$$\hat{\sigma}_{mn} = 2\mu\hat{\gamma}_{mn}, \tag{7.11.15}$$

$$\hat{\sigma}_{rr} = 3K\hat{\gamma}_{rr}. \tag{7.11.16}$$

Stress-strain relations in terms of μ and K

$$\gamma_{mn} = \frac{1}{2\mu}[\sigma_{mn} - \tfrac{1}{3}\delta_{mn}\sigma_{rr}] + \frac{1}{9K}\delta_{mn}\sigma_{rr}. \tag{7.11.17}$$

Relations between various elastic constants

$$\lambda = \frac{\nu E}{(1 + \nu)(1 - 2\nu)}, \tag{7.11.18}$$

$$\mu = \frac{E}{2(1 + \nu)}, \tag{7.11.19}$$

$$K = \frac{3\lambda + 2\mu}{3}, \tag{7.11.20}$$

$$3\lambda + 2\mu = \frac{E}{1 - 2\nu}. \tag{7.11.21}$$

Various forms of strain energy densities

$$U^* = \tfrac{1}{2}\sigma_{mn}\gamma_{mn}, \tag{7.11.22}$$

$$U^* = \frac{1}{2E}[(1 + \nu)\sigma_{mn}\sigma_{mn} - \nu(\sigma_{rr})^2], \tag{7.11.23}$$

$$U^* = \frac{1}{2E}[(\sigma_{11}^2 + \sigma_{22}^2 + \sigma_{33}^2) - 2\nu(\sigma_{11}\sigma_{22} + \sigma_{22}\sigma_{33} + \sigma_{33}\sigma_{11})$$
$$+ 2(1 + \nu)(\sigma_{12}^2 + \sigma_{23}^2 + \sigma_{31}^2)], \tag{7.11.24}$$

$$U^* = \tfrac{1}{2}\lambda(\gamma_{rr})^2 + \mu\gamma_{mn}\gamma_{mn}, \tag{7.11.25}$$

$$U^* = \tfrac{1}{2}[\lambda(\gamma_{11} + \gamma_{22} + \gamma_{33})^2 + \mu(\gamma_{11}^2 + \gamma_{22}^2 + \gamma_{33}^2)$$
$$+ 2\mu(\gamma_{12}^2 + \gamma_{23}^2 + \gamma_{31}^2)], \tag{7.11.26}$$

$$U^* = \frac{1}{4\mu}[\sigma_{mn}\sigma_{mn} - \tfrac{1}{3}(\sigma_{rr})^2] + \frac{1}{18K}(\sigma_{rr})^2. \tag{7.11.27}$$

Distortion energy density

$$\hat{U}^* = \frac{1}{4\mu}[\sigma_{mn}\sigma_{mn} - \tfrac{1}{3}(\sigma_{rr})^2]. \tag{7.11.28}$$

Dilatation energy density

$$\hat{U}^* = \frac{1}{18K}(\sigma_{rr})^2. \tag{7.11.29}$$

7.12 SUMMARY: THE EQUATIONS OF LINEAR ELASTICITY

We are now in a position to summarize the equations of the linear theory of elasticity. The unknown quantities are the linear strain tensor, ϵ_{mn}, the displacement tensor, u_m, and the stress tensor, σ_{mn}. These tensors are a function of the coordinates y_1, y_2, y_3 of the structure. Their tensor properties are denoted by the following transformation laws:

$$\tilde{\epsilon}_{mn}(\tilde{y}_1, \tilde{y}_2, \tilde{y}_3) = \epsilon_{rs}(y_1, y_2, y_3) l_{r\tilde{m}} l_{s\tilde{n}}, \tag{7.12.1}$$

$$\tilde{u}_m(\tilde{y}_1, \tilde{y}_2, \tilde{y}_3) = u_r(y_1, y_2, y_3) l_{\tilde{m}r}, \tag{7.12.2}$$

$$\tilde{\sigma}_{mn}(\tilde{y}_1, \tilde{y}_2, \tilde{y}_3) = \sigma_{rs}(y_1, y_2, y_3) l_{\tilde{m}r} l_{\tilde{n}s}. \tag{7.12.3}$$

The objective of solid mechanics is the complete specification of these three tensors throughout any given structure under a prescribed loading condition and/or a geometrical boundary condition. The equations which govern the behavior of ϵ_{mn}, u_m, and σ_{mn} are as follows:

Linear strain-displacement relations

$$\epsilon_{mn} = \frac{1}{2}\left(\frac{\partial u_m}{\partial y_n} + \frac{\partial u_n}{\partial y_m}\right). \tag{7.12.4}$$

Force equilibrium equations for linear strains

$$\frac{\partial \sigma_{mn}}{\partial y_m} + F_n = 0. \tag{7.12.5}$$

Constitutive relations for anisotropic materials (without temperature changes)

$$\sigma_{mn} = E_{mnpr}\epsilon_{pr}, \tag{7.12.6}$$

$$\epsilon_{pr} = S_{mnpr}\sigma_{pr}. \tag{7.12.7}$$

These last relations, which are different forms of the generalized Hooke's law, introduce the elasticity tensor E_{mnpr} and the compliance tensor S_{mnpr}. The components of the elasticity tensor are a property of the material, and are determined experimentally. Since most of the high-strength engineering materials are isotropic, the bulk of the literature in solid mechanics is devoted to this class of materials. The number of independent components of the elasticity tensor reduces to two for isotropy, and Eqs. (7.12.6) and (7.12.7) become much simpler.

The *stress-strain relations* for isotropic materials with changes in temperature are

$$\epsilon_{mn} = \frac{1}{E}\left[(1 + \nu)\sigma_{mn} - \nu\,\delta_{mn}\sigma_{rr}\right] + \alpha\,\delta_{mn}\,\Delta T, \tag{7.12.8}$$

or alternatively,

$$\sigma_{mn} = 2\mu\epsilon_{mn} + \lambda\delta_{mn}\epsilon_{rr} - (3\lambda + 2\mu)\,\delta_{mn}\alpha\,\Delta T, \qquad (7.12.9)$$

where

$$\alpha = \text{Coefficient of thermal expansion,}$$

$$E = \text{Young's modulus,}$$

$$\nu = \text{Poisson's ratio,}$$

$$\mu = \frac{E}{2(1 + \nu)}, \qquad (7.12.10)$$

$$\lambda = \frac{\nu E}{(1 + \nu)(1 - 2\nu)}. \qquad (7.12.11)$$

The *boundary conditions* which Eqs. (7.12.4), (7.12.5), and either (7.12.8) or (7.12.9) must satisfy are of two types:

(a) *Geometric constraints on a region A_1 of the surface.* These mean the prescription of

$$\mathbf{u} = \mathbf{u}^*, \qquad (7.12.12)$$

where $\mathbf{u}^*$ is a prescribed displacement. At the points of support the boundary displacements are zero.

(b) *Applied surface loads on a region A_2 of the surface.* This specification means that the interior stresses in region A_2 must be in equilibrium with the externally applied surface loads. This restriction is expressed as

$$\sigma_{mn}n_m\mathbf{i}_n = \boldsymbol{\sigma}^*, \qquad (7.12.13)$$

where $\boldsymbol{\sigma}^*$ is the prescribed stress vector acting on region A_2.

We note that there are a total of fifteen unknowns, consisting of six components of the strain tensor, three components of the displacement tensor, and six components of the stress tensor. Balancing these unknowns are fifteen equations consisting of three equations of equilibrium, six strain-displacement relations, and six stress-strain relations. A remark should be made here concerning the six strain-displacement relations. We have shown in Section 5.13 that the six strain-displacement relations for linear strain also lead to a set of six compatibility relations between the components of the strain:

$$\frac{\partial^2\epsilon_{nk}}{\partial y_m\,\partial y_l} + \frac{\partial^2\epsilon_{ml}}{\partial y_n\,\partial y_k} - \frac{\partial^2\epsilon_{nl}}{\partial y_m\,\partial y_k} - \frac{\partial^2\epsilon_{mk}}{\partial y_n\,\partial y_l} = 0. \qquad (7.12.14)$$

Thus instead of using the strain-displacement relations one may verify the efficacy of a solution for the strain components by determining whether the compatibility equations are satisfied. However, Eqs. (7.12.4) must be employed in order to determine the displacements of the body.

The exact solution of the complete set of equations has not been accomplished except for a very few simple cases. The real challenge of solid mechanics to the engineer lies, then, in devising suitably accurate approximate solutions. Great ingenuity is required, first to construct a suitable mathematical model of the structure, and second, to solve the simplified equations which describe the mathematical model. Finally, engineering judgment is required to evaluate the validity of the analysis in the light of past experience and new experiments.

Our further studies in solid mechanics will be devoted to the solutions of the equations of elasticity. There are many different mathematical tools which can be used, and these range over a broad spectrum of mathematics from numerical analysis through the calculus of variations to the complex variable. The development of these analytical tools and their application to elasticity is a fascinating experience.

7.13   SIMPLE EXAMPLES OF SOLUTIONS FOR EQUATIONS OF ELASTICITY

As we have observed, for a solid of general geometry and arbitrary boundary conditions it is not possible to obtain a close solution of the fifteen equations of elasticity and the corresponding boundary conditions listed in the preceding section. For some important structural components of simplified geometry and loading, however, the number of equations can be reduced. Here for purposes of illustration we will present two simple examples for which the fifteen equations are satisfied. In addition we will derive the expressions of strain energy for the simple structural elements.

(a) Tension of a prismatic bar. Let us consider a uniform prismatic bar under uniform tension stress, as shown in Fig. 7.8. If body forces are not present, the equations of equilibrium (Eq. 7.12.5) are satisfied by the following components of stress:

$$\sigma_{11} = S = \text{Constant},$$
$$\sigma_{22} = \sigma_{33} = \sigma_{12} = \sigma_{23} = \sigma_{31} = 0. \qquad (7.13.1)$$

It is seen that the lateral surface around the bar is free of stress. Since all stress

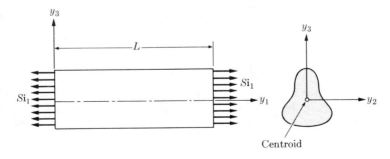

FIG. 7.8   Prismatic bar under uniaxial stress.

components except σ_{11} are zero, the boundary conditions given by Eq. (7.12.13) are satisfied. At the two ends, however, the boundary condition requires that the prescribed boundary stress vector $\boldsymbol{\sigma}^*$ be given by the normal stress $S i_1$.

Assuming that the material is isotropic, we can calculate the state of strain using Eq. (7.12.8). The result is

$$\epsilon_{11} = \sigma_{11}/E,$$
$$\epsilon_{22} = \epsilon_{33} = -\nu\sigma_{11}/E,$$
$$\epsilon_{12} = \epsilon_{23} = \epsilon_{31} = 0. \tag{7.13.2}$$

It is obvious that all the compatibility relations (7.12.14) are satisfied, since all strain components are constants. The strain-energy density U^* is simply [from Eq. (7.10.3)]

$$U^* = (1/2)\sigma_{11}\epsilon_{11}, \tag{7.13.3}$$

which can be written as either

$$U^* = (1/2E)\sigma_{11}^2, \tag{7.13.4}$$

or

$$U^* = (E/2)\epsilon_{11}^2. \tag{7.13.5}$$

For a bar of cross-sectional area A, length L, and total axial force P, we have

$$\dot{\sigma}_{11} = P/A, \tag{7.13.6}$$

and hence the total strain energy in the bar is

$$U = \iiint U^* \, dV = P^2 L/2AE. \tag{7.13.7}$$

If the elongation of the bar is Δ, we have

$$\epsilon_{11} = \Delta/L, \tag{7.13.8}$$

and

$$U = AE\Delta^2/2L. \tag{7.13.9}$$

(b) Pure bending of a uniform beam. Consider a uniform beam bent in its plane of symmetry by two equal and opposite bending moments M, as shown in Fig. 7.9. Let the y_1-axis lie along the centroid of the cross section. The bending moment as shown is about the y_2-axis. According to elementary bending theory the stress components are given by

$$\sigma_{11} = Cy_3,$$
$$\sigma_{22} = \sigma_{33} = \sigma_{12} = \sigma_{23} = \sigma_{31} = 0, \tag{7.13.10}$$

where C is a constant.

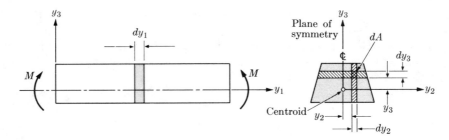

FIG. 7.9   Prismatic bar under pure bending.

We see that all the equations of equilibrium, Eq. (7.12.5), and the boundary conditions, Eq. (7.12.13), for the surface around the beam are satisfied. At the two ends of the beam the boundary condition requires that the prescribed boundary stress vector $\boldsymbol{\sigma}^*$ should consist only of normal stresses, $\sigma_{11}^* \mathbf{i}_1$. The variation of σ_{11}^* is also given by Cy_3, that is, it must be directly proportional to the distance from the y_2-axis. Since the prescribed boundary stresses σ_{11}^* must be equipollent to the applied bending moment M, we have

$$M = -\iint \sigma_{11}^* y_3 \, dA = -C \iint y_3^2 \, dA = -IC, \qquad (7.13.11)$$

where I is the cross-sectional moment of inertia with respect to the y_2-axis: $I = \iint y_3^2 \, dA$. In order to satisfy the other conditions of equilibrium we need to verify that the resulting axial force and moment about the y_3-axis are zero. This is indeed true:

$$\iint \sigma_{11}^* \, dA = C \iint y_3 \, dA = 0, \qquad (7.13.12)$$

and

$$\iint \sigma_{11}^* y_2 \, dy_2 \, dy_3 = C \iint y_2 y_3 \, dy_2 \, dy_3 = 0, \qquad (7.13.13)$$

because the y_2-axis passes through the centroid, and y_3 is the axis of symmetry of the cross section. We can now determine the constant C from Eq. (7.13.11),

$$C = -M/I, \qquad (7.13.14)$$

and substitute this value in Eq. (7.13.10).

$$\sigma_{11} = -My_3/I. \qquad (7.13.15)$$

For an isotropic material the strain components are given by

$$\epsilon_{11} = -My_3/EI, \qquad (7.13.16)$$

$$\epsilon_{22} = \epsilon_{33} = \nu My_3/EI, \qquad (7.13.17)$$

$$\epsilon_{12} = \epsilon_{23} = \epsilon_{31} = 0. \qquad (7.13.18)$$

The six compatibility equations are then satisfied, because the strain components are either zero or of only a first power in y_3.

Let us now consider the displacement of the beam under the condition of pure bending. The strain-displacement relations Eq. (7.12.4) yield the following differential equations:

$$\frac{\partial u_1}{\partial y_1} = \epsilon_{11} = -\frac{My_3}{EI}, \tag{7.13.19}$$

$$\frac{\partial u_2}{\partial y_2} = \epsilon_{22} = \nu\frac{My_3}{EI}, \tag{7.13.20}$$

$$\frac{\partial u_3}{\partial y_3} = \epsilon_{33} = \nu\frac{My_3}{EI}, \tag{7.13.21}$$

$$\frac{1}{2}\left(\frac{\partial u_1}{\partial y_2} + \frac{\partial u_2}{\partial y_1}\right) = \epsilon_{12} = 0, \tag{7.13.22}$$

$$\frac{1}{2}\left(\frac{\partial u_2}{\partial y_3} + \frac{\partial u_3}{\partial y_2}\right) = \epsilon_{23} = 0, \tag{7.13.23}$$

$$\frac{1}{2}\left(\frac{\partial u_1}{\partial y_3} + \frac{\partial u_3}{\partial y_1}\right) = \epsilon_{31} = 0. \tag{7.13.24}$$

The displacements u_1, u_2, and u_3 can be obtained by integrating these equations when the geometrical conditions, i.e., the constraints of the bar are given. Let us now focus our attention on the lateral deflection of the beam, i.e., the component u_3. From Eq. (7.13.19) we obtain

$$u_1 = -\frac{My_1y_3}{EI} + \bar{u}, \tag{7.13.25}$$

where $\bar{u}$ may be a function of y_2 and y_3. From Eq. (7.13.24) we obtain

$$\frac{\partial u_3}{\partial y_1} = -\frac{\partial u_1}{\partial y_3} = \frac{My_1}{EI} - \frac{\partial \bar{u}}{\partial y_3}. \tag{7.13.26}$$

Differentiation with respect to y_1 gives

$$\frac{\partial^2 u_3}{\partial y_1^2} = \frac{M}{EI}. \tag{7.13.27}$$

Since the bending moment is not a function of y_3, Eq. (7.13.27) yields the important result that the second derivative of lateral displacement for every longitudinal fiber of the beam is a constant. (A longitudinal fiber is an element of the beam with the same y_3-coordinate.) If we denote the lateral deflection of the axis of the beam by $w(y_1)$ instead of u_3, we have

$$\frac{d^2 w}{dy_1^2} = \frac{M}{EI}. \tag{7.13.28}$$

We recall from calculus that the curvature of a given curve, $y = f(x)$, is given by

$$\frac{1}{R} = \frac{d^2f/dx^2}{[1 + (df/dx)^2]^{3/2}},$$
(7.13.29)

where R is the radius of curvature. For a very flat curve for which df/dx is very small in comparison to unity, the curvature may be represented simply by the second derivative of the curve, that is,

$$\frac{1}{R} = \frac{d^2f}{dx^2}.$$
(7.13.30)

The deformed shapes of beams are such that the curvatures are small; thus d^2w/dy_1^2 is the curvature of the deformed beam. We also find that the normal strain ϵ_{11} is related to the curvature by

$$\epsilon_{11} = -y_3 \frac{d^2w}{dy_1^2}.$$
(7.13.31)

We consider next the strain energy dU for an element dy_1 of the beam. Since the only stress component is σ_{11}, we have

$$dU = \left(\frac{1}{2} \iint \frac{\sigma_{11}^2}{E} \, dy_2 \, dy_3\right) dy_1,$$
(7.13.32)

or

$$dU = \left(\frac{1}{2} \iint E\epsilon_{11}^2 \, dy_2 \, dy_3\right) dy_1.$$
(7.13.33)

Substituting Eq. (7.13.15) into Eq. (7.13.32), and Eq. (7.13.31) into Eq. (7.13.33), we obtain respectively

$$dU = \frac{M^2}{2EI} \, dy_1,$$
(7.13.34)

and

$$dU = \frac{EI}{2} \left(\frac{d^2w}{dy_1^2}\right)^2 dy_1.$$
(7.13.35)

7.14 ENGINEERING BEAM THEORY

At this point it is worth while to mention the so-called engineering beam theory which covers the nonuniform beam under general lateral loading conditions. In such a case, the bounding surface of the beam may not be free of stress, and/or the body force F_3 may not be zero. Also, in general, the shear force in each section is not zero, and hence the bending moment M is not constant along the beam. The engineering beam theory, however, also neglects the normal stresses σ_{22} and σ_{33}, because it can be shown that these components are of a much smaller order of magnitude than σ_{11}. The theory also assumes that for a beam, the deformation due to shear strain is negligible in comparison to that due to the normal strain ϵ_{11}. As a result, the normal stresses and strains may still be calculated using Eqs. (7.13.15) and (7.13.31), although the bending moment, curvature, and moment of inertia are no longer constant along y_1. This engineering beam theory also assumes that the moment-curvature relation, Eq. (7.13.28), will still hold, and that the strain energy of a complete beam can be calculated by integrating Eq. (7.13.35).

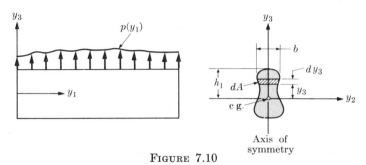

FIGURE 7.10

7.15 SUMMARY: ENGINEERING BEAM THEORY

The equations of engineering beam theory (see Figure 7.10 for coordinates and other nomenclature) are as follows:

Definition of equipollent transverse shear and bending moment

$$S = -\int_A \sigma_{13}\, dA, \tag{7.15.1}$$

$$M = -\int_A \sigma_{11} y_3\, dA. \tag{7.15.2}$$

Displacements

$$u_3(y_1, y_2, y_3) = w(y_1), \tag{7.15.3}$$

$$u_1(y_1, y_2, y_3) = -y_3 \frac{dw}{dy_1}. \tag{7.15.4}$$

Strain-displacement relation

$$\epsilon_{11} = -y_3 \frac{d^2 w}{dy_1^2}. \tag{7.15.5}$$

Moment-curvature relation

$$\frac{d^2 w}{dy_1^2} = \frac{M}{EI}. \tag{7.15.6}$$

Force equilibrium

$$\frac{dS}{dy_1} = p.$$ (7.15.7)

Moment equilibrium

$$\frac{dM}{dy_1} = S.$$ (7.15.8)

Stress-strain relation

$$\sigma_{11} = E\epsilon_{11}.$$ (7.15.9)

Bending stress versus moment relation

$$\sigma_{11} = -\frac{My_3}{I}.$$ (7.15.10)

Shear stress versus transverse shear relation

$$\sigma_{13} = \frac{SQ}{bI}.$$ (7.15.11)

Cross-sectional moment of inertia

$$I = \iint y_3^2 \, dA.$$ (7.15.12)

Static moment

$$Q = \iint_{y_3}^{h_1} y_3 \, dA.$$ (7.15.13)

PROBLEMS

7.1   (a) Thin flat panels (Fig. P.7.1) with external forces acting in the y_1y_2-plane of the panel are said to be in a *state of plane stress*, that is, $\sigma_{33} = \sigma_{13} = \sigma_{23} = 0$ (see Problem 6.4).   For an isotropic material verify the following expressions for γ_{11}, γ_{22}, and γ_{12} in terms of σ_{11}, σ_{22}, and σ_{12} and the elastic constants E and ν.

$$\gamma_{11} = \frac{1}{E}(\sigma_{11} - \nu\sigma_{22}), \qquad \gamma_{22} = \frac{1}{E}(\sigma_{22} - \nu\sigma_{11}), \qquad \gamma_{12} = \frac{1+\nu}{E}\sigma_{12}.$$

(b) Determine the following inverse stress-strain relations, i.e., the expression of σ_{11}, σ_{22}, and σ_{12} in terms of γ_{11}, γ_{22}, and γ_{12}:

$$\sigma_{11} = \frac{E}{1 - \nu^2}(\gamma_{11} + \nu\gamma_{22}), \qquad \sigma_{22} = \frac{E}{1 - \nu^2}(\gamma_{22} + \nu\gamma_{11}), \qquad \sigma_{12} = \frac{E}{1 + \nu}\gamma_{12}.$$

How are the elastic constants related to the elasticity tensor E_{mnpr}?

7.2   The fibers of a reinforced plastic panel are evenly distributed along three different preferred directions, AA, BB, and CC, which are 60° apart.   All are parallel to the face of the panel, as shown in Fig. P.7.2.   One of these directions is along the

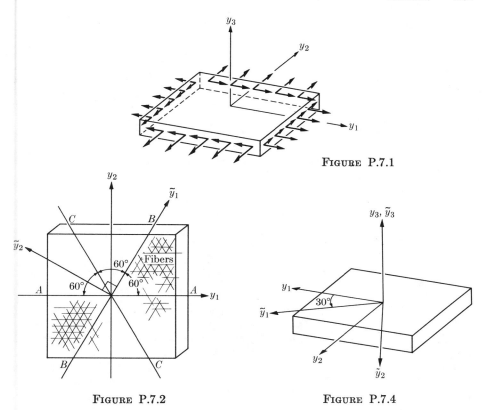

FIGURE P.7.1

FIGURE P.7.2   FIGURE P.7.4

y_1-axis. Show that the relation between stress and strain components in the y_1y_2-plane is governed by only two elastic constants, that is,

$$\gamma_{11} = S_1\sigma_{11} + S_2\sigma_{22}, \qquad \gamma_{22} = S_2\sigma_{11} + S_3\sigma_{22}, \qquad \gamma_{12} = S_4\sigma_{12},$$

where $S_3 = S_1$ and $S_4 = S_1 - S_2$. Thus the elastic properties are isotropic with respect to any direction normal to the y_3-axis. *Hint:* Prove $S_1 = S_3$ by considering the stress and strain components along the y_1y_2-coordinates for the case of uniaxial tension σ_{11}. Observe that the elastic constants for the y_1y_2-system and for $\tilde{y}_1\tilde{y}_2$-system are the same. Prove that $S_4 = S_1 - S_2$ by a similar consideration for the case of pure shear stress σ_{12}.

7.3   A hexagonal material has five independent elastic constants. (See Section 7.9 for example of hexagonal materials and the form of the elastic coefficient array.) Use symmetry arguments to reduce the elasticity matrix for orthotropic material (nine constants) to the one for hexagonal material.

7.4   Figure P.7.4 shows a panel made of orthotropic material with y_1-, y_2-, and y_3-axes parallel respectively to the axis of symmetry of the material. (a) Show that under plane stress conditions ($\sigma_{33} = \sigma_{13} = \sigma_{23} = 0$) the stress-strain relations for this panel can be reduced to the following matrix form which involves only four independent constants:

$$\begin{bmatrix} \sigma_{11} \\ \sigma_{22} \\ \sigma_{12} \end{bmatrix} = \begin{bmatrix} G_{11} & G_{12} & 0 \\ G_{12} & G_{22} & 0 \\ 0 & 0 & H_{12} \end{bmatrix} \begin{bmatrix} \gamma_{11} \\ \gamma_{22} \\ \gamma_{12} \end{bmatrix}.$$

(b) Derive expressions for G_{11}, G_{22}, G_{12}, and H_{12} in terms of the elasticity tensor E_{mnpr}.

(c) The inverse stress-strain relations may be written as

$$\begin{bmatrix} \gamma_{11} \\ \gamma_{22} \\ \gamma_{12} \end{bmatrix} = \begin{bmatrix} \dfrac{1}{E_1} & -\dfrac{\nu_{12}}{E_1} & 0 \\ -\dfrac{\nu_{12}}{E_1} & \dfrac{1}{E_2} & 0 \\ 0 & 0 & \dfrac{1}{H_{12}} \end{bmatrix} \begin{bmatrix} \sigma_{11} \\ \sigma_{22} \\ \sigma_{12} \end{bmatrix}.$$

Determine G_{11}, G_{22}, G_{12} in terms of E_1, E_2, and ν_{12}.

(d) Determine the stress-strain relations (strains in terms of stresses) for the $\tilde{y}_1$-, $\tilde{y}_2$-, $\tilde{y}_3$-coordinate system where $\tilde{y}_3$ and y_3 coincide and $\widehat{y_1 \tilde{y}_1} = 30°$. Note that shear stress $\tilde{\sigma}_{12}$ will cause normal strains $\tilde{\gamma}_{11}$ and $\tilde{\gamma}_{22}$, and normal stress $\tilde{\sigma}_{11}$ or $\tilde{\sigma}_{22}$ will cause shear strain $\tilde{\gamma}_{12}$. Thus, for an orthotropic material the principal directions of stress and strain do not in general coincide. But for an isotropic material they always coincide.

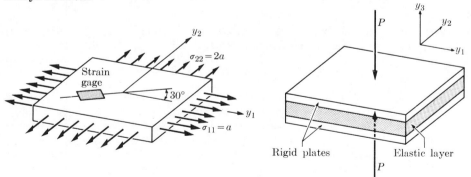

FIGURE P.7.5   FIGURE P.7.7

7.5 A rectangular plate of isotropic material is acted upon by boundary stresses $\sigma_{11} = a$, $\sigma_{22} = 2a$, $\sigma_{12} = 0$. The material properties are $E = 10^7$ psi, $\nu = 0.3$, and $\alpha = 10^{-6}/°F$. A strain gage is mounted at an angle of $30°$ from the y_1-axis, as shown in Fig. P.7.5. (a) What will the strain gage read under the above loading? (b) If the loading is kept constant and the temperature is raised $100°F$, what will be the new reading of the strain gage?

7.6   The state in which the strain components γ_{33}, γ_{13}, and γ_{23} are zero is called the *plane strain state*. For an isotropic material in plane strain verify the following expressions of γ_{11}, γ_{22}, and γ_{12} in terms of σ_{11}, σ_{22}, and σ_{12} and the elastic constants E and ν.

$$\gamma_{11} = \frac{1+\nu}{E}[(1-\nu)\sigma_{11} - \nu\sigma_{22}],$$

$$\gamma_{22} = \frac{1+\nu}{E}[(1-\nu)\sigma_{22} - \nu\sigma_{11}],$$

$$\gamma_{12} = \frac{1+\nu}{E}\sigma_{12}.$$

7.7   An elastic layer is sandwiched between two rigid plates to which it is bonded, as shown in Fig. P.7.7. The layer is compressed, i.e., the normal stress σ_{33} is negative. Supposing that the bonding to the plates prevents any lateral strain ϵ_{11} or ϵ_{22}, find the ratio $\sigma_{33}/\epsilon_{33}$ in terms of E and ν of the elastic material.

7.8   A material has *equal* strength properties along two axes oblique to each other at an angle of 35°. The third axis is perpendicular to the plane of the other two axes, and properties along the third axis are different. How many elasticity components are required to describe the elastic behavior of this material?

7.9   (a) An isotropic panel of thickness h is in a state of plane stress. Derive the expression for the strain energy of an element of volume $dy_1\,dy_2$ in terms of the stress components σ_{11}, σ_{22}, and σ_{12}. (b) Derive the expression of the volume element for the strain energy in terms of the strain components γ_{11}, γ_{22}, and γ_{12}.

7.10   Derive the expression for the strain-energy density U^* in terms of σ_{11}, σ_{22}, and σ_{12} under the condition of plane strain (see Problem 7.6).

7.11   A long rectangular rod hangs under gravity in a pool of water as shown in Fig. P.7.11. The water exerts only a *normal* pressure $p = \rho_w g y_1$ on the surfaces of the rod. The rod itself has a density ρ_r and length l. (a) State the boundary conditions on *all* six faces of the rod (use an averaged boundary condition on the face $y_1 = 0$). (b) The assumed solution for the stress σ_{mn} given below satisfies all boundary conditions and the fifteen equations of elasticity. Evaluate the three unknowns C_1, C_0, and the end load P in terms of the given quantities ρ_w, ρ_r, l, b, h and g. (c) Determine the deflection u_1 everywhere in the rod. Assume that $u_1 = 0$ at $y_1 = 0$.

Assumed solution:

$$\sigma_{11} = C_1 y_1 + C_0, \qquad \sigma_{12} = 0,$$
$$\sigma_{22} = -\rho_w g y_1, \qquad \sigma_{23} = 0,$$
$$\sigma_{33} = -\rho_w g y_1, \qquad \sigma_{31} = 0,$$

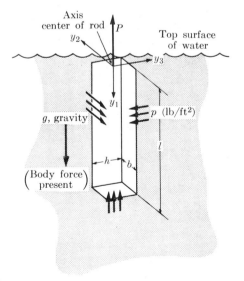

FIGURE P.7.11

and C_1, C_0, P are constants.

$$\text{Density of water,} \quad \rho_w \ (\text{slugs/ft}^3);$$
$$\text{Density of rod,} \quad \rho_r \ (\text{slugs/ft}^3);$$
$$\text{Isotropic material,} \quad E, \ \nu.$$

7.12 Given a cantilever beam loaded by a tip load P (Fig. P.7.12). By integrating the moment curvature relation (that is, $M = EI \ (d^2w/dy_1^2)$ with the prescribed displacement boundary conditions, determine the vertical deflection $w(y_1)$ of the beam at any point, and the maximum deflection at the tip $w(a)$. What is the slope of the beam at $y_1 = a$? [Neglect the weight of the beam.]

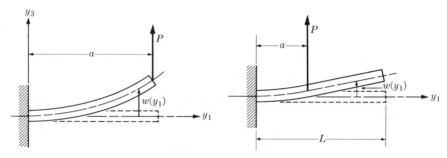

FIGURE P.7.12 FIGURE P.7.13

7.13 A cantilever beam of length L is loaded by a vertical load P at a distance a from the clamped end (Fig. P.7.13). Determine the tip deflection $w(L)$. *Hint:* Make use of the result of the previous problem. What is the curvature of the beam $y_1 > a$?

7.14 Many problems in plane stress are formulated in terms of the Airy stress function, which we encountered in Problem 6.11. We recall that the definition of the Airy stress function,

$$\sigma_{11} = \frac{\partial^2 \phi}{\partial y_2^2}, \qquad \sigma_{22} = \frac{\partial^2 \phi}{\partial y_1^2}, \qquad \sigma_{12} = -\frac{\partial^2 \phi}{\partial y_1 \, \partial y_2},$$

ensures that the equations of equilibrium will be satisfied. To complete the formulation it is necessary to replace all the equations of plane elasticity by a single equation in ϕ alone. The equations of plane elasticity are summarized as follows:

Equations of equilibrium,

$$\frac{\partial \sigma_{11}}{\partial y_1} + \frac{\partial \sigma_{21}}{\partial y_2} = 0, \qquad \frac{\partial \sigma_{12}}{\partial y_1} + \frac{\partial \sigma_{22}}{\partial y_2} = 0.$$

Strain-displacement relations,

$$\epsilon_{11} = \frac{\partial u_1}{\partial y_1}, \qquad \epsilon_{22} = \frac{\partial u_2}{\partial y_2}, \qquad \epsilon_{12} = \frac{1}{2}\left(\frac{\partial u_1}{\partial y_2} + \frac{\partial u_2}{\partial y_1}\right).$$

Stress-strain relations,

$$\epsilon_{11} = \frac{1}{E}(\sigma_{11} - \nu\sigma_{22}), \qquad \epsilon_{22} = \frac{1}{E}(\sigma_{22} - \nu\sigma_{11}), \qquad \epsilon_{12} = \frac{1+\nu}{E}\sigma_{12}.$$

(a) Show that combination of the stress-strain relations, the definition of ϕ, and the strain-displacement relations (in the form of a compatibility condition) leads to the biharmonic equation of plane elasticity,

$$\frac{\partial^2 \phi}{\partial y_1^4} + 2\frac{\partial^4 \phi}{\partial y_1^2 \partial y_2^2} + \frac{\partial^4 \phi}{\partial y_2^4} = 0.$$

(b) A certain problem in plane stress is solved via the biharmonic equation, and it is found that

$$\phi(y_1, y_2) = k(y_1^4 - 6y_1^2 y_2^2 + y_2^4),$$

where k is a constant. Calculate the components of stress and strain in the structure, including the strain ϵ_{zz}.

7.15   Derive a formula for the strain-energy density in terms of the stress function ϕ of the structure under the condition of plane stress.

$$8$$

Plastic Behavior of Solids

8.1 INTRODUCTION

In Chapter 7 the simplest kind of constitutive relation was described and formulated, namely, that of a linear elastic stress-strain relation. However, the engineering use of materials requires not only a knowledge of elastic behavior, but also of inelastic or plastic behavior. For example, plastic deformations in structures are usually permitted under emergency or sometimes even under normal operating conditions. In fact, the metal-forming processes involve essentially a plastic flow of materials. For the former problem the plastic strains are usually small; for the latter, the plastic strains may be of the order of several hundred percent. Thus, in some cases, the elastic strain may be completely neglected. In this chapter we shall consider the relations between stress and strain in the range of small plastic strain.

Much knowledge has been gained during the past two decades about how structural defects in the orderly arrangement of the atoms cause the observed mechanical behavior of materials. Much further research must be done before all portions of the ordinary stress-strain curve can be predicted or explained. Notwithstanding its humble nature, the ordinary stress-strain curve of the common engineering materials is amazingly reproducible, and contains most of the material data necessary for design. The response of materials to stress and strain from a microscopic and atomistic point of view is a fascinating study and the subject of many recent books. In this first course on solid mechanics, however, we shall adopt a phenomenological point of view and content ourselves with an engineering approach to plasticity. However, we must be aware of those developments in the field of physical metallurgy and solid-state physics which will aid our engineering approach.

8.2 STRESS-STRAIN RELATIONS UNDER UNIAXIAL LOADING CONDITIONS

The simplest description of the plastic behavior of solids is the stress-strain curve under the condition of uniaxial stress. This curve is usually obtained by a simple tensile test which involves a slender specimen pulled in a direction parallel to its axis. Figure 8.1 shows a narrow strip cut from a thin sheet. The center part of the strip was machined so that a uniform cross-sectional area could be maintained. When the specimen is pulled apart, the elongation between previously indented gage marks is measured. The ratio between the elongation Δl and the original gage length l is the strain in the axial direction; the applied load divided by the original area of the specimen yields the uniaxial stress. Figure 8.2 shows stress-strain curves for several typical engineering materials. It is worth while to discuss briefly some of the mechanical properties defined by the tensile stress-strain curves.

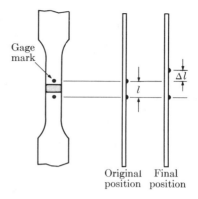

FIG. 8.1   Tension specimen.

The ultimate tensile stress is defined by the maximum load the specimen can sustain divided by the original area of the cross section. For the brittle material shown in Fig. 8.2(c), the ultimate tensile stress is actually the fracture stress of the material. For a ductile material, the ultimate tensile stress is the stress at which local necking of a specimen occurs. Thus, although the so-called engineering stress (i.e., load divided by the original area) decreases after necking begins, the true stress (defined by the tensile load divided by the instantaneous minimum area) actually increases. The specimen finally fractures at a maximum value of true stress.

In Fig. 8.2(a) and (b) the stress-strain curves corresponding to the necking region are drawn as dashed lines. For this part of the curve, the cross section of the specimen is no longer uniform between the two gage points. Consequently, the strain is nonuniform, and indeed for some materials the regions of large strain are concentrated in small bands. The average value of strain obtained by dividing the elongation by the original gage length thus loses its value as a description of the process of deformation. It is obvious that this value will be different when the original gage length is different.

The yielding, or the onset of plastic flow, for certain ductile materials such as low carbon steel, can be clearly identified as shown in Fig. 8.2(a). In fact, for this case upper and lower yield points exist. The upper yield point exists because interstitial solute atoms serve as obstacles to the movement of dislocations. When the stress becomes sufficiently high to tear the obstacles loose, the dislocations can move at a lower value of applied stress. For a low carbon steel this lower yield stress remains constant for a certain range of strain until the phenomenon of strain hardening becomes significant.

For most ductile materials strain hardening follows plastic flow immediately, hence there is no clear-cut indication of the onset of plastic flow. We normally define the yield stress as the stress which produces a very small, but easily measurable, amount of plastic deformation. The most commonly used yield

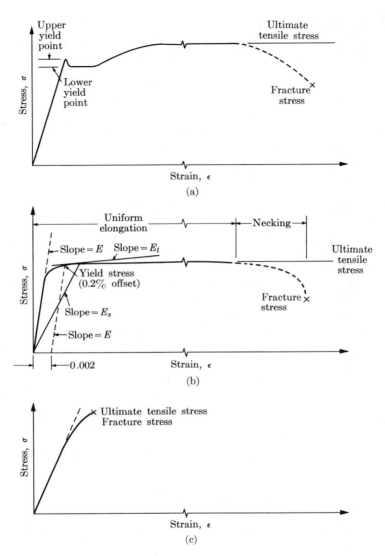

Fɪɢ. 8.2   The nomenclature of uniaxial stress-strain diagrams.

stress is defined by the intersection of a line which passes through the strain axis at a strain of 0.002, and is parallel to the elastic line, as shown in Fig. 8.2(b). When a tensile specimen is loaded up to this stress and then is unloaded, the unloading curve is, in general, parallel to the initial straight portion of the stress-strain curve. Thus, when the stress has returned to zero, there will remain a 0.2% permanent strain, which is often called the *permanent offset*. The above arbitrary definition of yield stress is often called 0.2% *offset yield stress*.

The *proportional limit* is the stress at which strain ceases to be proportional to stress. This limit is very difficult to obtain experimentally. In fact, the more precise the technique of measurement, the lower the proportional limit will be found. It is thus customary to adopt an arbitrary definition by taking the proportional limit as the stress which would produce a plastic strain of 0.0001 in./in.

The *modulus of elasticity* (or *Young's modulus*) is the slope of the initial straight-line portion of the stress-strain curve. At high stresses, when the relation between stress and strain ceases to be linear, two moduli are often used. They are: (1) the tangent modulus, E_t, defined by the slope of the tangent to the stress-strain curve, $E_t = d\sigma/d\epsilon$; and (2) the secant modulus, E_s, defined by the slope of a line from the origin to a point on the stress-strain curve, $E_s = \sigma/\epsilon$. (see Fig. 8.2b)

For most materials the available stress-strain data are limited to tensile properties because of the simplicity of conducting a tensile test. However, it should be mentioned that the stress-strain curve for compression is generally not the same as that for tension. Figure 8.3 shows typical stress-strain curves of a 2024-T3 aluminum alloy sheet. The curves show clearly that there is an appreciable difference between the curves in tension and in compression. This figure also shows that although Young's modulus for the rolled aluminum alloy sheet remains the same both for the direction with the grain and across the grain, the properties in the plastic range are certainly nonisotropic.

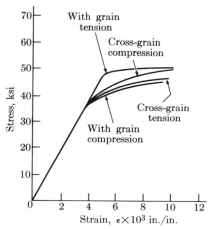

FIG. 8.3   Uniaxial stress-strain diagrams for 2024-T3 aluminum alloy (rolled sheet).

For many structural materials it has been found that the experimental stress-strain curve can be closely fitted over a fairly large range by the following relatively simple polynomial relations known as the *Ramberg-Osgood* relations (Ref. 8.1):

$$\epsilon = \frac{\sigma}{E} + \left(\frac{\sigma}{B}\right)^n. \tag{8.2.1}$$

The first term on the right-hand side represents the elastic strain, and the second term represents the plastic strain. Typical values for the three constants E, B, and n for various materials are given in Table 8.1.

It should be remarked that for materials with clearly defined yield points, such as low carbon steel (Fig. 8.2a), the effect of the slight rise of the upper yield point may be neglected. Thus for the range of small plastic strains, the stress remains practically constant. In this case, the stress-strain curve may be

TABLE 8.1   *Typical Stress-Strain Constants (Approximate)*

Material	Ultimate tensile stress, psi	E, psi $(\times 10^6)$	B, psi	n
Alloy steel	100,000	29	122,400	25
Alloy steel	180,000	29	202,000	30
Aluminum alloy 2024-T3	65,000	10.5	72,300	10
Aluminum alloy 7075-T6	83,000	10.5	101,200	20
Magnesium alloy	39,000	6.5	47,500	10

idealized by two straight lines, one with a slope of E, representing the elastic range, the other with zero slope representing the plastic range (Fig. 8.4a). This idealized model is often called the elastic perfectly plastic material. It is interesting to point out that in applying Eq. (8.2.1) to this limiting case, the constant n should approach infinity while the constant B is equal to the yield stress of the material.

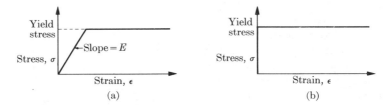

FIG. 8.4   Idealized uniaxial stress-strain diagrams: (a) elastic perfectly plastic material; (b) rigid perfectly plastic material.

Another idealized model is called the rigid perfectly plastic material (Fig. 8.4b). For a problem in which the plastic strain is much larger than the elastic strain, it is justifiable to neglect the latter. In this case, the solid will be considered as rigid when the applied stress is less than the yield stress, and the material will flow whenever the yield stress is reached.

8.3   YIELD CONDITION UNDER GENERAL STATE OF STRESS FOR ISOTROPIC MATERIALS

In the previous section we discussed the plastic behavior of a slender bar under uniaxial loading. For materials such as low carbon steel it was shown that plastic flow will occur when the normal stress reaches the yield stress, or yield point, of the material. In fact, for all materials plastic strain begins to appear when the uniaxial stress reaches a certain value. The question now arises as to what is the condition under which plastic deformation or yielding occurs under a general state of combined stress, i.e., two-dimensional or three-dimensional stress. To be more specific, the yield condition under combined stress can be expressed as a certain function of the stress components. When the value of this function reaches a certain level, plastic flow will occur. In the present discussion we shall limit ourselves to isotropic materials.

The plastic deformations of crystalline materials are the result of slip along atomic planes or of twin-glide displacement. In either case, the deformation does not involve permanent changes in the distance between atomic lattice planes. In fact, in both cases the deformation is the result of shearing action. It can thus be concluded that there will be no volume change in the solid due to plastic deformation, and that a state of stress which involves no shear component will not cause plastic deformation, no matter how high the stress magnitude may be. Examples of this type of stress state are *hydrostatic pressure*, for which the three principal stresses are of equal magnitude in compression, and the so-called *hydrostatic tension*, for which the three principal stresses are of equal magnitude in tension.

Since plastic deformation is directly associated with shearing action, it is logical to assume that the material will yield when the maximum shear stress reaches a critical value. This critical value is the maximum shear stress corresponding to the yielding of a simple tensile specimen; hence we speak of the *maximum shear-stress criterion for yielding*. This criterion is also called the *Tresca yield condition*.

Under a uniaxial stress condition, the maximum shear stress $\tau_{\max}$ is equal to half the applied normal stress:

$$(\tau_{\max})_{\text{yield}} = \frac{\sigma_0}{2}, \tag{8.3.1}$$

where $\sigma_0 =$ yield stress in simple tension. Under a three-dimensional stress condition, the maximum shear stresses are given by (see Section 6.9)

$$\tfrac{1}{2}|\sigma_I - \sigma_{II}|, \qquad \tfrac{1}{2}|\sigma_{II} - \sigma_{III}|, \qquad \tfrac{1}{2}|\sigma_{III} - \sigma_I|,$$

where σ_I, σ_{II}, and σ_{III} are the three principal stresses. The maximum shear stress to be used in Eq. (8.3.1) is the largest of these three values. If, for example, we assume that $\sigma_I < \sigma_{II} < \sigma_{III}$, then the Tresca yield condition is written as

$$\sigma_{III} - \sigma_I = \sigma_0. \tag{8.3.2}$$

A complete description of the Tresca yield condition is given in Table 8.2.

The yield condition under three given principal stresses is represented by a surface in three-dimensional space with the principal stresses as the coordinates. The yield surface under the maximum shear-stress criterion is the continuous

TABLE 8.2

Stress state	Tresca condition
$\sigma_I < \sigma_{II} < \sigma_{III}$	$\sigma_{III} - \sigma_I = \sigma_0$
$\sigma_{III} < \sigma_{II} < \sigma_I$	$\sigma_I - \sigma_{III} = \sigma_0$
$\sigma_I < \sigma_{III} < \sigma_{II}$	$\sigma_{II} - \sigma_I = \sigma_0$
$\sigma_{II} < \sigma_{III} < \sigma_I$	$\sigma_I - \sigma_{II} = \sigma_0$
$\sigma_{II} < \sigma_I < \sigma_{III}$	$\sigma_{III} - \sigma_{II} = \sigma_0$
$\sigma_{III} < \sigma_I < \sigma_{II}$	$\sigma_{II} - \sigma_{III} = \sigma_0$

surface formed by the six planes given by the equations of Table 8.2. It can be seen that the first two planes are parallel to the σ_{II}-axis and inclined at a 45° angle to the $\sigma_I\sigma_{II}$-plane, the third and fourth planes are parallel to the σ_{III}-axis and inclined at a 45° angle to the $\sigma_{II}\sigma_{III}$-plane, and the last two planes are parallel to the σ_I-axis and inclined at a 45° angle to the $\sigma_I\sigma_{III}$-plane. These six planes form a hexagonal prism which is extended to infinity in both directions, as shown in Fig. 8.5. The axis of this prism intersects the three coordinate axes at the same angle. The surface of this prism is the Tresca yield surface. When a point which represents the three principal stresses falls within this surface, the material is in the elastic range, and when the point lies on this surface, yielding will occur.

Another logical criterion for yielding is associated with the strain-energy content of the material at yielding. We have just pointed out that hydrostatic pressure or tension, which causes only volume change without distortion, will not produce any plastic deformation. This implies that the yield condition will depend only on the distortion energy and be independent of the dilatation energy. The distortion-energy criterion for yielding can thus be described by the following

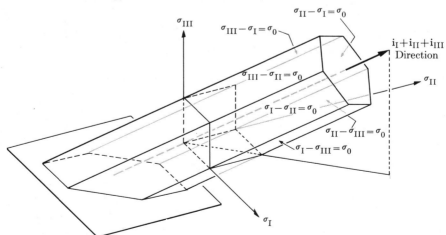

FIG. 8.5 Representation of the Tresca criterion as a surface in principal stress space.

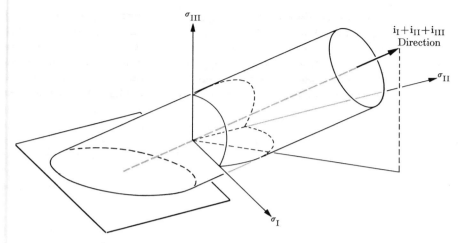

FIG. 8.6   The Mises-Hencky criterion shown as a surface in principal stress space.

statement: *Plastic flow will occur when the distortion-energy density in the material reaches the value corresponding to the yielding of a simple tensile specimen.* This criterion is often called the *Huber-von Mises-Hencky** or *Mises-Hencky Theory.*

The distortion-energy density under three-dimensional stress was given in Section 7.10:

$$U^* = \frac{1}{12G}[(\sigma_I - \sigma_{II})^2 + (\sigma_{II} - \sigma_{III})^2 + (\sigma_{III} - \sigma_I)^2].$$

Under uniaxial stress conditions the distortion energy becomes

$$U^* = \frac{1}{6G}(\sigma_0)^2.$$

The distortion-energy criterion can thus be stated as

$$\frac{1}{12G}[(\sigma_I - \sigma_{II})^2 + (\sigma_{II} - \sigma_{III})^2 + (\sigma_{III} - \sigma_I)^2] = \frac{1}{6G}(\sigma_0)^2,$$

or

$$(\sigma_I - \sigma_{II})^2 + (\sigma_{II} - \sigma_{III})^2 + (\sigma_{III} - \sigma_I)^2 = 2(\sigma_0)^2. \qquad (8.3.3)$$

We can see that, if the three principal stresses are chosen as the coordinate axes, the Mises-Hencky yield condition will be represented by a circular cylindrical surface whose axis coincides with the line which intersects the origin at equal angles with the three coordinate axes. The radius of this cylinder is equal to $\sqrt{2/3}\,\sigma_0$. This yield surface extends to infinity in both directions, (see Fig. 8.6).

---

* The maximum distortion energy criterion was discussed in two German articles by von Mises (Ref. 7) and Hencky (Ref. 8) in 1913 and 1925 respectively; the same yield condition had been suggested by Huber (Ref. 9) in 1904 in a Polish publication but did not attract general attention until nearly twenty years later.

Let us now discuss an important physical picture given by the two yield surfaces. The axis for either the hexagonal prism or the circular cylinder makes equal angles with the three principal stress axes. Hence it represents the state of hydrostatic tension or hydrostatic pressure. Thus both yield conditions are in agreement with our previous conclusion that hydrostatic tension or pressure, no matter how large in magnitude, will not produce yielding. Another way of interpreting this is by considering two points a and a' (Fig. 8.6) which are on a line parallel to the axis of the yield surface. Point a' represents the superposition of a hydrostatic pressure on the state of stress given by point a. If point a is not on the yield surface, a superposition of a hydrostatic pressure on this state of stress should not produce yielding in the solid. This fact has been verified by Bridgeman by conducting tension tests under very high hydrostatic pressures (Ref. 8.2).

Next let us consider the mathematical interpretations of the two yield conditions developed above. We have seen that a yield condition can be expressed in terms of stress components. For an isotropic material it should be independent of the choice of reference axes. Thus, we should expect the yield condition to be a function of the stress invariants. In fact, since it is not affected by the hydrostatic component of stress, we expect it to be dependent only on the deviatoric stress invariants $\hat{I}_2$ and $\hat{I}_3$. The yield condition hence will take the general form

$$f(\hat{I}_2, \hat{I}_3) = 0. \tag{8.3.4}$$

By referring to Eqs. (6.8.18), (6.8.19) and (6.10.6) we see that we can write

$$\hat{I}_2 = -\tfrac{1}{6}[(\sigma_{\mathrm{I}} - \sigma_{\mathrm{II}})^2 + (\sigma_{\mathrm{II}} - \sigma_{\mathrm{III}})^2 + (\sigma_{\mathrm{III}} - \sigma_{\mathrm{I}})^2].$$

The Mises-Hencky criterion then becomes

$$\hat{I}_2 + \frac{(\sigma_0)^2}{3} = 0. \tag{8.3.5}$$

The Tresca yield condition given by Table 8.2 can be represented by the following equation:

$$[(\sigma_{\mathrm{I}} - \sigma_{\mathrm{II}})^2 - (\sigma_0)^2][(\sigma_{\mathrm{II}} - \sigma_{\mathrm{III}})^2 - (\sigma_0)^2][(\sigma_{\mathrm{III}} - \sigma_{\mathrm{I}})^2 - (\sigma_0)^2] = 0. \tag{8.3.6}$$

In terms of the deviatoric stress invariants this equation becomes

$$4\hat{I}_2^3 - 27\hat{I}_3^2 - 9(\sigma_0)^2\hat{I}_2^2 + 6(\sigma_0)^4\hat{I}_2 - (\sigma_0)^6 = 0. \tag{8.3.7}$$

Another way of interpreting the Mises-Hencky criterion is to associate the yield condition with the octahedral shear stress τ_{oct}. We showed in Chapter 6 (see Problem 6.8) that

$$\tau_{\text{oct}} = \tfrac{1}{3}\sqrt{(\sigma_{\text{I}} - \sigma_{\text{II}})^2 + (\sigma_{\text{II}} - \sigma_{\text{III}})^2 + (\sigma_{\text{III}} - \sigma_{\text{I}})^2} = \sqrt{-(\tfrac{2}{3})\hat{I}_2}.$$

$$(8.3.8)$$

Thus the Mises-Hencky criterion given by $\hat{I}_2 = \text{const}$ can also be written as $\tau_{\text{oct}} = \text{const}$. This interpretation was given by Nadai (Ref. 8.3).

For convenience, the yield condition under combined stress is usually defined by an equivalent stress $\bar{\sigma}$. When the equivalent stress $\bar{\sigma}$ is equal to σ_0, the yield stress under uniaxial stress conditions, plastic flow will begin. Thus under the maximum shear-stress criterion, $\bar{\sigma}$ is equal to twice the maximum shear stress. Under the distortion energy criterion, we have

$$\bar{\sigma} = \sqrt{\tfrac{1}{2}[(\sigma_{\text{I}} - \sigma_{\text{II}})^2 + (\sigma_{\text{II}} - \sigma_{\text{III}})^2 + (\sigma_{\text{III}} - \sigma_{\text{I}})^2]}. \qquad (8.3.9)$$

Experimental verifications of the yield criteria have been conducted by many researchers during the last several decades (see Ref. 3, Chap. 17). Most of these studies were concerned with two-dimensional states of stress obtained by testing thin-walled tubes under combined internal pressure and axial load, or under combined torsion and tension.

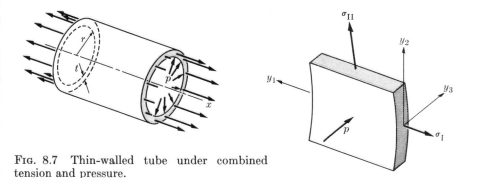

FIG. 8.7 Thin-walled tube under combined tension and pressure.

Let us consider a thin-walled tube under internal pressure and axial tension or compression loads, as shown in Fig. 8.7. The three principal stresses for an element shown in the lower half of the figure are σ_{I}, the axial stress; σ_{II}, the circumferential stress; and σ_{III}, the radial stress. It can be seen that the radial stress σ_{III} varies from $-p$ to 0 from the inner surface to the outer surface, while the circumferential stress resulting from hoop tension is equal to pr/t. For thin-walled tubes where r/t is much larger than unity, the radial stress can be ignored, and the problem may be considered a two-dimensional problem with the principal stress $\sigma_{\text{III}} = 0$.

In applying the maximum shear criterion to this case, we see that when both σ_I and σ_{II} are positive, the yield condition is determined by

$$\sigma_I - \sigma_{III} = \sigma_I = \sigma_0, \quad \text{when} \quad \sigma_I > \sigma_{II};$$

or

$$\sigma_{II} - \sigma_{III} = \sigma_{II} = \sigma_0, \quad \text{when} \quad \sigma_{II} > \sigma_I.$$

These equations are represented by the vertical line AB and the horizontal line CB in Fig. 8.8. When both σ_I and σ_{II} are negative, the yield condition will be given by

$$-\sigma_I = \sigma_0, \quad \text{when } \sigma_I < \sigma_{II};$$

$$-\sigma_{II} = \sigma_0, \quad \text{when } \sigma_{II} < \sigma_I.$$

These equations are represented by lines DE and EF, respectively. However, when σ_I and σ_{II} are of opposite sign, the yield condition must be determined by

$$\sigma_I - \sigma_{II} = \pm\sigma_0.$$

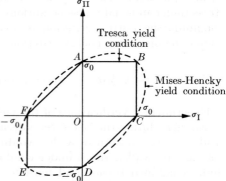

Fɪɢ. 8.8   Yield conditions, under biaxial stress.

These equations are represented by the lines AF and CD. The complete Tresca yield condition is thus given by the closed hexagon in Fig. 8.8.

In applying the distortion energy criterion to this case we see that Eq. (8.3.3) is reduced to

$$(\sigma_I)^2 - 2\sigma_I\sigma_{II} + (\sigma_{II})^2 + (\sigma_I)^2 + (\sigma_{II})^2 = 2(\sigma_0)^2,$$

or

$$(\sigma_I)^2 - \sigma_I\sigma_{II} + (\sigma_{II})^2 = (\sigma_0)^2. \tag{8.3.10}$$

This equation represents an ellipse which passes through points A, B, C, D, E, and F as shown in Fig. 8.8.

Experimental results by Lode (Ref. 4) and by Ros and Eichinger (Ref. 5) are illustrated in Fig. 8.9. The materials used in these experiments included copper, nickel, and steel. It can be seen that the experimental results correlate better with the distortion-energy theory.

When a thin-walled tube is acted on by combined tension and twisting, the principal stresses σ_I and σ_{II} can be obtained by using the Mohr circle for stress:

$$\sigma_I = \sigma/2 + \sqrt{(\sigma/2)^2 + (\tau)^2}, \quad \sigma_{II} = \sigma/2 - \sqrt{(\sigma/2)^2 + (\tau)^2},$$

where σ = the axial tensile stress, and τ = shear stress due to torsion. When we apply the maximum shear-stress theory, we substitute the above equations into $\sigma_I - \sigma_{II} = \sigma_0$, from which we obtain

$$2\sqrt{(\sigma/2)^2 + (\tau)^2} = \sigma_0$$

or

$$(\sigma)^2 + 4(\tau)^2 = (\sigma_0)^2. \qquad (8.3.11)$$

When we apply the distortion-energy theory, we substitute the same equations into

$$(\sigma_I)^2 - (\sigma_I \sigma_{II}) + (\sigma_{II})^2 = (\sigma_0)^2,$$

and obtain

$$(\sigma)^2 + 3(\tau)^2 = (\sigma_0)^2. \qquad (8.3.12)$$

Equations (8.3.11) and (8.3.12) represent the two ellipses shown in Fig. 8.10. Experimental results obtained by Taylor and Quinney (Ref. 8.6) using copper, aluminum, and mild steel tubes are plotted for comparison. In their experiments they first strain-hardened each tube by tension load, and then after partially removing the load they applied a gradually increas-

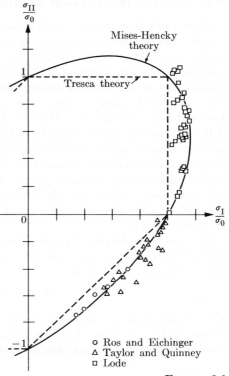

FIGURE 8.9

ing torque until yielding occurred. Thus for these tests the onset of plastic flow could be distinctly observed in these materials, although in their virgin states there had been no clear-cut indication of such phenomena. Taylor and Quinney's results again substantiate the distortion-energy yield criterion.

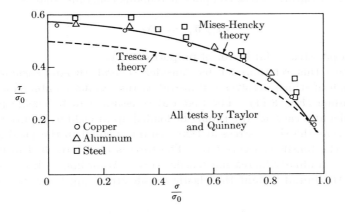

FIGURE 8.10

8.4 PLASTIC STRESS-STRAIN RELATIONS

In the previous section we discussed only the state of combined stress that would cause a material to begin to deform plastically. For an elastic perfectly plastic material, if the equivalent stress $\bar{\sigma}$ reaches the yield stress, the material will flow. But the problem is, "How will the material flow and what are the magnitudes of the various strain components when the material does flow?" For a uniaxial loading condition, the value of plastic strain for a perfectly plastic material is indeterminate. Thus in the case of a combined loading condition, if the equivalent stress is maintained at the yield stress, the magnitudes of the plastic strain components are also indeterminate. However, it is possible to determine the relations among the various components of strain. This problem on the plastic flow of elastic plastic material may also be approached in the following way: "What are the various components of stress when an elastic perfectly plastic material is deformed into the plastic range with known components of strain?"

For materials deformed beyond the elastic range, it is convenient to express the total strain ϵ in terms of two components, the elastic strain ϵ^e and the plastic strain ϵ^p, such that

$$\epsilon_{mn} = \epsilon_{mn}^e + \epsilon_{mn}^p. \tag{8.4.1}$$

The elastic-strain components can be related to the components of stress by means of Eq. (7.7.28). For the plastic-strain part, we note that the volume remains unchanged during plastic deformation, that is,

$$\epsilon_{ii}^p = 0. \tag{8.4.2}$$

From Eq. (5.11.2) we conclude that the total plastic-strain tensor ϵ_{ij}^p is identical to the deviatoric plastic-strain tensor $\hat{\epsilon}_{ij}^p$.

A simple theory for plastic flow consistent with the Mises-Hencky yield condition is that the ratio of the increment of each plastic strain component to its corresponding deviatoric stress component remains constant, that is,

$$d\epsilon_{ij}^p/\hat{\sigma}_{ij} = d\lambda, \tag{8.4.3}$$

where $d\lambda$ is a positive scalar factor of proportionality.

To illustrate this relation, let us consider several two-dimensional stress problems, each of which involves a thin-walled tube under combined axial and torsional loading (Fig. 8.11). The material is assumed to be elastic perfectly plastic. In the first case, let the tubing be loaded up to yield by a pure torsional moment, so that the shear stress at that instant is equal to the yielding shear stress, while the tensile stress is zero. The tube is then stretched in the axial direction by a gradually increasing tensile stress. According to Eq. (8.4.3), we find that at the initial instant the plastic tensile strain must also be zero, and

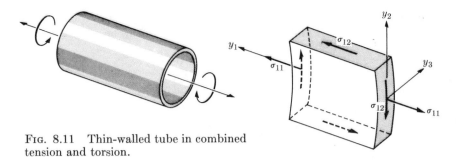

Fig. 8.11   Thin-walled tube in combined
tension and torsion.

the initial tensile strain will consist of only an elastic component. The initial
slope of the tensile stress-strain curve must then remain equal to the elastic
modulus E. In the second case, let the yield condition be reached by combined
shear and tensile stresses. When the tube is stretched beyond yielding both
elastic and plas⁺ic strains will occur, and the slope of the tensile stress-strain
curve must be less than E. In the limiting case where the initial yielding was
the result of pure tension, the slope of the tensile stress-strain curve beyond this
point will certainly be zero, because no further tensile stress can be added beyond
this point.

We can extend this example further by calculating the complete tensile stress-
strain relation of the thin-walled tube for the period preceding the initial
yielding. Let us consider the case for which the yielding has been attained by
pure shear, and the twisting angle is maintained constant while the tensile
strain is applied.

The increments in total shear strain and total tensile strain are, respectively,

$$d\epsilon_{12} = d\epsilon_{12}^e + d\epsilon_{12}^p, \tag{8.4.4}$$

$$d\epsilon_{11} = d\epsilon_{11}^e + d\epsilon_{11}^p. \tag{8.4.5}$$

Substituting the relations

$$d\epsilon_{12}^e = d\sigma_{12}/2G, \tag{8.4.6}$$

$$d\epsilon_{11}^e = d\sigma_{11}/E, \tag{8.4.7}$$

$$d\epsilon_{12}^p = \hat{\sigma}_{12}\, d\lambda = \sigma_{12}\, d\lambda, \tag{8.4.8}$$

$$d\epsilon_{11}^p = \hat{\sigma}_{11}\, d\lambda = \tfrac{2}{3}\sigma_{11}\, d\lambda, \tag{8.4.9}$$

into Eqs. (8.4.4) and (8.4.5), we obtain

$$d\epsilon_{12} = \frac{d\sigma_{12}}{2G} + \sigma_{12}\, d\lambda, \tag{8.4.10}$$

$$d\epsilon_{11} = \tfrac{2}{3}\sigma_{11}\, d\lambda + \frac{d\sigma_{11}}{E}. \tag{8.4.11}$$

We see that $d\lambda$ can be eliminated from these equations. By substituting $d\epsilon_{12} = 0$ for our present problem and introducing the yield condition (Eq. 8.3.12),

$$(\sigma_{11})^2 + 3(\sigma_{12})^2 = (\sigma_0)^2, \tag{8.4.12}$$

we obtain the following relation between ϵ_{11} and σ_{11}:

$$d\epsilon_{11} = \frac{(E/3 - G)(\sigma_{11})^2 + G(\sigma_0)^2}{EG[(\sigma_0)^2 - (\sigma_{11})^2]} \, d\sigma_{11}. \tag{8.4.13}$$

Integrating, we obtain

$$\epsilon_{11} = \int_0^{\epsilon_{11}} d\epsilon_{11} = \int_0^{\sigma_{11}} \frac{(E/3 - G)(\sigma_{11})^2 + G(\sigma_0)^2}{EG[(\sigma_0)^2 - (\sigma_{11})^2]} \, d\sigma_{11}, \tag{8.4.14}$$

or

$$\epsilon_{11} = \frac{G - (E/3)}{EG} \sigma_{11} + \frac{\sigma_0}{6G} \ln \left(\frac{\sigma_0 + \sigma_{11}}{\sigma_0 - \sigma_{11}} \right). \tag{8.4.15}$$

For a typical engineering material the stress-strain relation given by Eq. (8.4.15) is shown in Fig. 8.12. It is seen that the initial slope of this curve is equal to E.

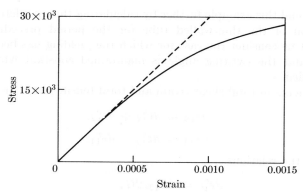

FIG. 8.12   A plot of Eq. (8.4.4) for steel.

The next problem to be considered is the plastic stress-strain relation for materials which are characterized by strain-hardening in the plastic range. Suppose we conduct a simple tensile test and load the specimen into the plastic range to a point a, as shown in Fig. 8.13. If we then release the load and reload again, we find that the yield stress has increased to the value σ_a because of the previous plastic deformation.

We now raise the question, "If, after the release of the tensile load, a combined loading is applied, what will be the yield condition?" According to a

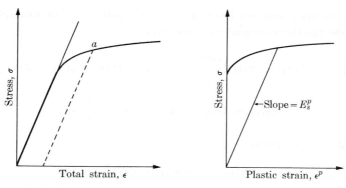

FIG. 8.13   The plastic secant modulus.

simple theory of isotropic strain-hardening, yielding will occur when the equivalent stress $\bar{\sigma}$ determined by, say, the distortion energy theory, has reached the
value σ_a. The question to be answered next is, "What is the relation between the
plastic strain and the equivalent stress?" In this case we are also interested in
the relations among the various strain components. Of course, the latter relations are the same as those for the perfectly plastic material, given by Eq. (8.4.3),
i.e., the ratio between each incremental plastic strain and the corresponding
deviatoric stress remains a constant value $d\lambda$. However, for materials which
are characterized by strain hardening the value $d\lambda$ can be uniquely determined
if the stress-strain ($\bar{\sigma}$ vs. $\bar{\epsilon}$) curve under uniaxial loading is given. Under uniaxial
tension, the incremental plastic strain is $d\bar{\epsilon}^p$ while the deviatoric stress is $\frac{2}{3}\bar{\sigma}$.
Thus

$$d\lambda = \frac{3}{2}\frac{d\bar{\epsilon}^p}{\bar{\sigma}}. \tag{8.4.16}$$

Substituting this equation into Eq. (8.4.3), we obtain

$$d\epsilon_{11}^p = [\sigma_{11} - \tfrac{1}{2}(\sigma_{22} + \sigma_{33})]\frac{d\bar{\epsilon}^p}{\bar{\sigma}}, \qquad d\epsilon_{12}^p = \tfrac{3}{2}\sigma_{12}\frac{d\bar{\epsilon}^p}{\bar{\sigma}},$$

$$d\epsilon_{22}^p = [\sigma_{22} - \tfrac{1}{2}(\sigma_{11} + \sigma_{33})]\frac{d\bar{\epsilon}^p}{\bar{\sigma}}, \qquad d\epsilon_{23}^p = \tfrac{3}{2}\sigma_{23}\frac{d\bar{\epsilon}^p}{\bar{\sigma}},$$

$$d\epsilon_{33}^p = [\sigma_{33} - \tfrac{1}{2}(\sigma_{11} + \sigma_{22})]\frac{d\bar{\epsilon}^p}{\bar{\sigma}}, \qquad d\epsilon_{31}^p = \tfrac{3}{2}\sigma_{31}\frac{d\bar{\epsilon}^p}{\bar{\sigma}}; \tag{8.4.17}$$

or

$$d\epsilon_{ij}^p = \tfrac{3}{2}\hat{\sigma}_{ij}\frac{d\bar{\epsilon}^p}{\bar{\sigma}}. \tag{8.4.18}$$

We can eliminate stresses from the above equation. For example, if the reference axes are the principal axes, we have

$$d\epsilon_I^p = [\sigma_I - \tfrac{1}{2}(\sigma_{II} + \sigma_{III})]\frac{d\bar\epsilon^p}{\bar\sigma}, \qquad d\epsilon_{II}^p = [\sigma_{II} - \tfrac{1}{2}(\sigma_I + \sigma_{III})]\frac{d\bar\epsilon^p}{\bar\sigma},$$

$$d\epsilon_{III}^p = [\sigma_{III} - \tfrac{1}{2}(\sigma_I + \sigma_{II})]\frac{d\bar\epsilon^p}{\bar\sigma}.$$

(8.4.19)

From the above we obtain

$$(d\epsilon_I^p - d\epsilon_{II}^p)^2 + (d\epsilon_{II}^p - d\epsilon_{III}^p)^2 + (d\epsilon_{III}^p - d\epsilon_I^p)^2$$

$$= \tfrac{9}{4}[(\sigma_I - \sigma_{II})^2 + (\sigma_{II} - \sigma_{III})^2 + (\sigma_{III} - \sigma_I)^2]\left(\frac{d\bar\epsilon^p}{\bar\sigma}\right)^2 \qquad (8.4.20)$$

$$= \tfrac{3}{2}(d\bar\epsilon^p)^2,$$

or

$$d\bar\epsilon^p = \sqrt{\tfrac{2}{9}[(d\epsilon_I^p - d\epsilon_{II}^p)^2 + (d\epsilon_{II}^p - d\epsilon_{III}^p)^2 + (d\epsilon_{III}^p - d\epsilon_I^p)^2]}. \qquad (8.4.21)$$

Knowing the incremental values $d\bar\epsilon^p$, we can determine by integration the total equivalent plastic strain $\bar\epsilon^p$.

Under the present theory, which is often called the *incremental strain theory* or *flow theory*, the equivalent stress $\bar\sigma$ and equivalent strain $\bar\epsilon$ are related in the *same* manner *as* the tensile stress-strain curve of the material. It should be noted that the individual strain components such as ϵ_I^p, ϵ_{II}^p, and ϵ_{III}^p, which are obtainable by integration, are usually not unique functions of the stress components. Under the incremental strain theory the final state of strain is a function of the loading path. Thus if the same state of strain is reached through two different paths, the resulting states of stress will be different.

We shall consider now a particular type of combined loading under which the proportions of the various stress components remain the same. This means that the ratio $\hat\sigma_{ij}/\bar\sigma$ remains constant. We can thus integrate Eq. (8.4.18) and write

$$\epsilon_{ij}^p = \frac{3}{2}\frac{\hat\sigma_{ij}}{\bar\sigma}\bar\epsilon^p, \qquad (8.4.22)$$

or

$$\epsilon_{ij}^p = \frac{3}{2}\frac{\hat\sigma_{ij}}{E_s^p}, \qquad (8.4.23)$$

where E_s^p is the secant modulus referred to the plastic component of strain, as shown in Fig. 8.13. It is seen that this stress-strain relation is expressed in terms of final strain. Thus the stress-strain relation given by Eq. (8.4.12) is often referred to as the total strain or deformation theory of plasticity. In a strict sense, this theory is applicable only under proportional loading conditions.

Illustrative Examples

(a) Shear stress-strain relations. The stress-strain curve obtained by a simple tensile test of an isotropic material can be expressed as

$$\epsilon = \frac{\sigma}{E} + \left(\frac{\sigma}{B}\right)^n. \tag{8.4.24}$$

It is required to determine the shear stress-shear strain relation for the same material. The shear modulus in the elastic range is G.

For the shear stress and strain we use the engineering notation τ and γ (see Sections 5.5 and 7.7). We have

$$\gamma = \frac{\tau}{G} + \gamma^p, \tag{8.4.25}$$

$$\frac{\gamma^p/2}{\tau} = \frac{3}{2}\frac{1}{E_s^p}, \tag{8.4.26}$$

$$E_s^p = \frac{\bar{\sigma}}{\bar{\epsilon}^p} = \frac{\bar{\sigma}}{(\bar{\sigma}/B)^n}. \tag{8.4.27}$$

Under pure shear the equivalent stress given by the distortion-energy theory is

$$\bar{\sigma} = \sqrt{3}\,\tau. \tag{8.4.28}$$

Thus, by combining the above equations, we obtain

$$\gamma = \frac{\tau}{G} + 3^{(n+1)/2}\,B^{-n}\tau^n. \tag{8.4.29}$$

We can then write

$$\gamma = \frac{\tau}{G} + \left(\frac{\tau}{A}\right)^n, \tag{8.4.30}$$

where

$$A = \frac{B}{3^{(1/2+1/2n)}}. \tag{8.4.31}$$

(b) Plate under varying biaxial stresses. Consider a thin plate under biaxial tension stresses σ_I and σ_{II}, as shown in Fig. 8.14. The material is isotropic, and its plastic behavior may be idealized as rigid-plastic with linear strain-hardening. Under uniaxial tension the stress-strain diagram is as shown in Fig. 8.15. The tangent modulus E_t is a constant. Now let the principal stresses σ_I and σ_{II} be varying with respect to time. We represent this condition as

$$\sigma_I = f(t), \tag{8.4.32}$$

and

$$\sigma_{II} = g(t). \tag{8.4.33}$$

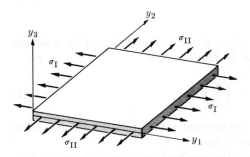

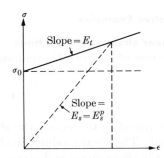

Fig. 8.14 Thin plate under biaxial tension stresses.

Fig. 8.15 Stress-strain diagram: rigid-plastic with linear strain-hardening.

Our objective is to calculate the time history of the three principal strains ϵ_I, ϵ_{II}, and ϵ_{III}, based on the incremental strain theory.

Since the material is rigid-plastic, the total strains ϵ_I, ϵ_{II}, and ϵ_{III} are the same as the plastic strains. Thus, based on Eq. (8.4.19), the increments of the principal strains can be expressed as

$$d\epsilon_I = d\epsilon_I^p = (\sigma_I - \tfrac{1}{2}\sigma_{II}) \frac{d\bar{\epsilon}^p}{\bar{\sigma}}, \tag{8.4.34}$$

$$d\epsilon_{II} = d\epsilon_{II}^p = (\sigma_{II} - \tfrac{1}{2}\sigma_I) \frac{d\bar{\epsilon}^p}{\bar{\sigma}}, \tag{8.4.35}$$

$$d\epsilon_{III} = d\epsilon_{III}^p = -\tfrac{1}{2}(\sigma_I + \sigma_{II}) \frac{d\bar{\epsilon}^p}{\bar{\sigma}}, \tag{8.4.36}$$

where the equivalent stress $\bar{\sigma}$ is given by

$$\bar{\sigma} = (\sigma_I^2 - \sigma_I\sigma_{II} + \sigma_{II}^2)^{1/2} = (f^2 - fg + g^2)^{1/2}. \tag{8.4.37}$$

For the present linear strain-hardening behavior,

$$\frac{d\bar{\epsilon}^p}{d\bar{\sigma}} = \frac{1}{E_t} = \text{const.} \tag{8.4.38}$$

Let us rewrite Eq. (8.4.34) in the form

$$d\epsilon_I = \frac{(\sigma_I - (1/2)\sigma_{II})}{\bar{\sigma}} \frac{d\bar{\epsilon}^p}{d\bar{\sigma}} d\bar{\sigma}. \tag{8.4.39}$$

By substituting Eqs. (8.4.32), (8.4.33), (8.4.37), (8.4.38), and the following one,

$$d\bar{\sigma} = \tfrac{1}{2}(f^2 - fg + g^2)^{-1/2}[f'(2f - g) + g'(2g - f)], \tag{8.4.40}$$

into Eq. (8.4.39), we obtain

$$d\epsilon_I = \frac{1}{4E_t} \frac{(2f - g)[f'(2f - g) + g'(2g - f)]}{f^2 - fg + g^2} dt. \qquad (8.4.41)$$

Thus

$$\epsilon_I = \frac{1}{4E_t} \int_{t_0}^{t} \frac{(2f - g)[f'(2f - g) + g'(2g - f)]}{f^2 - fg + g^2} dt, \qquad (8.4.42)$$

where t_0 is the instant of time when the material begins to yield under biaxial stresses, i.e., when

$$[f(t)]^2 - f(t)g(t) + [g(t)]^2 = \sigma_0^2. \qquad (8.4.43)$$

Similarly, for the other two principal strains, we obtain

$$\epsilon_{II} = \frac{1}{4E_t} \int_{t_0}^{t} \frac{(2g - f)[f'(2f - g) + g'(2g - f)]}{f^2 - fg + g^2} dt \qquad (8.4.44)$$

and

$$\epsilon_{III} = -\frac{1}{4E_t} \int_{t_0}^{t} \frac{(f + g)[f'(2f - g) + g'(2g - f)]}{f^2 - fg + g^2} dt. \qquad (8.4.45)$$

Thus based on the incremental strain theory, the state of strain is a function of the complete loading history.

It is of interest to note that under the deformation theory the state of strain at a given instant of time is given by Eqs. (8.4.23):

$$\epsilon_I = \epsilon_I^p = \frac{\sigma_I - \frac{1}{2}\sigma_{II}}{E_s^p}, \qquad \epsilon_{II} = \epsilon_{II}^p = \frac{\sigma_{II} - \frac{1}{2}\sigma_I}{E_s^p},$$

$$\epsilon_{III} = \epsilon_{III}^p = -\frac{1}{2}\left(\frac{\sigma_I + \sigma_{II}}{E_s^p}\right), \qquad (8.4.46)$$

where for a given value of the equivalent stress $\bar{\sigma}$ the secant modulus is

$$E_s^p = \frac{\bar{\sigma}E_t}{\bar{\sigma} - \sigma_0}. \qquad (8.4.47)$$

Under the deformation theory the state of strain is only a function of the instantaneous state of stress.

8.5 SUMMARY

This chapter presents the two most important considerations in defining the plastic behavior of materials, the state of combined stresses which will cause the material to yield, and the multiaxial stress-strain relations in the plastic range. For simplicity this discussion has been limited to isotropic materials.

Yield conditions

The Tresca or maximum shear-stress condition. Yielding occurs whenever the maximum shear stress is equal to one-half the yield stress in simple tension. Under the state of biaxial stress when $\sigma_{III} = 0$, the yield boundary is a closed hexagon defined by the following lines:

$$\sigma_I = \pm\sigma_0,$$

$$\sigma_{II} = \pm\sigma_0, \tag{8.5.1}$$

and

$$\sigma_I - \sigma_{II} = \pm\sigma_0.$$

The Mises-Hencky or distortion-energy criterion: yielding occurs whenever the distortion energy is equal to $(1/6G)\sigma_0^2$ or when the following condition for the principal stresses is satisfied:

$$(\sigma_I - \sigma_{II})^2 + (\sigma_{II} - \sigma_{III})^2 + (\sigma_I - \sigma_{III})^2 = 2\sigma_0^2. \tag{8.5.2}$$

The biaxial stress yield condition is given by an ellipse:

$$\sigma_I^2 - \sigma_I\sigma_{II} + \sigma_{II}^2 = \sigma_0^2. \tag{8.5.3}$$

Plastic stress-strain relations for elastic perfectly plastic materials

$$\epsilon_{mn} = \epsilon_{mn}^e + \epsilon_{mn}^p. \tag{8.5.4}$$

The elastic-strain components are related to the stress components by the generalized Hooke's law:

$$\epsilon_{mn}^e = (1/E)[(1 + \nu)\sigma_{mn} - \nu\,\delta_{mn}\sigma_{rr}]. \tag{8.5.5}$$

The plastic strain components are characterized by

$$\epsilon_{rr}^p = 0 \quad \text{and} \quad \frac{d\epsilon_{mn}^p}{\hat{\sigma}_{mn}} = d\lambda, \tag{8.5.6}$$

where $d\lambda$ is an indication of plastic flow.

Plastic stress-strain relation for materials which are characterized by strain hardening. The plastic behavior is governed by an equivalent stress $\bar{\sigma}$ and equivalent plastic strain $\bar{\epsilon}^p$, where $\bar{\sigma}$ is defined by

$$\bar{\sigma} = \sqrt{\tfrac{1}{2}[(\sigma_I - \sigma_{II})^2 + (\sigma_{II} - \sigma_{III})^2 + (\sigma_I - \sigma_{III})^2]} \tag{8.5.7}$$

and $\bar{\epsilon}^p$ is the integral of $d\bar{\epsilon}^p$ defined by

$$d\bar{\epsilon}^p = \sqrt{\tfrac{2}{9}[(d\epsilon_I^p - d\epsilon_{II}^p)^2 + (d\epsilon_{II}^p - d\epsilon_{III}^p)^2 + (d\epsilon_I^p - d\epsilon_{III}^p)^2]},$$

$$\tag{8.5.8}$$

where

$$d\epsilon_I^p = \frac{[\sigma_I - \frac{1}{2}(\sigma_{II} + \sigma_{III})]\, d\bar{\epsilon}^p}{\bar{\sigma}},$$

$$d\epsilon_{II}^p = \frac{[\sigma_{II} - \frac{1}{2}(\sigma_I + \sigma_{III})]\, d\bar{\epsilon}^p}{\bar{\sigma}}, \qquad (8.5.9)$$

$$d\epsilon_{III}^p = \frac{[\sigma_{III} - \frac{1}{2}(\sigma_I + \sigma_{II})]\, d\bar{\epsilon}^p}{\bar{\sigma}}.$$

PROBLEMS

8.1 (a) Plot stress-strain diagrams for the following materials: steel alloy, $\sigma_{ult} =$ 180,000 psi; aluminum alloy, 2024-T3; and magnesium alloy. Use the data given in Table 8.1. (b) Find graphically the 0.2% offset yield stress for each of the above materials. (c) From the graphs obtained in part (a), draw graphs of the tangent and secant moduli versus total strain for each material.

8.2 The following analytical expression has been used to describe the stress-strain relation of a certain material:

$$\epsilon = \frac{\sigma}{B} + \frac{\sigma}{\sigma_{ult} - \sigma},$$

where B and σ_{ult} are constants. From a uniaxial tensile test of this material the following test data were obtained:

σ(psi)	500	1000	10,000
ϵ(in./in.)	0.0051	0.0104	0.1500

(a) Determine the values of B, σ_{ult}, and the elastic modulus E. (b) Let σ_0 be the 0.2% offset yield stress. Calculate the numerical value of σ_0 from the given test data. (c) Derive an expression for the relation between shear stress and engineering strain for the material, using (i) the distortion energy theory, and (ii) the maximum shear stress theory.

8.3 A two-material bar is built into rigid walls at both ends, as shown in Fig. P.8.3(a). Both bars have the same cross-sectional area and coefficient of thermal expansion, $\alpha = 10^{-4}/°F$. Bar A may be treated as a rigid perfectly plastic material, and bar B as elastic and linear strain-hardening. Figure P.8.3(b) illustrates the material properties

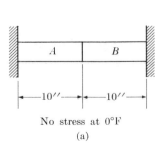

No stress at 0°F

(a)

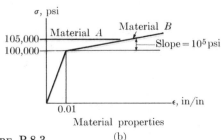

Material properties

(b)

FIGURE P.8.3

of bars A and B. It is assumed that the material properties are independent of temperature. (a) Describe the deformation of the system as the temperature is increased. (b) Calculate the temperature at which materials A and B start to yield. Neglect three-dimensional effects by assuming that $\nu = 0$ in both the elastic and the plastic ranges. Describe the deformation and stress histories of the system as the temperature is increased further.

8.4   A sheet of material having a yield stress $\sigma_0 = 50{,}000$ psi is loaded in a state of plane stress by the stresses $\sigma_{11} = 20{,}000$ psi, $\sigma_{22} = -20{,}000$ psi, and σ_{12}. Compute the value of σ_{12} which is required to cause the material to yield under the assumptions of (a) the Tresca Theory and (b) the Distortion-Energy Theory.

8.5 A thin plate is in the state of uniform plane stress (Fig. P.8.5). The stress components are in the following proportions:

$$\sigma_{11} : \sigma_{22} : \sigma_{12} = 1 : +\tfrac{1}{2} : +\tfrac{1}{2}.$$

Determine the plane along which slip will occur when the stress is increasing proportionally. Assume slip will occur along planes of maximum shear stress.

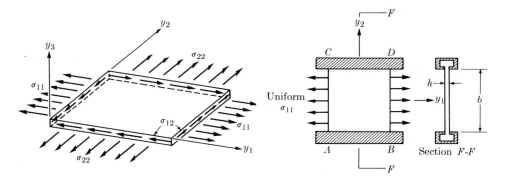

FIGURE P.8.5   FIGURE P.8.6

8.6   The rectangular plate shown in Fig. P.8.6 is constrained against deformation in the y_2-direction by rigid tracks at the edges AB and CD. The material is elastic perfectly plastic, with $E = 10{,}000{,}000$ psi, $\nu = 0.3$, and $\sigma_0 = 50{,}000$ psi for uniaxial tension. Using (a) the Mises-Hencky criterion, and (b) the Tresca criterion, calculate the values of σ_{11} and ϵ_{11} at which the constrained plate begins to yield.

8.7   A square plate is subjected to uniform tension in both the y_1- and y_2-directions, such that σ_{11} is always twice the value of σ_{22} (see Fig. P.8.7a). The material is isotropic and can be idealized as elastic and linear strain-hardening, as shown in Fig. P.8.7(b). Poisson's ratio is equal to $\tfrac{1}{3}$ in the elastic range. Calculate and plot the strains ϵ_{11} and ϵ_{22} as functions of σ_{11} for the range $0 \leq \sigma_{11} \leq 2\sigma_0$.

8.8   For the plate in the previous problem, if the value of σ_{22} remains always $\tfrac{1}{2}\sigma_0$ while σ_{11} increases gradually from zero, calculate and plot the strains ϵ_{11} and ϵ_{22} as functions of σ_{11} for the range $0 \leq \sigma_{11} \leq 2\sigma_0$.

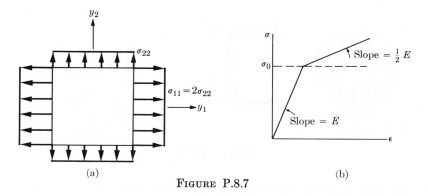

(a)

(b)

FIGURE P.8.7

8.9   The plate illustrated by Figs. 8.14 and 8.15 is strained by biaxial tension. The time histories of the two principal strains are given by $\epsilon_I = F(t)$ and $\epsilon_{II} = G(t)$. Develop the equations which are required for determining the time histories of the two principal stresses.

8.10   A thin-wall tube with closed ends as shown in Fig. P.8.10 is acted upon by a constant torque of 12,560 in -lb. The shear stress σ_{12} due to T is given by $\sigma_{12} = T/2\pi r^2 t$, where t = thickness of the tube = 0.05 in. and r = radius of the tube = 2 in. As shown in the sketch, y_1, y_2, and y_3 are along the longitudinal, circumferential, and radial directions, respectively.

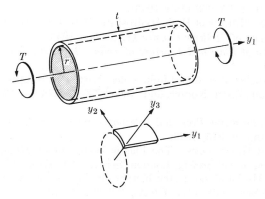

FIGURE P.8.10

The torque is kept constant while the internal pressure is gradually increased. Express the percentage increase in tube diameter as a function of the internal pressure p for each of the following sets of material properties:

(a) elastic linear strain-hardening with elastic modulus = E = 10^7 psi, yield stress under uniaxial tension = σ_0 = 40,000 psi, tangent modulus = E_t = 2×10^6 psi;

(b) stress-strain relation in tension given by

$$\epsilon = \frac{\sigma}{10^7} + \left(\frac{\sigma}{10^6}\right)^2.$$

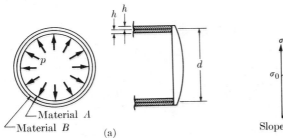
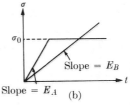

FIGURE P.8.11

8.11   The wall of a long closed-end tube consists of two layers as shown in Fig. P.8.11(a). The thickness for each layer is h which is small in comparison to the diameter d of the tube. The elastic moduli for the two materials A and B are E_A and E_B, respectively, and Poisson's ratio for both materials is ν. Material A becomes perfectly plastic at σ_0, while material B has a very high elastic limit (see Fig. P.8.11b). Derive the equations required for evaluating the elastic and plastic deformation of the tube under increasing internal pressure.

REFERENCES

(1) W. RAMBERG and W. R. OSGOOD, "Description of Stress-Strain Curves by Three Parameters," *NACA TN* 902 (July 1943).

(2) P. W. BRIDGEMAN, "Effects of High Hydrostatic Pressure on the Plastic Properties of Metals," *Reviews of Modern Physics* **17**, 3 (Jan. 1945).

(3) A. NADAI, *Theory of Flow and Fracture of Solids*, 2nd Ed., Vol. I, McGraw-Hill, New York, 1950.

(4) W. LODE, "Versuche über den Einfluss der Mittleren Hauptspannung auf das Fliessen der Metalle Eisen, Kupfer, und Nickel," *Zeitschrift für Physik* **36**, 913–939 (1926).

(5) M. ROS and A. EICHINGER, *Proc. 2nd International Congress for Applied Mechanics*, Zurich, Switzerland, p. 315, 1926.

(6) G. I. TAYLOR and H. QUINNEY, "The Plastic Distortion of Metals," *Philosophical Transactions of the Royal Society*, London, Ser. A. **230**, 323–362 (1931).

(7) R. VON MISES, "Mechanik der festen Körper im plastischen Deformationzustand", *Göttinger Nachrichten, math.-phys. Klasse*, 582–592 (1913).

(8) H. HENCKY, "Über das Wesen der plastischen Verformung," *Z. Verendeut. Ing.* **69**, 695 (1925).

(9) M. T. HUBER, "The Work of Deformation as a Proper Measure of Material Failure" (in Polish), *Czasopismò technicznè*, Lwov (Lemberg) **22**, 81 (1904).

9

Energy Principles in Solid Continuum

9.1 INTRODUCTION

The preceding seven chapters have laid the foundation for the statics of deformable bodies. We have seen that equations of static equilibrium, consistent deformation, and constitutive relationships are required for solving many problems. These equations for linearly elastic problems form a system of fifteen partial differential equations. The prospect of tackling these equations head on is somewhat frightening, and the chances of success are not very bright. Further along in our study of statics we shall discuss the possibility of reducing the general equations of three-dimensional elasticity to a considerably simplified form for problems involving the bending and twisting of beams and the bending of plates, and for problems which involve only two of the three space coordinates. We shall also discuss the direct approach of finding solutions to the simplified equations.

The direct approach of finding solutions to these fifteen equations is often difficult. An alternative method derives from considering the energy of the structural system. This direct approach may be loosely classified as a "vector" method, whereas the energy approach is often described as "scalar." The vector method examines in detail the stress vectors and displacement vectors along predetermined directions. Energy is an invariant, i.e., the energy is independent of the coordinate system used to describe the structure. The application of the energy approach, in general, obviates the task of writing some vector equations—a task often very difficult to accomplish.

This chapter will derive the various energy theorems for a general solid continuum. However, applications of these theorems will be confined to simple structural forms such as cables, rods, beams, and trusses. Applications of the energy methods to more complex structural forms will be investigated at a later time.

We must keep in mind that the various energy approaches represent alternative methods of solving the elasticity equations. For example, some energy approaches may be interpreted as a replacement of the equations of static equilibrium, while others can be used to derive the conditions of consistent deformation. There are many names attached to these theorems. The theorems are all related, as indeed they must be, since they are based on the same set of equations, i.e., the equations of elasticity. The various energy theorems merely attack the problem from different quarters. Some are better adapted for the calculation of deflections, others for the calculation of stresses. With slightly different re-interpretations, the same theorem can handle a variety of problems. The cardinal feature, however, is that the energy theorems enable us to circumvent a frontal attack on the fifteen partial differential equations of elasticity.

9.2 WORK AND INTERNAL ENERGY

In Chapter 7 we stated, on the basis of the First Law of Thermodynamics, that the work performed by the loads applied externally to a structure is equal to the energy stored in the structure by the internal stresses and strains, if we neglect the very small temperature changes which occur. Let us elaborate on this theme in order to develop a terminology and methodology which we can use in developing the energy theorems.

Consider a deformable structure which is fixed in space. Such a structure may be schematically represented in Fig. 9.1 as a planar body supported at points A and B, although the results to be derived will be applicable to three-dimensional bodies. The structure is acted upon by a system of Q-forces, as shown in the figure. There are N such forces. Note that Q_1 shown acting at point 1 is a force in the usual sense of the word, but that Q_3 shown acting at point 3 is a couple force or moment. The term "generalized forces" is used to include both forces and moments; henceforth, unless specifically stated otherwise, the symbol Q_n will represent a *generalized force*.

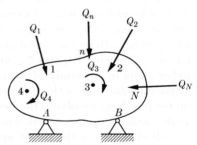

Fig. 9.1   Body loaded at N points.

There is defined at each of the N points a *generalized displacement*, q_n, representing the deformation of the body at that point. For Q_2, the generalized displacement is a linear deflection along the line of action of Q_2, and for the couple force (moment) Q_3, it is an angular rotation about a line perpendicular to the plane of the couple force Q_3. These two typical generalized displacements are shown in Fig. 9.2, where the generalized force Q_3 is depicted as a couple force in order to indicate more clearly the angular rotation nature of q_3. Thus q_3 represents the rotation of a small line segment at point 3 as the body is deformed. In general, therefore, the product of a generalized force and its associated

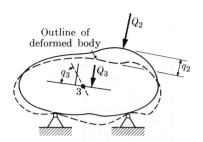
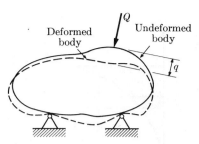

Fig. 9.2   Two types of generalized forces and generalized displacements.

Fig. 9.3   Body under a single force.

generalized displacement represents work. The positive sense of the generalized displacement is the same as that of the corresponding force.

In order to develop some fundamental concepts, let us focus our attention on a body which is acted upon by a single generalized force (see Fig. 9.3). The dashed line represents to an exaggerated degree the deformed shape of the body after Q has reached its final magnitude. The relationship between the force Q and the displacement q is shown as a force-displacement curve in Fig. 9.4. Note that we have not as yet postulated a linear relationship. For an infinitesimal increment in q, the increment in work done by Q is

$$dW = Q\,dq, \qquad (9.2.1)$$

which is the crosshatched area in Fig. 9.4. By applying the First Law of Thermodynamics (neglecting temperature changes), we find that the work done by Q is stored as internal energy in the structure, provided the supports are unyielding. This may be expressed as

$$dU = dW, \qquad (9.2.2)$$

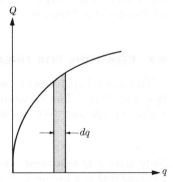

where U denotes the strain energy (see Chapter 7). The total strain energy is given by the integral

Fig. 9.4   General force-displacement relationship.

$$U = \int_0^q Q'\,dq', \qquad (9.2.3)$$

where Q and q represent the final values, and the primes merely indicate dummy integration variables.

Let us also define at this time the complementary strain energy or, as it is usually called, the *complementary energy*. It is given by a relation which is conjugate to Eq. (9.2.3),

$$U_c = \int_0^Q q'\,dQ', \qquad (9.2.4)$$

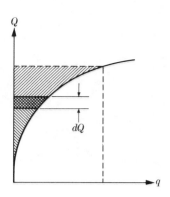

FIG. 9.5 Definition of complementary energy.

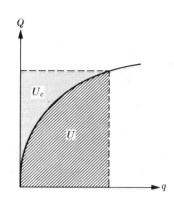

FIG. 9.6 Strain energy and complementary energy.

and is represented by the crosshatched area in Fig. 9.5. If Eq. (9.2.4) is integrated by parts, we get

$$U_c = Qq - U, \tag{9.2.5}$$

a result geometrically verified by Fig. 9.6, in which the product Qq is seen to be the area of a rectangle. The relation given by Eq. (9.2.5) is sometimes known as *Friederich's Theorem*.

9.3 RELATIONS FOR LINEARLY ELASTIC SYSTEMS

The force-displacement curve for a linearly elastic structure is a straight line (see Fig. 9.7). The constant slope of the curve, denoted by K, is called the *stiffness coefficient*. We can write

$$Q = Kq. \tag{9.3.1}$$

Note that if Q is a force, then K has the dimensions of force per unit length, whereas if Q is a moment (couple force) then K has the units of force-length per radian (or just force-length, since the radian measure of an angle is dimensionless). The determination of K, as we discovered in Chapters 4 and 7, requires the solution of the equations of elasticity. For the present we will assume that the value of K is known.

With the linear relation of Eq. (9.3.1), the strain energy given by Eq. (9.2.3) becomes

$$U = \int_0^q Kq' \, dq' = \tfrac{1}{2}Kq^2. \tag{9.3.2}$$

If the final value of the force is factored out of Eq. (9.3.2), we find that

$$U = \tfrac{1}{2}Qq. \tag{9.3.3}$$

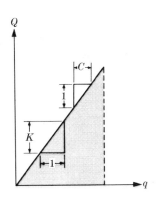

FIG. 9.7 Stiffness (K) and compliance (C) coefficients.

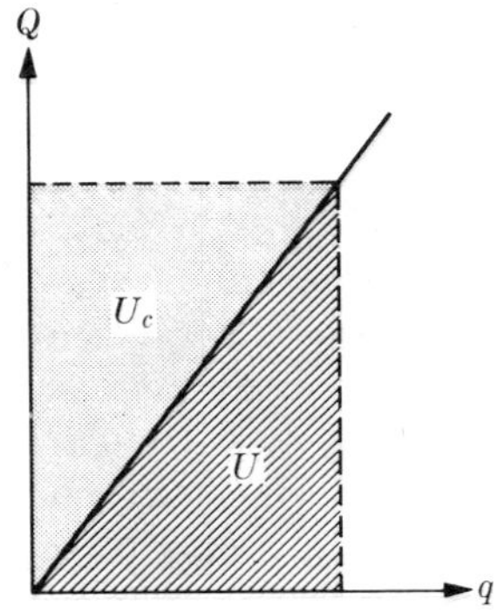

FIG. 9.8   Strain energy and complementary energy—linear system.

The preceding formula which states that the strain energy is equal to one-half the product of the force and the corresponding displacement is known as *Clapeyron's Law*. Geometrically, Eq. (9.3.3) is the area of the shaded triangle in Fig. 9.7. We will state Clapeyron's Law more rigorously at a later stage (cf. Eq. 9.3.21).

The relation conjugate to Eq. (9.3.1) is

$$q = CQ, \tag{9.3.4}$$

where C is called the *flexibility* or *compliance coefficient*. It has the dimensions of length per unit force if q is a linear displacement, and radian per unit force-length if q is an angular displacement.

By substituting Eq. (9.3.4) into Eq. (9.2.4), we obtain the complementary energy of an elastic system,

$$U_c = \int CQ' \, dQ' = \tfrac{1}{2}CQ^2. \tag{9.3.5}$$

However, it is an obvious but very useful fact that for a *linearly elastic system* the complementary energy is numerically equal to the strain energy (see Fig. 9.8), or

$$U = U_c. \tag{9.3.6}$$

For this illustrative example in which only a single force is acting on the body, we see that

$$C = K^{-1}, \tag{9.3.7}$$

that is, the compliance or flexibility coefficient is equal to the inverse of the stiffness coefficient.

Let us extend these results to a linearly elastic structure which is acted upon by a system of N (a specific number) generalized forces. Then the increment in strain energy is given by

$$dU = \sum_{m=1}^{N} Q_m \, dq_m. \qquad (9.3.8)$$

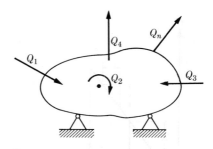

It is evident, for example, that the force Q_3 (see Fig. 9.9) will cause deformations throughout the body and specifically at the N points of particular interest. Thus the displacement at point n, for instance, is due not only to the load at

Fig. 9.9   Body loaded at N points.

point n, but to all the other $N - 1$ generalized forces. We may therefore write

$$q_n = q_n(Q_1, Q_2, \ldots, Q_N), \qquad (9.3.9)$$

to indicate that q_n is a function of all the generalized forces which are acting on the structure. The total differential of q_n is

$$dq_n = \frac{\partial q_n}{\partial Q_1} dQ_1 + \frac{\partial q_n}{\partial Q_2} dQ_2 + \cdots + \frac{\partial q_n}{\partial Q_N} dQ_N. \qquad (9.3.10)$$

Since the structure is linearly elastic, the partial derivatives are constants. We will call them the flexibility or compliance influence coefficients:

$$C_{nm} = \frac{\partial q_n}{\partial Q_m}. \qquad (9.3.11)$$

It is seen that C_{nm} is the generalized displacement at point n in the structure, caused by a unit value of the generalized force at point m. This is illustrated in Fig. 9.10 by means of a cantilever beam. We will later prove that the flexibility influence coefficients are symmetric, that is,

$$C_{nm} = C_{mn}. \qquad (9.3.12)$$

The linear nature of the problem permits us to integrate Eq. (9.3.10) to obtain

$$q_n = \sum_{m=1}^{N} C_{nm}Q_m, \qquad n = 1, 2, \ldots, N. \qquad (9.3.13)$$

We will assume that the relation given by Eq. (9.3.9) can be inverted to yield

$$Q_n = Q_n(q_1, q_2, \ldots, q_N). \qquad (9.3.14)$$

By a line of reasoning parallel to that used for C_{nm} we can develop the equation

$$Q_m = \sum_{n=1}^{N} K_{mn}q_n, \qquad (9.3.15)$$

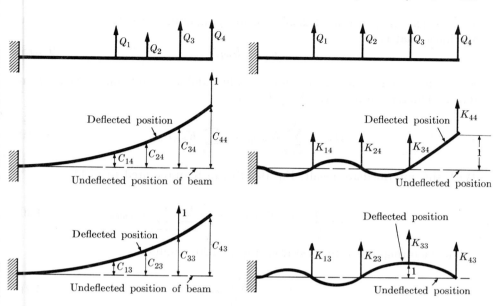

FIG. 9.10 Illustration of flexibility influence coefficients.

FIG. 9.11 Illustration of stiffness influence coefficients.

where the K_{mn}, called the *stiffness influence coefficients*, are given by

$$K_{mn} = \frac{\partial Q_m}{\partial q_n}. \tag{9.3.16}$$

The stiffness influence coefficients are somewhat more difficult to visualize. The coefficient K_{nm} is the generalized force Q_n required at point n when the generalized displacement q_m has a unit value while all the other generalized displacements are zero. We again use a cantilever beam (shown in Fig. 9.11) to illustrate the meaning of the stiffness influence coefficient. We will later show that the stiffness influence coefficients are also symmetric, that is,

$$K_{nm} = K_{mn}. \tag{9.3.17}$$

In order to evaluate the total strain energy by integrating Eq. (9.3.8), we must become cognizant of the conservative nature of our linearly elastic system. The strain energy stored in the body is dependent only on the final state of the forces and displacements, and not on the process whereby the final state is attained. This means that the sequence of load applications, that is, Q_1 first, then Q_2, etc., or any order whatever, is unimportant. Mathematically, it is said that the "path" of integration is unimportant. In evaluating the strain energy it is therefore permissible and helpful in the integration process to assume that the generalized displacements q_n maintain the same proportion at all times, that is,

$$q_n = B_n q, \tag{9.3.18}$$

where the B_n's are factors of proportionality and q is a function varying from 0 to 1. The generalized force Q_m is then given by

$$Q_m = \sum_{n=1}^{N} K_{mn} q_n = \sum_{n=1}^{N} K_{mn} B_n q. \tag{9.3.19}$$

With this intermediate result, Eq. (9.3.8) can be integrated to yield

$$U = \frac{1}{2} \sum_{m=1}^{N} \sum_{n=1}^{N} K_{mn} B_m B_n q^2 = \frac{1}{2} \sum_{m=1}^{N} \sum_{n=1}^{N} K_{mn} q_m q_n. \tag{9.3.20}$$

The factoring of Eq. (9.3.15) from Eq. (9.3.20) results in

$$U = \frac{1}{2} \sum_{n=1}^{N} Q_n q_n. \tag{9.3.21}$$

This formula is Clapeyron's Law which can be stated as follows:

The work done by a system of forces acting on a linearly elastic structure is independent of the rate at which the forces increase and of the order in which the forces are applied. The work done, which is stored as strain energy, is equal to one-half the product of the final magnitudes of the generalized force and its corresponding generalized displacement. The structure must be initially stress free and not subjected to temperature changes.

By using an argument similar to that used for U, the complementary energy relation can be derived:

$$U_c = \frac{1}{2} \sum_{m=1}^{N} \sum_{n=1}^{N} C_{mn} Q_m Q_n. \tag{9.3.22}$$

The relationship between the flexibility influence coefficients and the stiffness influence coefficients can be derived from Eqs. (9.3.13) and (9.3.15), which are two sets of simultaneous linear algebraic equations. They can be written in matrix notation (see Appendix, Section A-20):

$$\begin{bmatrix} q_1 \\ q_2 \\ \vdots \\ q_N \end{bmatrix} = \begin{bmatrix} C_{11} & C_{12} & \cdots & C_{1N} \\ C_{21} & \cdots & \cdots & C_{2N} \\ \vdots & & & \vdots \\ C_{N1} & \cdots & \cdots & C_{NN} \end{bmatrix} \begin{bmatrix} Q_1 \\ Q_2 \\ \vdots \\ Q_N \end{bmatrix}, \tag{9.3.23}$$

and

$$\begin{bmatrix} Q_1 \\ Q_2 \\ \vdots \\ Q_N \end{bmatrix} = \begin{bmatrix} K_{11} & K_{12} & \ldots & K_{1N} \\ K_{21} & \ldots & \ldots & K_{2N} \\ \vdots & & & \vdots \\ K_{N1} & \ldots & \ldots & K_{NN} \end{bmatrix} \begin{bmatrix} q_1 \\ q_2 \\ \vdots \\ q_N \end{bmatrix}. \tag{9.3.24}$$

These may be written compactly as

$$\{q\} = [C]\{Q\} \tag{9.3.25}$$

and

$$\{Q\} = [K]\{q\}, \tag{9.3.26}$$

where $[C]$ and $[K]$ are, respectively, the matrices of flexibility and stiffness influence coefficients. By premultiplying Eq. (9.3.25) by $[C]^{-1}$, the inverse of the matrix of the flexibility influence coefficients (see Appendix, Section A–26), we obtain

$$[C]^{-1}\{q\} = [C]^{-1}[C]\{Q\} = \{Q\}. \tag{9.3.27}$$

Comparing with Eq. (9.3.26) we can conclude that

$$[K] = [C]^{-1}. \tag{9.3.28}$$

This says that the matrix of stiffness influence coefficients is the inverse of the matrix of flexibility influence coefficients. We can also show that

$$[C] = [K]^{-1}. \tag{9.3.29}$$

9.4 PRINCIPLE OF VIRTUAL WORK

We are now in position to develop the Principle of Virtual Work. This energy principle will be used as the foundation for other energy principles and theorems. It is simply stated, but the many different possible interpretations not only lead to new theorems, but give the principle a great deal of versatility.

Let us consider again a structure acted on by a system of generalized forces Q_m (see Fig. 9.12). The generalized forces induce stresses and deformations which satisfy the differential equations of equilibrium (Eq. 7.12.5), strain-displacement relations (Eq. 7.12.4), and the stress-strain relations. The last relations are not necessarily elastic. We can gain information about the behavior of this particular structure by deliberately perturbing (or provoking) the structure. The cause of the perturbations will, in general, be entirely unrelated to the Q forces already acting on the body, and will depend on the type of information desired. For example, the virtual deformations may be caused by another system of forces acting on the body or by a temperature change. In the development of the principle our only concern is that the perturbations

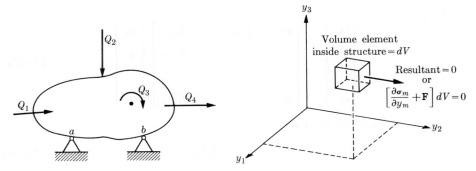

FIG. 9.12 Body under a system of gen-   FIG. 9.13 Equilibrium of an elementary
eralized forces.   volume.

manifest themselves in what will be called a virtual distortion of the body. The
word "virtual" is used to signify that the distortion is independent of the Q-
forces. In the interior of the structure the virtual distortion will be denoted by
the vector $\delta\mathbf{u}$ (y_1, y_2, y_3), and at the points of application of the Q forces, by
δq_m. The virtual displacement $\delta\mathbf{u}$ must be consistent with the geometrical
constraints of the structure, e.g., in Fig. 9.12, we know that $\delta\mathbf{u}$ must be zero at
points a and b. Additionally, the virtual displacement cannot affect the equilibrium of the Q-force system. Consequently, the magnitudes of $\delta\mathbf{u}$ are usually
such that the lines of action of the Q-forces and the Q-induced internal stresses
can be assumed to be unaffected by the distortion.

The Principle of Virtual Work is stated mathematically as

$$\delta U = Q_m \, \delta q_m, \tag{9.4.1}$$

where δU represents the work done internally by the Q-induced stresses under
the strains engendered by the virtual deformation $\delta\mathbf{u}$. In words, the Principle
may be stated as follows:

*If a body is in equilibrium and remains in equilibrium while it is subjected to
a virtual distortion compatible with the geometrical constraints, the virtual work
done by the external forces is equal to the virtual work done by the internal stresses.*

Let us now present a formal proof of Eq. (9.4.1). The resultant force acting
on a volume element is given by Eq. (6.6.3), and because the volume element is
in equilibrium, this resultant force must be zero. These results are shown
schematically in Fig. 9.13. The point of application of the force vector shown
in this figure is subjected to the small virtual displacement $\delta\mathbf{u}$, and hence,
although work is done by each of the components, the net result is zero. We
write this as the integral of the scalar product of force and displacement:

$$\iiint \left\{ \left(\frac{\partial \boldsymbol{\sigma}_m}{\partial y_m} + \mathbf{F} \right) \cdot \delta\mathbf{u} \right\} dV = 0. \tag{9.4.2}$$

A slight rearrangement of Eq. (9.4.2) leads to the form

$$\iiint \left\{ \frac{\partial(\boldsymbol{\sigma}_m \cdot \delta \mathbf{u})}{\partial y_m} + \mathbf{F} \cdot \delta \mathbf{u} \right\} dV = \iiint \left\{ \boldsymbol{\sigma}_m \cdot \frac{\partial(\delta \mathbf{u})}{\partial y_m} \right\} dV. \qquad (9.4.3)$$

With the aid of the Divergence Theorem, or Green's Theorem (see Section A-19), the first term on the left-hand side of the preceding equation can be converted to a surface integral:

$$\iiint \frac{\partial(\boldsymbol{\sigma}_m \cdot \delta \mathbf{u})}{\partial y_m} dV = \iint \boldsymbol{\sigma}_m \cdot \delta \mathbf{u} \, n_m \, dA, \qquad (9.4.4)$$

where the n_m are the components of the unit outward normal vector which locates the orientation of the element of area dA on the surface of the body (see Chapter 6).

The surface integral in Eq. (9.4.4) is recognizable as the virtual work of the externally applied surface forces, since $\boldsymbol{\sigma}_m$ will be zero except at those places where forces are applied. Thus, if only discrete forces are applied to the surface of the structure as in Fig. (9.12), then

$$Q_n \, \delta q_n = \iint \{\boldsymbol{\sigma}_m \cdot \delta \mathbf{u} \, n_m \, dA\}. \qquad (9.4.5)$$

The right-hand side of Eq. (9.4.3) can also be simplified and interpreted in a meaningful fashion. By introducing Eq. (6.3.4) we can change the integral into the form

$$\boldsymbol{\sigma}_m \cdot \frac{\partial(\delta \mathbf{u})}{\partial y_m} = \sigma_{mn} \mathbf{i}_n \cdot \frac{\partial(\delta \mathbf{u})}{\partial y_m}. \qquad (9.4.6)$$

Since the stress tensor σ_{mn} is symmetric, the terms can be further rearranged:

$$\sigma_{mn} \mathbf{i}_n \cdot \frac{\partial(\delta \mathbf{u})}{\partial y_m} = \sigma_{mn} \left\{ \frac{1}{2} \left[\mathbf{i}_n \cdot \frac{\partial(\delta \mathbf{u})}{\partial y_m} + \mathbf{i}_m \cdot \frac{\partial(\delta \mathbf{u})}{\partial y_n} \right] \right\}. \qquad (9.4.7)$$

The geometrical meaning of the expression inside the braces is recognized as the virtual strain tensor (compare with Eq. 5.9.1):

$$\delta \epsilon_{mn} = \frac{1}{2} \left[\mathbf{i}_n \cdot \frac{\partial(\delta \mathbf{u})}{\partial y_m} + \mathbf{i}_m \cdot \frac{\partial(\delta \mathbf{u})}{\partial y_n} \right]. \qquad (9.4.8)$$

If we write the virtual displacement in component form,

$$\delta \mathbf{u} = \delta u_m \mathbf{i}_m, \qquad (9.4.9)$$

then

$$\delta \epsilon_{mn} = \frac{1}{2} \left[\frac{\partial \delta u_m}{\partial y_n} + \frac{\partial \delta u_n}{\partial y_m} \right]. \qquad (9.4.10)$$

All these expressions [Eqs. (9.4.8) through (9.4.10)] are of the same form as those used in Chapter 5. The only difference is that instead of using **u** to describe the deformation of the structure we are now using $\delta\mathbf{u}$ to describe the virtual deformation.

The virtual-work equation which started out as Eq. (9.4.2) now assumes a form which can be directly applied to structural problems:

$$\iint \boldsymbol{\sigma}_m \cdot \delta\mathbf{u} n_m \, dA + \iiint \mathbf{F} \cdot \delta\mathbf{u} \, dV = \iiint \sigma_{mn} \, \delta\epsilon_{mn} \, dV. \qquad (9.4.11)$$

The left-hand side has a physical interpretation as the virtual work of the external forces, and the right-hand side is the virtual work of the internal stresses (Eq. 9.4.1). It will be helpful to write Eq. (9.4.11) as

$$\delta W = \delta U, \qquad (9.4.12)$$

with

$$\delta W = \iint \boldsymbol{\sigma}_m \cdot \delta\mathbf{u} n_m \, dA + \iiint \mathbf{F} \cdot \delta\mathbf{u} \, dV, \qquad (9.4.13)$$

and

$$\delta U = \iiint \sigma_{mn} \, \delta\epsilon_{mn} \, dV. \qquad (9.4.14)$$

If, as has already been remarked, the external forces consist only of generalized forces of a discrete nature, then Eq. (9.4.12) has the special form given by Eq. (9.4.1). The semantics of the word "external" may be a little troublesome since the force **F** is a volume force and appears in the interior of the body. However, "external" is used to mean the applied loads or the cause of the stresses and strains.

Let us present several examples to illustrate the application of the Principle of Virtual Work.

(a) Some simple examples for δU. The virtual work of the internal stresses takes on simple forms for the simpler types of structures. For a member in a state of uniaxial stress (such as a rod or the members of a truss) there is only one stress component and one strain component. Let us consider the situation shown in Fig. 9.14. A rod of uniform cross-sectional area A and length L is loaded at its free end by a load Q. If we let the virtual deformation of the rod be represented by the end elongation δq, the virtual strain $\delta\epsilon$ in the rod is given by $\delta q/L$. The axial stress caused by Q is

$$\sigma^{(Q)} = Q/A. \qquad (9.4.15)$$

Hence the virtual work of the internal stress is

$$\delta U = \int_{\text{vol}} \sigma^{(Q)} \, \delta\epsilon \, dV = Q \, \delta q. \qquad (9.4.16)$$

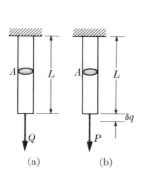

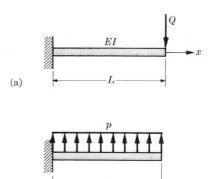

FIG. 9.14   Prismatic bar under uniaxial   FIG. 9.15   Cantilever beam under two
tension.   systems of loading.

If we consider the virtual deformation as that caused by a load P (Fig. 9.14b)
we have

$$\delta q = PL/AE, \tag{9.4.17}$$

and

$$\delta U = QPL/AE. \tag{9.4.18}$$

Consider next a cantilever beam as shown in Fig. 9.15. The stresses caused by
Q (see Section 7.13) are

$$\sigma^{(Q)} = -\frac{M^{(Q)}(x)z}{I}, \tag{9.4.19}$$

where I is the cross-sectional moment of inertia and $M^{(Q)}(x)$ is the bending-
moment function caused by the Q-forces. We shall now consider the virtual
deformation as that due to a uniformly distributed load p. Again, according
to Chapter 7, the virtual strain distribution in the beam corresponding to this
virtual deformation is given by

$$\delta\epsilon = -\frac{M^{(p)}(x)z}{EI}, \tag{9.4.20}$$

where $M^{(p)}(x)$ is the bending-moment function caused by the distributed load p
on the beam. The virtual work of the internal stresses for the bending of a
beam is

$$\delta U = \int_{\text{vol}} \sigma^{(Q)} \, \delta\epsilon \, dV = \int_0^L \frac{M^{(Q)}(x)M^{(p)}(x)}{EI} \, dx. \tag{9.4.21}$$

Note the similarity in form of Eqs. (9.4.18) and (9.4.21): one represents the
internal virtual work of a rod (which could be a truss member) and the other,
the internal virtual work of a beam. Nevertheless, both contain a product of
two generalized forces, one of which is associated with the Q-force and the other
with the forces which cause distortion.

(b) Solution for simple statically indeterminate trusses. Let us consider next the problem of a plane truss with three members acted on by a single load Q, as shown in Fig. 9.16. This statically indeterminate problem was treated previously in Chapter 4 by solving simultaneously the equations of equilibrium, the equations of consistent deformation, and the constitutive equations. The solutions for the three member forces were given in Eqs. (4.4.8) and (4.4.9).

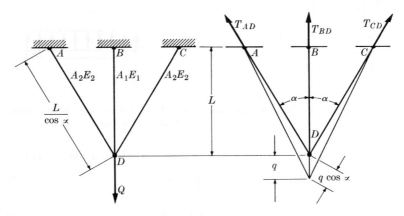

FIG. 9.16   Statically indeterminate planar truss.

For the present problem the only external force is Q. If we let the vertical displacement at the point D be denoted by q, then the elongation of the member BD is q, while for geometrical compatibility the elongation of member AD or CD is $q \cos \alpha$. The internal forces and elongation of the truss members are given by the constitutive relations as follows:

$$T_{BD} = q A_1 E_1 / L, \tag{9.4.22}$$

$$T_{AD} = T_{CD} = q A_2 E_2 \cos^2 \alpha / L. \tag{9.4.23}$$

Let us now prescribe a virtual distortion such that the vertical deflection at D is δq. This means that the elongation of BD is δq, while the geometric compatibility requires that the elongation of AD or BD be $\delta q \cos \alpha$. According to Eq. (9.4.16), the work done internally by the Q-induced stresses under this prescribed virtual displacement is

$$\delta U = \frac{q A_1 E_1 \, \delta q}{L} + \frac{2q A_2 E_2 \cos^3 \alpha}{L} \, \delta q. \tag{9.4.24}$$

According to the principle of virtual work we have

$$\delta U = Q \, \delta_q. \tag{9.4.25}$$

Substituting δU from Eq. (9.4.24) and simplifying, we obtain

$$q = \frac{QL}{A_1E_1[1 + 2(A_2E_2/A_1E_1)\cos^3\alpha]}. \tag{9.4.26}$$

Substituting into Eqs. (9.4.22) and (9.4.23), we find that

$$T_{BD} = \frac{Q}{1 + 2A_2E_2\cos^3\alpha/A_1E_1}, \tag{9.4.27}$$

$$T_{AD} = T_{CD} = \frac{Q\sec\alpha}{2 + A_1E_1/A_2E_2\cos^3\alpha}. \tag{9.4.28}$$

These agree with the result obtained previously in Chapter 4. It is seen that in evaluating the change in strain energy the geometrical compatibility condition for the elongation of the three bars was used. However, the vector equation of equilibrium was not introduced. The conditions of equilibrium are thus implied in the energy principle.

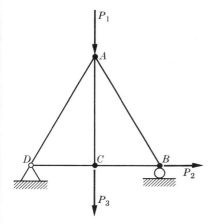

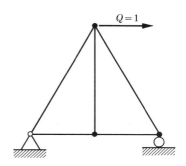

FIG. 9.17   Statically determinate planar truss.

FIG. 9.18   Planar truss under a dummy load.

(c) Virtual Work applied to deflection of a truss. The Principle of Virtual Work is well suited to the calculation of deflections. Consider a simple truss loaded as shown in Fig. 9.17. We wish to calculate the horizontal deflection of joint A caused by the system of P-forces. This can be accomplished by an adroit choice of a Q-force system consisting of a unit horizontal load applied to joint A (see Fig. 9.18). When the deformation by the system of P-forces is considered as the virtual deformation, the virtual work equation becomes

$$Q\,\delta q = (1)\,\Delta_h = \delta U, \tag{9.4.29}$$

where Δ_h is the desired horizontal displacement of joint A. Each bar of a truss is in a state of uniaxial stress, and so for each bar there is an expression for the virtual work of the internal stresses in the form given by Eq. (9.4.18). We can represent this calculation as

$$(1) \quad \Delta_h = \sum \frac{F_i^{(Q)} F_i L_i}{A_i E_i}, \qquad (9.4.30)$$

where $F_i^{(Q)}$ and F_i are the axial loads in the ith member due to, respectively, the unit load Q and the P-forces. Here L_i, A_i, and E_i are, respectively, the length, area, and elastic modulus of the ith member. Thus we have obtained the deflection of a truss by a clever choice of the Q-force system and by a calculation of the virtual work of the internal strain energy. Note that the cause of distortion, namely the P-forces, is the major item of interest, and that the Q-force system is merely a vehicle to the desired result. The deflection calculation is reduced to two stress analyses, one under the P-forces and the other under the single Q-force. Hence the integration of strain-displacement relations has been circumvented.

(d) Virtual Work applied to deflection of a beam. Consider the cantilever beam loaded in some arbitrary manner (see Fig. 9.19). It is desired to determine the deflection at the tip of the beam. As for the truss, we choose as the Q-force system a unit load on the point at which, and in the direction along which the deflection is desired. The virtual work equation becomes, with the aid of Eq. (9.4.21),

$$(1) \quad \Delta_z = \int_0^L \frac{M^{(Q)}(x) M(x)}{EI} \, dx, \qquad (9.4.31)$$

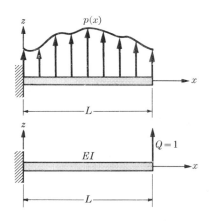

where $M(x)$ is the bending moment due to the distributed load $p(x)$, $M^{(Q)}(x)$ is the bending moment due to the unit load, and Δ_z is the desired displacement. We are able to calculate the deflection at a point of a beam merely by calculating the bending-moment function for the cause of distortion, and using a specially chosen Q-force system. Again the integration of strain-displacement relations has been avoided.

Fig. 9.19   Cantilever beam under two systems of loading.

Because the approach we have used in these examples involves the introduction of a unit load at the point where the structure deflection is required, it is sometimes called the *unit load method*. In the latter part of this chapter we will see that the equation obtained by the unit load method is identical with that developed by the dummy load method which is derived from the Principle of Minimum Complementary Energy (see Section 9.10).

9.5 BETTI'S LAW AND MAXWELL'S LAW

The Principle of Virtual Work can be used to formulate *Betti's Law*. Let us consider our all-purpose example structure which will be loaded in turn by two different systems of generalized forces. One system will be called the Q- and the other the P-system (see Fig. 9.20). The structure is assumed to be linearly elastic. There are L number of Q-forces, and K number of P-forces. Let the displacements of the points of application of the Q-forces caused by the P-forces be denoted by $\Delta_l^{(P)}$. Similarly the deflections of the points of application of the P-forces caused by the Q-forces will be denoted by $\Delta_k^{(Q)}$.

The virtual work of the Q-system of forces and the internal stresses, $\sigma_{mn}^{(Q)}$ under the displacements caused by the P-forces is given by

$$\sum_{l=1}^{L} Q_l \, \Delta_l^{(P)} = \iiint \sigma_{mn}^{(Q)} \, \delta\epsilon_{mn}^{(P)} \, dV, \tag{9.5.1}$$

where $\delta\epsilon_{mn}^{(P)}$ are the internal virtual strains caused by the P-forces. Similarly, the virtual work of the P-system of forces and internal stresses $\sigma_{mn}^{(P)}$ under the displacements caused by the Q-forces is given by

$$\sum_{k=1}^{K} P_k \, \Delta_k^{(Q)} = \iiint \sigma_{mn}^{(P)} \, \delta\epsilon_{mn}^{(Q)} \, dV. \tag{9.5.2}$$

The virtual strains $\delta\epsilon_{mn}^{(P)}$ are caused by the P-induced stresses $\sigma_{mn}^{(P)}$. Thus for a linearly elastic body we can write (see Eq. 7.2.35)

$$\delta\epsilon_{mn}^{(P)} = S_{mnrs}\sigma_{rs}^{(P)}. \tag{9.5.3}$$

Similarly,

$$\delta\epsilon_{mn}^{(Q)} = S_{mnrs}\sigma_{rs}^{(Q)}. \tag{9.5.4}$$

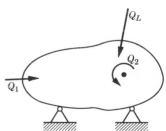

The two virtual-work equations, (9.5.1) and (9.5.2), become

$$\sum_{l=1}^{L} Q_l \, \Delta_l^{(P)} = \iiint \sigma_{mn}^{(Q)} \, S_{mnrs}\sigma_{rs}^{(P)} \, dV, \tag{9.5.5}$$

$$\sum_{k=1}^{K} P_k \, \Delta_k^{(Q)} = \iiint \sigma_{mn}^{(P)} \, S_{mnrs}\sigma_{rs}^{(Q)} \, dV. \tag{9.5.6}$$

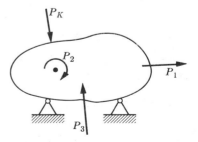

It is observed that the integrands are the same because of the symmetry property

FIG. 9.20 Body under two systems of generalized forces.

$S_{mnrs} = S_{rsmn}$. We obtain

$$\sum_{l=1}^{L} Q_l \, \Delta_l^{(P)} = \sum_{k=1}^{K} P_k \, \Delta_k^{(Q)}, \qquad (9.5.7)$$

which is the mathematical statement of Betti's Law:

In any structure which is linearly elastic, the virtual work done by a system of Q-forces under a distortion caused by a system of P-forces is equal to the virtual work done by the system of P-forces under a distortion caused by the system of Q-forces.

Maxwell's Law of Reciprocal Deflections follows directly from, and is a special case of Betti's Law. We consider just one Q-force and one P-force which are *numerically* equal (see Fig. 9.21). Then according to Betti's Law, we have

$$Q \, \Delta_1^{(P)} = Q \, \Delta_2^{(Q)}. \qquad (9.5.8)$$

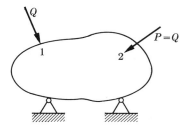

Further, let us denote the deflection at point m caused by a load at point n by Δ_{mn}, and the deflection at point n caused by the same magnitude of load at point m, by Δ_{nm}. Then Eq. (9.5.8) reads, in general terms,

FIG. 9.21   Body under two numerically equal forces.

$$\Delta_{mn} = \Delta_{nm}, \qquad (9.5.9)$$

which is a general formula for Maxwell's Law of Reciprocal Deflections:

In any structure which is linearly elastic the generalized deflection at point m caused by a generalized load Q at point n is numerically equal to the generalized deflection at point n caused by the same magnitude of generalized force Q at point m.

It is a small step from Eq. (9.5.9) to prove Eq. (9.3.12), which states that the flexibility-influence coefficients, previously defined, are symmetric.

9.6   PRINCIPLE OF MINIMUM POTENTIAL ENERGY

The Principle of Minimum Potential Energy may be looked upon as a specialization of the Principle of Virtual Work. We now use the symbol δ (in Eqs. 9.4.13 and 9.4.14) to mean the variation of a function according to the Calculus of Variations (see Section A.18). At a given point (y_1, y_2, y_3) of a body, the state of stress and strain is represented by σ_{mn} and ϵ_{mn}. Then $\delta\epsilon_{mn}$ represents any small change of the function $\epsilon_{mn}(y_1, y_2, y_3)$, subject to the condition that the compatibility equations and boundary constraints be satisfied. As shown in Section A-18, the use of the symbol δ is in many respects analogous to the use of the differential symbol.

For linear isotropic elastic behavior, the stress-strain relation and the expression of the strain-energy density U^* are given by Eqs. (7.7.14) and (7.10.7), respectively. These are rewritten by replacing γ_{mn} by ϵ_{mn} as follows:

$$\sigma_{mn} = 2\mu\epsilon_{mn} + \lambda\,\delta_{mn}\epsilon_{rr}, \tag{9.6.1}$$

and

$$U^* = \mu\epsilon_{mn}\epsilon_{mn} + \tfrac{1}{2}\lambda(\epsilon_{rr})^2. \tag{9.6.2}$$

It is seen that the variation in internal strain energy U is†

$$\delta U = \delta \iiint [\mu\epsilon_{mn}\epsilon_{mn} + \tfrac{1}{2}\lambda(\epsilon_{rr})^2]\, dV$$

$$= \iiint [2\mu\epsilon_{mn} + \lambda\,\delta_{mn}\epsilon_{rr}]\,\delta\epsilon_{mn}\, dV. \tag{9.6.3}$$

By substituting Eq. (9.6.1) into Eq. (9.6.3), we obtain

$$\delta U = \iiint \sigma_{mn}\,\delta\epsilon_{mn}\, dV. \tag{9.6.4}$$

We see that, according to the Principle of Virtual Work, we may now treat δU in Eq. (9.4.1) as the variation in the internal strain energy.

The stage is now set to rewrite the Principle of Virtual Work in the form

$$\delta\pi_p = 0, \tag{9.6.5}$$

where

$$\pi_p = U(\epsilon_{mn}) - W, \tag{9.6.6}$$

$$U = \iiint \tfrac{1}{2}[2\mu\epsilon_{mn}\epsilon_{mn} + \lambda(\epsilon_{rr})^2]\, dV, \tag{9.6.7}$$

and

$$W = \iint \sigma_m \cdot \mathbf{u}\, n_m\, dA + \iiint \mathbf{F} \cdot \mathbf{u}\, dV. \tag{9.6.8}$$

The scalar π_p is called the *total potential energy* (or just potential energy) of the structural system, and W, as before, is the work done by the applied body and surface forces.

It should be carefully noted that in order to satisfy the condition of continuity of displacements over the entire body, the function ϵ_{mn} must first be

---

† Care should be exercised in applying the δ- or d-operation to equations written in summation notation. The best approach is to write out first the equations in explicit form and then to perform the desired operation.

expressed in terms of the displacement functions u_m by

$$\epsilon_{mn} = \frac{1}{2}\left(\frac{\partial u_m}{\partial y_n} + \frac{\partial u_n}{\partial y_m}\right), \tag{9.6.9}$$

and the variation of the strain energy U must then be carried out with respect to the displacement functions.

The result, succinctly summarized by Eq. (9.6.5) and subject to Eqs. (9.6.6), (9.6.7), (9.6.8), and (9.6.9), is known as the Principle of Minimum Potential Energy and is stated as follows:

Among all the geometrically possible states of displacement, the correct state of displacement is that which makes the potential energy a minimum.

From a physical argument one can conclude that for a stable equilibrium condition the total potential energy of a system must be a minimum. The fact that π_p is a minimum can also be proved more rigorously by comparing the potential energy π_p at the equilibrium state with that at a neighboring state, π_p'. The state at π_p' is characterized by strains, $\epsilon_{mn} + \delta\epsilon_{mn}$, and the corresponding displacements, $\mathbf{u} + \delta\mathbf{u}$. We find that

$$\pi_p' = \iiint \left\{ \tfrac{1}{2}[2\mu(\epsilon_{mn} + \delta\epsilon_{mn})(\epsilon_{mn} + \delta\epsilon_{mn}) + \lambda(\epsilon_{rr} + \delta\epsilon_{rr})^2] \right.$$
$$\left. -\mathbf{F} \cdot (\mathbf{u} + \delta\mathbf{u}) \right\} dV - \int \sigma_m \cdot (\mathbf{u} + \delta\mathbf{u})n_m \, dA. \tag{9.6.10}$$

The right-hand side of this equation can be regrouped so that

$$\pi_p' = \iiint \left\{ \tfrac{1}{2}[2\mu\epsilon_{mn}\epsilon_{mn} + \lambda(\epsilon_{rr})^2] - \mathbf{F}\cdot\mathbf{u} \right\} dV - \int \sigma_m \cdot \mathbf{u} \, n_m \, dA$$
$$+ \iiint \left\{ (2\mu\epsilon_{mn} + \lambda\,\delta_{mn}\epsilon_{rr})\,\delta\epsilon_{mn} - \mathbf{F}\cdot\delta\mathbf{u} \right\} dV - \int \sigma_m \cdot \delta\mathbf{u} \, n_m \, dA$$
$$+ \iiint [2\mu(\delta\epsilon_{mn})(\delta\epsilon_{mn}) + \lambda(\delta\epsilon_{rr})(\delta\epsilon_{rr})] \, dV. \tag{9.6.11}$$

It is seen that the first group of terms is π_p, while the second group is $\delta\pi_p$ and hence is zero according to Eq. (9.6.5). The last group is a positive quadratic in $\delta\epsilon_{mn}$, and is therefore nonnegative for any arbitrary variation of ϵ_{mn}. We have thus shown $\pi_p' - \pi_p$ to be always nonnegative. This means that for stable equilibrium the potential energy must be a minimum.

A remark should be made here that the strain energy U is also a characteristic of materials which exhibit nonlinear elastic behavior. For example, in the simple case of the uniform bar under uniformly distributed stress shown in Fig. 9.22, the uniaxial stress-strain relation can be represented by

$$\sigma = f(\epsilon). \tag{9.6.12}$$

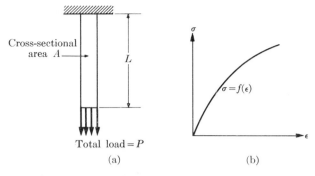

FIG. 9.22   Strain energy of a prismatic bar—nonlinear stress-strain relation.

The strain energy in the bar is

$$U = \iiint \left[\int_0^\epsilon \sigma \, d\epsilon \right] dV, \tag{9.6.13}$$

or

$$U = AL \int_0^\epsilon f(\epsilon) \, d\epsilon, \tag{9.6.14}$$

where A is the cross-sectional area and L is the length of the bar. The variation in strain energy is thus

$$\delta U = AL f(\epsilon) \, \delta\epsilon = \iiint \sigma \, \delta\epsilon \, dV, \tag{9.6.15}$$

which is the work done by the internal stresses due to the virtual deformation $\delta\epsilon$.

The Principle of Minimum Potential Energy is most advantageously used in conjunction with the calculus of variation. We will more fully exploit this principle at a later time. At this point we will be content to make a few qualitative observations. The statement of the principle contains the words "the correct state," and "correct" as used in this context means that the rigorous application of $\delta\pi_p = 0$ would lead back to the three differential equations of force equilibrium. However, the words "the correct state" can be changed to mean "the most nearly correct state" or "the best state" if we are willing to accept approximate solutions. Herein lies the greatest utility of the Principle of Minimum Potential Energy, namely, the development of approximate equations of equilibrium to replace the three coupled partial differential equations.

9.7   CASTIGLIANO'S FIRST THEOREM

The Principle of Minimum Potential Energy leads naturally and easily to another theorem adapted for structures which are loaded discretely. Let us again consider a structure which is acted upon by a system of generalized forces Q_m. Since the body forces are zero, the work expression, Eq. (9.6.8), becomes simply

$$W = \sum_{m=1}^{N} Q_m q_m, \qquad (9.7.1)$$

where q_m are the displacements of the points of application of the generalized forces. Then the potential energy of the system is written as

$$\pi_p = U(q_m) - \sum_{m=1}^{N} Q_m q_m, \qquad (9.7.2)$$

where the internal energy is expressed in terms of the generalized displacements. The Principle of Minimum Potential Energy requires that

$$\delta \pi_p = \sum_{m=1}^{N} \frac{\partial \pi_p}{\partial q_m} \delta q_m = 0, \qquad (9.7.3)$$

and since the set of displacements q_m is independent, we have

$$\frac{\partial \pi_p}{\partial q_m} = 0, \qquad (9.7.4)$$

or, from Eq. (9.7.2),

$$Q_m = \frac{\partial U}{\partial q_m}. \qquad (9.7.5)$$

This result which relates Q_m to the rate of change of strain energy with respect to the corresponding displacement is known as Castigliano's First Theorem. Let us consider several simple applications of Eqs. (9.7.5).

(a) Verification of Eq. (9.3.15). The strain energy U is expressed in terms of the stiffness influence coefficients K_{mn} in Eq. (9.3.20). We see that the application of Eq. (9.7.5) to Eq. (9.3.20) leads to

$$Q_m = \sum_{n=1}^{N} K_{mn} q_n, \qquad (9.7.6)$$

which is a verification of Eq. (9.3.15).

(b) Statically indeterminate plane truss. Let us consider again the statically indeterminate truss shown in Fig. 9.16. The strain energy in the three members can be expressed in terms of the vertical displacement q at D as

$$
\begin{aligned}
U &= \frac{1}{2} \frac{A_1 E_1}{L} q^2 + 2 \cdot \frac{1}{2} \frac{A_2 E_2}{L/\cos \alpha} (q \cos \alpha)^2 \\
&= \frac{1}{2} \frac{A_1 E_1}{L} q^2 \left(1 + \frac{2 A_2 E_2}{A_1 E_1} \cos^3 \alpha \right).
\end{aligned} \qquad (9.7.7)
$$

By means of Eq. (9.7.5) we obtain

$$Q = \frac{dU}{dq} = \frac{A_1 E_1}{L}\left(1 + \frac{2 A_2 E_2}{A_1 E_1} \cos^3 \alpha\right) q, \qquad (9.7.8)$$

which agrees with the result previously derived in Eq. (9.4.26).

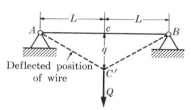

FIG. 9.23   Wire under a single load.   Wire area $= A$
Elastic modulus $= E$

(c) Deflection of a wire.   In the two previous examples the resulting force-displacement relations are linear. Let us now consider an example in which the resulting force displacement relation is nonlinear. Figure 9.23 shows a wire which is held fixed at both ends, and is acted on by a downward force Q at the mid-span point. The deformed shape of the wire is shown by the dashed line. The strain energy stored in the stretched wire is given by

$$U = 2 \cdot \frac{1}{2} \frac{EA}{L} \Delta^2, \qquad (9.7.9)$$

where E is the elastic modulus, A the cross-sectional area, and Δ the elongation of the segment AC. The original length of the wire is $2L$. We have

$$\Delta = (L^2 + q^2)^{1/2} - L, \qquad (9.7.10)$$

where q is the deflection at mid-span. When we restrict our discussion to small deflections, (i.e. such that q is small in comparison to L), we obtain

$$\Delta = L\left[1 + \frac{(q/L)^2}{2} + \cdots\right] - L,$$

or approximately,

$$\Delta = \frac{q^2}{2L}. \qquad (9.7.11)$$

Thus the strain energy is

$$U = EAq^4/4L^3, \qquad (9.7.12)$$

and, from Castigliano's First Theorem,

$$Q = \partial U/\partial q = EAq^3/L^3. \qquad (9.7.13)$$

(d) Deflection of a simply supported beam. As a final example, let us calculate the deflection of a simply supported uniform beam acted upon by a uniformly distributed load (see Fig. 9.24). We will first integrate the differential equations of beam theory to obtain a close form solution, and then we will use the Principle of Minimum Potential Energy to obtain an approximate solution of those same beam equations. Let us derive the differential equation governing w, the deflection of the beam. We have from Eq. (7.13.28)

$$\frac{d^2 w}{dy_1^2} = \frac{M}{EI},$$ (9.7.14)

where I is the cross-sectional moment of inertia, and E is the elastic modulus of the beam. We also have the relation between the bending moment and the distributed load p (see Chapter 3):

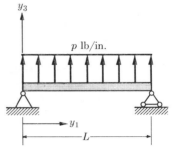

$$\frac{dS}{dy_1} = -p,$$ (9.7.15)

$$\frac{dM}{dy_1} = -S.$$ (9.7.16)

Equations (9.7.15) and (9.7.16) can be combined to yield

$$\frac{d^2 M}{dy_1^2} = p.$$ (9.7.17)

Fig. 9.24   Simply supported beam under uniform lateral loading.

By differentiating Eq. (9.7.14) twice and substituting into Eq. (9.7.17), we obtain

$$EI \frac{d^4 w}{dy_1^4} = p.$$ (9.7.18)

This is a fourth-order differential equation with four boundary conditions as follows:

$$w(0) = 0,$$ (9.7.19)

$$w(L) = 0,$$ (9.7.20)

$$M(0) = EI \frac{d^2 w}{dy_1^2}\bigg|_{y_1=0} = 0, \quad \text{or} \quad \frac{d^2 w}{dy_1^2}\bigg|_{y_1=0} = 0,$$ (9.7.21)

$$M(L) = 0, \quad \text{or} \quad \frac{d^2 w}{dy_1^2}\bigg|_{y_1=L} = 0.$$ (9.7.22)

The solution is

$$w(y_1) = \frac{py_1}{24EI} (y_1^3 - 2Ly_1^2 + L^3),$$ (9.7.23)

and the maximum deflection which occurs at mid-span is

$$w_{\max} = w\left(\frac{L}{2}\right) = \frac{5}{384} \frac{pL^4}{EI} = 0.013020 \frac{pL^4}{EI}.$$ (9.7.24)

The results shown in Eqs. (9.7.23) and (9.7.29) are considered as exact within the framework of the engineering theory of beams. Now let us proceed to an approximate solution by applying the Principle of Minimum Potential Energy. We assume that the deflection $w(y_1)$ can be represented by a Fourier series,

$$w(y_1) = \sum_{i=1}^{\infty} w_i \sin \frac{i\pi y_1}{L}.$$ (9.7.25)

It is seen that in each term the deflections at the two ends are zero. The assumed deflection $w(y_1)$ thus satisfies the displacement boundary conditions.† The strain energy in the beam is given by Eq. (7.13.35):

$$U = \frac{1}{2} \int_0^L EI \left(\frac{d^2w}{dy_1^2}\right)^2 dy_1$$

$$= \frac{EI}{2} \int_0^L \left[\sum_{i=1}^{\infty} \left(\frac{i\pi}{L}\right)^2 w_i \sin \frac{i\pi y_1}{L}\right]^2 dy_1.$$ (9.7.26)

Introducing the orthogonality property of the Fourier series, that is,

$$\int_0^L \sin \frac{i\pi y_1}{L} \sin \frac{j\pi y_1}{L} dy_1 = \begin{cases} 0, & i \neq j, \\ \frac{L}{2}, & i = j, \end{cases}$$ (9.7.27)

we obtain

$$U = \tfrac{1}{4}EIL \sum_{i=1}^{\infty} \left(\frac{i\pi}{L}\right)^4 w_i^2.$$ (9.7.28)

The potential energy due to the distributed load is

$$-W = \int_0^L (-pw) dy_1 = -p \int_0^L \sum_{i=1}^{\infty} w_i \sin \frac{i\pi y_1}{L} dy_1 = -p \sum_{i=1,3,5,\dots}^{\infty} \frac{2}{\pi} \frac{w_i}{i}.$$ (9.7.29)

---

† In the present case it happens the moment boundary conditions are also satisfied in each term. This is, however, not required in the application of the Principle of Minimum Potential Energy.

Thus

$$\pi_p = \tfrac{1}{4}EI \sum_{i=1}^{\infty} \frac{(i\pi)^4}{L^3} w_i^2 - p \sum_{i=1,3,5,\dots}^{\infty} \frac{2}{\pi} \frac{w_i}{i}. \tag{9.7.30}$$

Here π_p is expressed in terms of w_i. For π_p to be a minimum we must set

$$\frac{\partial \pi_p}{\partial w_i} = 0 \qquad i = 1, 2, \dots, \infty,$$

which yields

$$\frac{\partial \pi_p}{\partial w_1} = 0; \qquad \tfrac{1}{2}EI\left(\frac{\pi}{L}\right)^4 w_1 = \frac{2}{\pi} p, \qquad \text{or} \qquad w_1 = \frac{4pL^4}{\pi^5 EI};$$

$$\frac{\partial \pi_p}{\partial w_2} = 0; \qquad \tfrac{1}{2}EI(16)\left(\frac{\pi}{L}\right)^4 w_2 = 0, \qquad \text{or} \qquad w_2 = 0; \tag{9.7.31}$$

$$\frac{\partial \pi_p}{\partial w_3} = 0; \qquad \tfrac{1}{2}EI(81)\left(\frac{\pi}{L}\right)^4 w_3 = \frac{2}{\pi} \frac{p}{3}, \qquad \text{or} \qquad w_3 = \frac{4pL^4}{\pi^5 EI} \frac{1}{(81)(3)};$$

$$\vdots$$

Substituting the values of w_i into Eq. (9.7.25), we obtain

$$w(y_1) = \frac{4}{\pi^5} \frac{pL^4}{EI} \sum_{i=1,3,5,\dots}^{\infty} \frac{1}{i^5} \sin \frac{i\pi y_1}{L}. \tag{9.7.32}$$

It can be shown that this is the Fourier expansion for the deflection shape given by Eq. (9.7.23). The infinite series can, of course, be considered also as a close form solution. However, its utility lies in the fact that approximate results of acceptable accuracy can be obtained with only a few terms. For example, if we use only the leading term of the series (see Eq. 9.7.25), we obtain

$$w_1 = \frac{4}{\pi^5} \frac{pL^4}{EI}. \tag{9.7.33}$$

Thus the maximum deflection of the beam is

$$w\left(\frac{L}{2}\right) = w_1 = 0.013071 \frac{pL^4}{EI}. \tag{9.7.34}$$

It differs from the exact solution, Eq. (9.7.24), by only 0.04 percent.

9.8 PRINCIPLE OF VIRTUAL COMPLEMENTARY WORK

In the earlier sections of this chapter, the notions of strain energy and complementary energy were developed. There was seen to be a certain duality or conjugate nature between many of the ideas. We will pursue this duality by developing the Principle of Virtual Complementary Work which can, in a certain sense, be considered a dual or conjugate to the Principle of Virtual Work. Instead of considering a virtual deformation, let us consider a virtual change $\delta\sigma_{mn}$ in the stress tensor. These changes in stresses are to satisfy the equations of equilibrium,

$$\frac{\partial}{\partial y_m}(\delta\sigma_{mn}) = 0. \tag{9.8.1}$$

From Eq. (9.8.1) we conclude that the following integral also vanishes:

$$\iiint \frac{\partial\delta\sigma_{mn}}{\partial y_m} u_n \, dV = 0, \tag{9.8.2}$$

where u_m is the displacement caused by the stresses σ_{mn}. Equation (9.8.2) can also be written in terms of vector quantities,

$$\iiint \frac{\partial\delta\boldsymbol{\sigma}_m}{\partial y_m} \cdot \mathbf{u} \, dV = 0 \tag{9.8.3}$$

when the following vector relations are recognized:

$$\delta\boldsymbol{\sigma}_m = \delta\sigma_{mn}\mathbf{i}_n, \tag{9.8.4}$$

and

$$\mathbf{u} = u_n\mathbf{i}_n. \tag{9.8.5}$$

Equation (9.8.3) can be rearranged to read

$$\iiint \frac{\partial}{\partial y_m}(\delta\boldsymbol{\sigma}_m \cdot \mathbf{u}) \, dV = \iiint \delta\boldsymbol{\sigma}_m \cdot \frac{\partial\mathbf{u}}{\partial y_m} \, dV. \tag{9.8.6}$$

By making use of the Divergence Theorem, we can convert the left-hand side of Eq. (9.8.6) to a surface integral:

$$\iiint \frac{\partial}{\partial y_m}(\delta\boldsymbol{\sigma}_m \cdot \mathbf{u}) \, dV = \iint (\delta\boldsymbol{\sigma}_m \cdot \mathbf{u})n_m \, dA. \tag{9.8.7}$$

This integral represents the work done by the virtual surface stresses $\delta\sigma_{mn}^* n_m$, which obviously must be consistent with the changes in internal stress $\delta\sigma_{mn}$.

By making use of Eqs. (9.8.6) and (5.9.1) and the fact that σ_{mn} is symmetric, we can express the right-hand side of Eq. (9.8.6) as

$$\iiint \delta\boldsymbol{\sigma}_m \cdot \frac{\partial\mathbf{u}}{\partial y_m} \, dV = \iiint \epsilon_{mn} \, \delta\sigma_{mn} \, dV. \tag{9.8.8}$$

This is clearly the internal complementary work done under the actual state of strain ϵ_{mn}, and a virtual change $\delta\sigma_{mn}$ of the stress tensor. Equation (9.8.6) thus becomes

$$\iiint \epsilon_{mn} \, \delta\sigma_{mn} \, dV = \iint [\delta\boldsymbol{\sigma}_m \cdot \mathbf{u}]n_m \, dA. \tag{9.8.9}$$

This is the mathematical summary of the Principle of Virtual Complementary Work, which may be stated as follows:

> *The virtual complementary work done under the actual state of strain and virtual stress change $\delta\sigma_{mn}$, which satisfies the differential equations of equilibrium, is equal to the complementary work done by the virtual surface forces.*

More advanced study will show that a rigorous solution of the variational problem represented by Eq. (9.8.9) leads to the strain-displacement relations and the displacement-boundary conditions at the surface of the body. At this stage, however, there is no restriction on the type of stress-strain relation, i.e., these may be nonlinear. The principle summarized by Eq. (9.8.9) is complementary to the Principle of Virtual Work given by Eq. (9.4.11). Each represents an alternative approach to the small-deflection equations of elasticity. The application of these principles to the same problem will yield the same answer if the problem can be solved exactly, but if approximate answers are desired, the different approaches will, in many instances, yield slightly different answers.

9.9 PRINCIPLE OF MINIMUM COMPLEMENTARY ENERGY

The Principle of Minimum Complementary Energy represents a specialization of the principle derived in the previous section. As in the development of the Principle of Minimum Potential Energy (Section 9.6), we again consider the virtual stress change $\delta\sigma_{mn}$ as a variation of the stress σ_{mn}. The meaning of the integral $\iiint \epsilon_{mn}\,\delta\sigma_{mn}\,dV$ is, thus, clearly the change in complementary energy δU_c.

For a linearly elastic and isotropic material the relation between the strain and stress tensors has been given by Eq. (7.7.24). By replacing γ_{mn} by ϵ_{mn} the equation is rewritten in the following way:

$$\epsilon_{mn} = (1/E)[(1 + \nu)\sigma_{mn} - \nu\delta_{mn}\sigma_{rr}]. \tag{9.9.1}$$

For a linearly elastic material the value of strain energy U is identical to that of the complementary energy U_c. The expression of the strain energy density, U^*, in terms of stresses was presented previously in Eq. (7.10.5). Thus we may also write the complementary energy as

$$U_c = \iiint (1/2E)[(1 + \nu)\sigma_{mn}\sigma_{mn} - \nu(\sigma_{rr})^2]\,dV. \tag{9.9.2}$$

Clearly, the variation in internal complementary strain energy δU_c can be written as

$$\delta U_c = \delta\iiint (1/2E)[(1 + \nu)\sigma_{mn}\sigma_{mn} - \nu(\sigma_{rr})^2]\,dV$$

$$= \iiint (1/E)[(1 + \nu)\sigma_{mn} - \nu\,\delta_{mn}\sigma_{rr}]\delta\sigma_{mn}\,dV,$$

or

$$\delta U_c = \iiint \epsilon_{mn}\,\delta\sigma_{mn}\,dV. \tag{9.9.3}$$

If the total complementary energy is now defined as

$$\pi_c = U_c - \iint [\boldsymbol{\sigma}_m \cdot \mathbf{u}] n_m \, dA, \tag{9.9.4}$$

then the Principle of Virtual Complementary Work becomes the Principle of Minimum Complementary Energy:

$$\delta \pi_c = 0, \tag{9.9.5}$$

which may be stated as follows:

Of all the states of stress which satisfy the equations of equilibrium, the actual state of stress is that which makes the complementary energy a minimum.

It should be emphasized again that in the application of this principle the quantities to be varied are the internal stresses σ_{mn} and the surface stresses $\boldsymbol{\sigma}_m n_m$. This should be contrasted with the Principle of Minimum Potential Energy.

We shall discuss the application of the Principle of Minimum Complementary Energy in the following two sections which are concerned with Castigliano's Second Theorem and the Theorem of Least Work. Both theorems can be derived from the Principle of Minimum Complementary Energy.

9.10 CASTIGLIANO'S SECOND THEOREM

Let us consider again our all-purpose body as it is acted upon by a system of N generalized forces Q_m, and assume that the internal stresses have been explicitly expressed in terms of these generalized forces. Then for a structure whose material is linearly elastic the total complementary energy can be written as

$$\pi_c = U(Q_1, Q_2, \ldots, Q_N) - \sum_{m=1}^{N} q_m Q_m, \tag{9.10.1}$$

where q_m is the generalized displacement corresponding to Q_m. The Principle of Minimum Complementary Energy can thus be expressed as

$$\delta \pi_c (Q_1, Q_2, \ldots, Q_N) = 0. \tag{9.10.2}$$

This equation can be expanded to yield

$$\sum_{m=1}^{N} \frac{\partial \pi_c}{\partial Q_m} \delta Q_m = 0. \tag{9.10.3}$$

Each of the variations δQ_m is independent of the others; hence the condition set forth by Eq. (9.10.3) yields N independent equations

$$\frac{\partial \pi_c}{\partial Q_m} = 0, \qquad m = 1, 2, \ldots, N. \tag{9.10.4}$$

The application of Eq. (9.10.4) to Eq. (9.10.1) leads to the result known as Castigliano's Second Theorem,

$$q_m = \frac{\partial U}{\partial Q_m}.$$ (9.10.5)

In words, the result shown in Eq. (9.10.5) states:

In a linearly elastic structure, the partial derivative of the strain energy with respect to an externally applied generalized force is equal to the displacement corresponding to that force.

Castigliano's Second Theorem can be applied in calculating the deflections of structures. Illustrative examples are given in the following subsections.

(a) Deflections in terms of flexibility—influence coefficients. The complementary energy U has been expressed in terms of the flexibility influence coefficients C_{mn} in Eq. (9.3.22). Castigliano's Second Theorem (since $U = U_c$) then yields

$$q_m = \sum_{n=1}^{N} C_{mn}Q_n,$$ (9.10.6)

a result shown in Eq. (9.3.13) which was obtained by another argument.

(b) Deflection of a truss at a loading point. Let us consider the simple statically determinate truss which is reproduced here in Fig. 9.25 with specific information about the dimensions and loadings. The cross-sectional areas for members AD and AB are 2 in², while those for members DC, CB, and AC are $\frac{1}{2}$ in². All bars are made of aluminum alloy whose modulus of elasticity is 10^7 psi. A single downward load P of 10,000 lb is acting at the joint C. It is required to determine the vertical deflection at C.

Through the application of equations of statics, the axial loads in all members can be calculated in terms of the applied load Q. These loads are listed in Table 9.1. This table also contains the calculation of the total strain energy U, which is given by

$$U = \frac{1}{2}\sum \frac{F_i^2 L_i}{A_i E_i},$$ (9.10.7)

where L_i, A_i, E_i are respectively the length, area, and elastic modulus of the ith member, and F_i is the axial force in the ith member due to given applied loads. In the present problem only a single load Q is involved. As shown in the calculation given in Table 9.1,

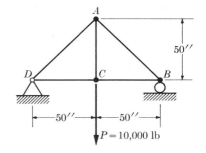

FIG. 9.25 Statically determinate planar truss.

TABLE 9.1 *Calculation of Truss Deflection*

Member	AB	AD	AC	DC	CB	Σ
Area (in²) A_i	2	2	$\frac{1}{2}$	$\frac{1}{2}$	$\frac{1}{2}$	—
Length (in.) L_i	$50\sqrt{2}$	$50\sqrt{2}$	50	50	50	—
Member load due to Q (Fig. 9.25): F_i	$-(\sqrt{2}/2)P$	$-(\sqrt{2}/2)P$	$+P$	$+P/2$	$+P/2$	—
$\dfrac{F_i^2 L_i}{A_i E_i}$	$12.5\sqrt{2}P^2/E$	$12.5\sqrt{2}P^2/E$	$100P^2/E$	$25P^2/E$	$25P^2/E$	$185.4P^2/E$
Member load due to $Y=1$ (Fig. 9.26): f_i	$-\sqrt{2}/2$	$+\sqrt{2}/2$	0	$+\frac{1}{2}$	$+\frac{1}{2}$	—
$\dfrac{F_i f_i L_i}{A_i E_i}$	$12.5\sqrt{2}P/E$	$-12.5\sqrt{2}P/E$	0	$25P/E$	$25P/E$	$50P/E$

the strain energy is

$$U = \frac{185.4}{2}\frac{P^2}{E}.$$

Hence the deflection at C is

$$q = \frac{\partial U}{\partial P} = 185.4\frac{P}{E} = 0.1854 \text{ in.}$$

(c) Calculation of structural deflection by Dummy Load method. As another example, let us suppose that we wish to compute the deflection of a truss at any joint which may or may not have an external load. In such a case we place a dummy load Y at the joint pointing toward the direction of the deflection to be determined. The strain energy of the truss due to the given applied loads Q_i and the dummy load Y is

$$U = \frac{1}{2}\sum \frac{(F_i + f_i Y)^2 L_i}{A_i E_i}, \tag{9.10.8}$$

where F_i, as before, is the load in the ith member of the truss due to the applied loads, while f_i is the load in the ith member due to a unit dummy load ($Y = 1$). According to Castigliano's Second Theorem, the deflection at this joint is

$\partial U/\partial Y$. Here, since Y is only a fictitious load, this deflection can be written as

$$\Delta = \lim_{Y \to 0} \left(\frac{\partial U}{\partial Y} \right) = \lim_{Y \to 0} \sum \frac{(F_i + f_i Y) f_i L_i}{A_i E_i},$$

or

$$\Delta = \sum \frac{F_i f_i L_i}{A_i E_i}. \tag{9.10.9}$$

This equation is the same as Eq. (9.4.30) which was obtained from the consideration of virtual work (with a slight change in notation: $f_i \equiv F_i^{(Q)}$).

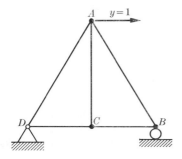
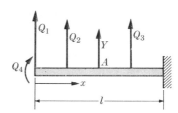

FIG. 9.26 Statically determinate planar truss under a dummy load.

FIG. 9.27 Statically determinate beam under external loads and a dummy load.

Let us refer again to the truss shown in Fig. 9.25. This time, however, it is required to determine the horizontal deflection at the joint A. In using the dummy load method a unit horizontal load is applied at A as shown in Fig. 9.26. The loads f_i in all members are then calculated by equations of statics. The last two rows in Table 9.1 also list the calculations for the horizontal deflection at A, that is,

$$\Delta = 50 \frac{P}{E} = 0.05 \text{ in.}$$

(d) Deflection of a simple beam. The dummy load method for deflection analysis can be applied to other structures in general. For example, it is required to compute the vertical deflection at A for a simple beam (Fig. 9.27) under a number of external forces $Q_1, \ldots, Q_n$, or a continuously distributed load $p(x)$. At A we apply a dummy vertical load Y. We see that the internal bending-moment distribution can be expressed in terms of the external forces and the dummy force Y. Let

$$M(x) = \text{bending moment due to } Q_1, \ldots, Q_n,$$
$$m(x) = \text{bending moment due to a unit load at } A.$$

We can then write the internal strain energy as

$$U = \frac{1}{2} \int_0^l \frac{(M + Ym)^2}{EI}\, dx. \tag{9.10.10}$$

According to Castigliano's Second Theorem, the vertical deflection at A is

$$\Delta_A = \lim_{Y \to 0} \frac{dU}{dY} = \lim_{Y \to 0} \int_0^l \frac{(M + Ym)m}{EI}\, dx = \int_0^l \frac{Mm}{EI}\, dx. \tag{9.10.11}$$

This equation is the same as Eq. (9.4.31), which was obtained previously from a different consideration (with a slight change in notation: $m \equiv M^{(Q)}$).

9.11 THEOREM OF LEAST WORK

The strain energy, U, for a statically indeterminate structure cannot be immediately written only in terms of the prescribed external forces. However, the strain energy can be expressed in terms of the applied external forces and a number of redundant internal forces or reactions. For example, the three-bar truss shown in Fig. 9.16 is a statically indeterminate structure, i.e., the member axial loads T_{AD} and T_{BD} cannot be determined by the equations of statics alone. However, if we let one of these, say T_{BD}, be the redundant force, we can express the other axial force T_{AD} in terms of T_{BD} and Q by applying the condition of equilibrium of internal forces at joint D.

In general, let a structure contain N independent redundant forces,

$$X_1, X_2, \ldots, X_N.$$

They are either internal forces or reactions forces. The latter are, of course, forces at the boundary; their corresponding generalized displacements, however, are zero. The expression for the complementary energy is

$$\pi_c = U(Q_1, Q_2, \ldots, Q_M, X_1, X_2, \ldots, X_N) - Q_m q_m, \tag{9.11.1}$$

where Q_1, Q_2, and Q_M are the prescribed external forces. The Principle of Minimum Complementary Energy thus consists of two groups of terms:

$$\partial \pi_c / \partial Q_m = 0, \tag{9.11.2}$$

and

$$\partial \pi_c / \partial X_n = 0. \tag{9.11.3}$$

The first set of equations again yields Castigliano's Second Theorem (Eq. 9.10.4); the second set, since the X_n redundants appear only in U, yields

$$\partial U / \partial X_n = 0. \tag{9.11.4}$$

This equation is referred to as the Theorem of Least Work, which may be stated in words as follows:

For a linear statically indeterminate structure, the derivative of the strain energy with respect to any redundant internal force or redundant reaction must be zero.

It can be concluded from Eq. (9.11.4) only that the redundant quantities are chosen such as to make the strain energy of the system either a maximum or a minimum. However, by calculating the second derivatives and showing that they are always positive for a stable system, it can be proved that the strain energy must be a minimum. Thus the theorem embracing Eq. (9.11.4) is known as the Theorem of Least Work.

It should be carefully observed that, whereas the Theorem of Minimum Complementary Energy applies to nonlinear as well as to linear systems, Castigliano's Theorem and the Theorem of Least Work apply only to systems in which the principle of superposition is valid, that is, to linear systems.

We shall consider several examples to illustrate the application of the Theorem of Least Work.

(a) Simple statically indeterminate truss. In the three-bar truss (Fig. 9.16), the strain energy is given by

$$U = \frac{1}{2} \frac{l}{A_1 E_1} T_{BD}^2 + \frac{l}{A_2 E_2 \cos \alpha} T_{AD}^2. \tag{9.11.5}$$

Here T_{BD} and T_{AD} are related by the condition of equilibrium at joint D, that is,

$$T_{BD} + 2 T_{AD} \cos \alpha = Q. \tag{9.11.6}$$

Let T_{BD} be the redundant internal force. We can write

$$T_{AD} = \frac{Q - T_{BD}}{2 \cos \alpha}. \tag{9.11.7}$$

Substitution into Eq. (9.11.5) yields

$$U = \frac{1}{2} \frac{l}{A_1 E_1} T_{BD}^2 + \frac{l}{A_2 E_2 \cos \alpha} \left(\frac{Q - T_{BD}}{2 \cos \alpha} \right)^2. \tag{9.11.8}$$

Application of the Theorem of Least Work thus leads to

$$\frac{\partial U}{\partial T_{BD}} = \frac{l}{A_1 E_1} T_{BD} + \frac{2l}{A_2 E_2 \cos \alpha} \left(\frac{Q - T_{BD}}{2 \cos \alpha} \right) \left(-\frac{1}{2 \cos \alpha} \right) = 0. \tag{9.11.9}$$

This equation can be solved for T_{BD}. The result is

$$T_{BD} = \frac{Q}{(1 + 2 A_2 E_2 \cos^3 \alpha) / A_1 E_1}. \tag{9.11.10}$$

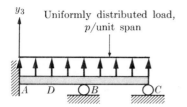
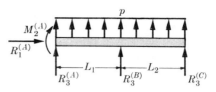

$$\text{Fig. 9.28}\quad\text{Statically indeterminate beam.}$$

(b) Statically indeterminate beam. Let us consider the beam shown in Fig. 9.28. There are five unknown reactions, but only three equilibrium equations of statics. The problem hence has two redundant forces. Of the five reactions, $R_1^{(A)}$ is found to be zero when the condition of force summation along the y_1-direction is imposed. However, any two of the remaining reactions may be used as the redundant forces. For example, if we let

$$X_1 = R_3^{(A)}, \qquad X_2 = M_3^{(A)}, \tag{9.11.11}$$

we can write

$$R_3^{(B)} + R_3^{(C)} = -p(L_1 + L_2) - X_1,$$

and

$$L_1 R_3^{(B)} + (L_1 + L_2)R_3^{(C)} = -p\frac{(L_1 + L_2)^2}{2} + X_2. \tag{9.11.12}$$

These two equations can be solved to obtain $R_3^{(B)}$ and $R_3^{(C)}$ in terms of X_1 and X_2.

We may also assign the redundant forces in the following manner: By figuratively cutting the beam apart at any station D (between stations A and B) and replacing the internal stresses by the equipollent shear force $S^{(D)}$ and bending moment $M^{(D)}$, we obtain two separate structures as shown in Fig. 9.29. It is clear that if $M^{(D)}$ and $S^{(D)}$ are assigned respectively as the two redundant forces X_1 and X_2, the bending-moment distribution in each part of the beam can be written in terms of X_1, X_2, and p. Thus, in turn, the strain energy U of the complete beam can be written as

$$U = U(X_1, X_2). \tag{9.11.13}$$

The two redundant forces can then be calculated by solving the two equations given by

$$\partial U/\partial X_1 = 0, \qquad \text{and} \qquad \partial U/\partial X_2 = 0.$$

$$\tag{9.11.14}$$

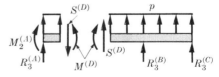

Fig. 9.29 Statically indeterminate beam and choice of redundant forces.

The Theorem of Least Work may also be interpreted as the result of Castigliano's Second Theorem. In order for the two parts illustrated in Fig. 9.29 to behave like the complete structure, the deformations at point D must be consistent, i.e., point D of the left portion must displace in exactly the same fashion as point D of the right portion. For example, Castigliano's Second Theorem can be applied to the bending moment $M^{(D)}$ acting on the left portion, yielding an angular displacement

$$\theta_{\text{left}} = \frac{\partial U_{\text{left}}}{\partial M^{(D)}}, \tag{9.11.15}$$

and also to the right portion yielding

$$\theta_{\text{right}} = \frac{\partial U_{\text{right}}}{\partial M^{(D)}}, \tag{9.11.16}$$

where θ is the rotation corresponding to the bending moment $M^{(D)}$. The relative rotation must be zero in order to satisfy the requirement of consistent deformations, that is,

$$\theta_{\text{left}} + \theta_{\text{right}} = 0, \tag{9.11.17}$$

or

$$\partial U/\partial M^{(D)} \equiv \partial U/\partial X_1 = 0, \tag{9.11.18}$$

because

$$U = U_{\text{left}} + U_{\text{right}}. \tag{9.11.19}$$

By an analogous treatment we can also obtain

$$\partial U/\partial S^{(D)} \equiv \partial U/\partial X_2 = 0. \tag{9.11.20}$$

When the results given by Eqs. (9.11.18) and (9.11.20) are interpreted in terms of the complete structure, they state that the relative generalized displacement of a pair of equal and opposite internal generalized forces is equal to zero. Such a result is the same as that derived from the Theorem of Least Work.

9.12 SUMMARY

The following energy theorems have been derived and illustrated:

Clapeyron's Law

The work done by a system of generalized forces Q_n acting on a linearly elastic structure is stored as strain energy and is equal to

$$U = \tfrac{1}{2} \sum_{n=1}^{N} Q_n q_n, \tag{9.12.1}$$

where q_n is the generalized displacement corresponding to Q_n.

Principle of Virtual Work

If a body is in equilibrium and remains in equilibrium while it is subjected to a virtual distortion compatible with the geometrical constraints, the virtual work done by the external forces is equal to the virtual work done by the internal stresses.

Betti's Law

In any structure which is linearly elastic, the virtual work done by a system of Q-forces under a distortion caused by a system of P-forces is equal to the virtual work done by the system of P-forces under a distortion caused by the system of Q-forces.

Maxwell's Law of Reciprocal Deflections

In any structure which is linearly elastic the generalized deflection at point m caused by a generalized load Q at point n is numerically equal to the generalized deflection at point n caused by the same magnitude of generalized force Q at point m.

Principle of Minimum Potential Energy

Among all the geometrically possible states of displacement, the correct state of displacement is that which makes the potential energy π_p a minimum, where $\pi_p = U - W$.

Castigliano's First Theorem

When the strain energy U can be expressed in terms of a system of generalized displacement q_m, then the generalized force is given by

$$Q_m = \frac{\partial U}{\partial q_m}. \tag{9.12.2}$$

Principle of Virtual Complementary Work

Due to virtual stress changes $\delta\sigma_{mn}$, which satisfy the differential equations of equilibrium, the virtual complementary work under the actual state of strain is equal to the work done by the virtual external forces.

Principle of Minimum Complementary Energy

Of all the states of stress which satisfy the equations of equilibrium, the actual state of stress is that which makes the complementary energy π_c a minimum, where $\pi_c = U_c - W$.

Castigliano's Second Theorem

For a linearly elastic structure, if the strain energy U can be expressed in terms of a system of generalized forces Q_m, then the generalized displacement is given by

$$q_m = \frac{\partial U}{\partial Q_m}. \tag{9.12.3}$$

Theorem of Least Work

For a linear statically indeterminate structure, the derivative of the strain energy with respect to any redundant internal force or redundant reaction must be zero.

PROBLEMS

9.1 A thin uniform rod with two concentrated masses is rotating about one of its ends at a constant velocity ω (Fig. P.9.1a). The cross-sectional area of the rod is A. The mass of the rod is negligibly small in comparison to that of the concentrated masses; hence the two centrifugal forces Q_1 and Q_2 are the generalized forces.

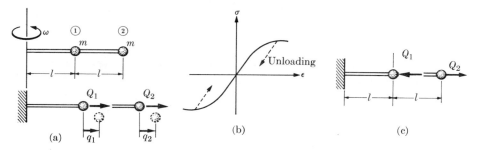

FIGURE P.9.1

(a) Let the rod be made of elastic material of Young's modulus E. Determine the strain energy stored in the rod by $U = \frac{1}{2}\int \epsilon\sigma\, dV$. Verify the strain energy expression by calculating the work done by the two concentrated forces. Will the order of application of the two forces affect your answer?

(b) Consider the rod strained into the plastic range. Let the stress-strain relation of the material be given by Fig. P.9.1(b). Determine the strain energy stored in the rod. Will the order of application of the two forces affect your answer?

(c) Now consider the problem illustrated by Fig. P.9.1(c). The forces Q_1 and Q_2 are in *opposite* directions, and $Q_2 = 4Q_1$. Will the order of application of the two forces affect the strain energy stored when the plastic behavior of the material is taken into consideration?

9.2 A truss is subject to the loading shown in Fig. P.9.2. Determine the bar forces and thereby determine the displacement of point d, at the mid-span, using the Principle of Virtual Work.

Bar	Area
A	5 in^2
B	20 in^2
C	10 in^2

$E = 30 \times 10^6$ psi

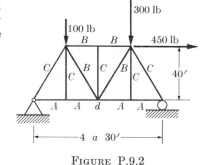

FIGURE P.9.2

9.3 A beam is subjected to a uniformly distributed load as shown in Fig. P.9.3. Determine by the Principle of Virtual Work, the displacement of a relative to the displacement of b. By a proper choice of a Q-force system, the solution can be obtained directly.

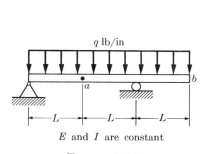

FIGURE P.9.3

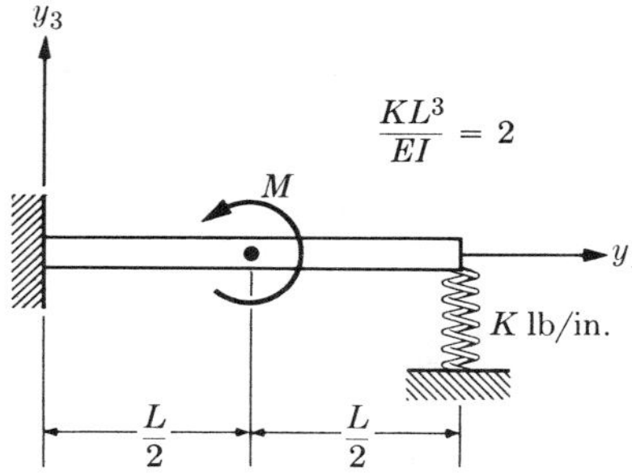

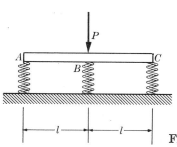

FIGURE P.9.5

9.4 Solve Problem 4.1 by the Principle of Minimum Potential Energy.

9.5 An elastic cantilever beam is restrained by a linear spring of spring constant k, and is acted upon by a couple M at the mid-span point (Fig. P.9.5). E and I are constant. By means of the Principle of Minimum Potential Energy determine approximately the vertical deflection at the end of the beam. Assume the beam has the following deflection shape,

$$w(y_1) = q_1 y_1^2 + q_2 y_1^3,$$

where q_1 and q_2 are the generalized displacements to be determined. Take $kL^3/EI = 2$.

9.6 Compute the forces in the bars of the system shown by Fig. P.9.6. All bars are of the same length, cross-sectional area, and material.

9.7 A beam of length $2l$, and bending stiffness EI is acted upon by a concentrated load P at the mid-span and is supported at three points A, B, and C by three springs each of which has a spring constant k lb/in. (Fig. P.9.7). Determine the deflection at the mid-span point B.

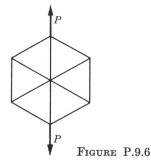

FIGURE P.9.6

FIGURE P.9.7

9.8 The beam AB is clamped at A and supported by a ball at B which is the end of another cantilever beam CB (Fig. P.9.8). Both beams have the same bending stiffness EI. The lengths are $2l$ and l, respectively. A vertical load P is acting at the mid-span of AB. Determine the vertical deflection at B.

9.9 Solve Problems 4.7 and 4.9 by an energy method.

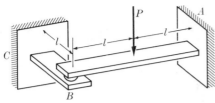

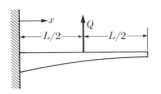

FIGURE P.9.8   FIGURE P.9.10

9.10   The cantilever beam shown in Fig. P.9.10 has a variable-section moment of inertia given by the following expression,

$$I(x) = I_0 L/(L + x),$$

where I_0 is the value at the clamped end. A concentrated load is acting at the mid-span point.   Determine the deflection curve by means of Castigliano's Second Theorem.

9.11   A truss is subjected to the loading as shown in Fig. P.9.11. For bars AE, BE, BD, and CD the area is 1 in^2. For bars AB, BC, and ED, the area is $\frac{1}{2}$ in^2. Determine the bar forces and the vertical deflection at B by Castigliano's Second Theorem. All the bars are constructed of the same material.

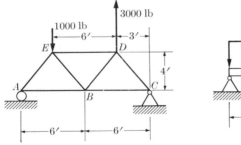

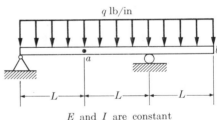

FIGURE P.9.11   FIGURE P.9.12

9.12   A beam is subjected to a uniformly distributed load as shown in Fig. P.9.12. Determine the deflections at points a and b. (a) Use Castigliano's Second Theorem to obtain the exact solution.   (b) Assume any reasonable deflection function and use the Principle of Minimum Potential Energy. Compare the result with that of (a).

9.13   Wire RS has a cross-sectional area A and modulus of elasticity E (Fig. P.9.13). A uniform load w lb/in. acts on the entire length of the beam PR which is also supported by a spring QT at mid-span and by a hinge at point P. The spring has a spring constant k lb/in. Calculate the deflection of point R, given that (a) the beam PR is a rigid body, (b) the beam PR has a bending stiffness EI.

9.14   The beam ABC is *clamped* at A and supported by a light truss for which the points $ABCD$ and E are *hinge joints*. The weight of the beam is 1 lb/in. of span, the cross-section moment of inertia is 10 in^4, and cross-sectional area is 5 in^2. The modulus

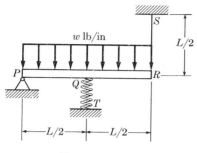

FIGURE P.9.13

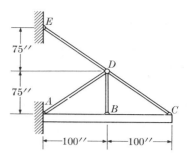

FIGURE P.9.14

of elasticity is $E_{\text{beam}} = 3 \times 10^7$ psi. The cross-sectional area of all the members of the truss is 0.1 in^2 and the modulus of elasticity is $E_{\text{truss member}} = 10^7$ psi. Determine the bending-moment distribution along the beam and the stresses in the truss members.

9.15 Determine the bending-moment distribution of the statically indeterminate beam shown in Fig. P.9.15. Determine the slope at the end of the beam and the deflection at $x = L/2$, using Castigliano's Second Theorem. Check your result with the approximate solution obtained in Problem 9.5.

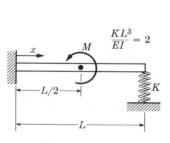

FIGURE P.9.15

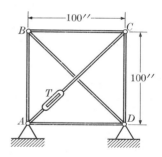

FIGURE P.9.16

9.16 The cross-sectional area for each member of the given truss (Fig. P.9.16) is 1 in^2, and the modulus is $E = 10^7$ psi. The turnbuckle T along AC is tightened such that the ends of the two bars at the turnbuckle are brought 0.2 in. closer to each other. Determine the stresses induced by this operation.

9.17 A frame composed of members rigidly joined together is acted upon by a uniform lateral load of 10 lb/in. along the vertical member CD as shown in Fig. P.9.17. The term "rigidly joined together at B" is used to denote the physical condition that the B-ends of members AB and BC undergo the same displacement and rotation under loads. Each member has a square cross section of dimensions 6 in. $\times$ 6 in. Determine the bending-moment distributions in

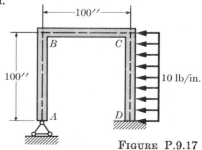

FIGURE P.9.17

the three members, and the horizontal deflection of the point C. (Neglect the strain energy due to axial loads in each member.) After solving the problem, calculate the strain energy due to axial loads and compare this result with the strain energy due to bending moment.

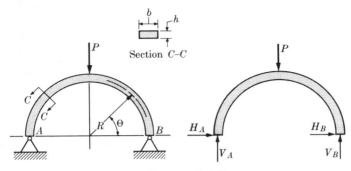

FIGURE P.9.18

9.18   The strain energy for a curved beam, with radius of curvature R much larger than its thickness h, can be obtained according to slender-beam analysis, i.e., according to the ordinary beam theory discussed in previous chapters. For example, if $M(\theta)$ represents the bending-moment distribution of the half-ring shown in Fig. P.9.18, then the total strain energy is

$$U = \int_0^\pi \frac{[M(\theta)]^2}{2EI} R \, d\theta.$$

The half-ring is acted upon by a concentrated load P as shown. Determine the bending moment at the section where the load P is applied. What is the deflection at this point? *Hint:* Remove the supports and replace with a shear force H and an axial force V, as shown. Note that because of symmetry, $H_A = -H_B$ and $V_A = V_B$.

Appendix

Mathematical Themes

A.1 INTRODUCTION

Many of the difficulties in the field of solid mechanics lie not in the basic ideas but arise, rather, as a result of the geometrical descriptions which must be made. Many variables must be sorted, and many equations derived. The sorting of the variables, i.e., the "cataloging" aspects will be minimized by adopting the summation and indicial conventions of tensor analysis. The derivation of the equations will become clearer and simplified by the use of ordinary vector analysis. In addition, the use of matrices will prove to be invaluable.

We will summarize in this Appendix some of the uses of mathematics which heretofore have not been presented in a first course in solid mechanics. Some of the topics included are covered in the first three semesters of calculus. The remaining topics are considered to be material for advanced calculus because of time limitations in the first courses, not because a great deal of mathematical sophistication is required.

A.2 THE SUMMATION AND INDICIAL NOTATION [1]*

The three-dimensional Euclidean space of solid mechanics will be described by the rectangular Cartesian coordinates y_1, y_2, and y_3, which are to be arranged in a right-handed system. A linear homogeneous function of the coordinates is usually written in the form

$$a_1 y_1 + a_2 y_2 + a_3 y_3 \equiv \sum_{m=1}^{3} a_m y_m, \tag{A.2.1}$$

where a_1, a_2, and a_3 are constants.

A great deal of writing and space can be saved by eliminating the $\sum$-sign with the adoption of the following two conventions:

(1) A lower-case subscript which is repeated in a term replaces the $\sum$-sign, i.e., a repeated lower-case subscript is to be summed from 1 to 3.
(2) A free, i.e., an unrepeated, lower-case subscript is to have the range of values 1, 2, and 3 unless otherwise noted.

---

* Numbers in brackets refer to references at the end of the Appendix.

Thus, Eq. (A.2.1) will now read

$$a_m y_m \equiv a_1 y_1 + a_2 y_2 + a_3 y_3. \tag{A.2.2}$$

A homogeneous quadratic function of the variables y_1, y_2, y_3 can be compactly written as

$$\begin{aligned}
a_{mn} y_m y_n \equiv\ & a_{11}(y_1)^2 + a_{12} y_1 y_2 + a_{13} y_1 y_3 \\
& + a_{21} y_2 y_1 + a_{22}(y_2)^2 + a_{23} y_2 y_3 \\
& + a_{31} y_3 y_1 + a_{32} y_3 y_2 + a_{33}(y_3)^2,
\end{aligned} \tag{A.2.3}$$

where the a_{mn}'s are again constants.

The repeated index is often called the "dummy" index because it is immaterial which index is used. Thus

$$a_m y_m \equiv a_r y_r, \qquad a_{mn} y_m y_n \equiv a_{rs} y_r y_s. \tag{A.2.4}$$

The system of linear equations

$$\begin{aligned}
b_1 &= a_{11} y_1 + a_{12} y_2 + a_{13} y_3, \\
b_2 &= a_{21} y_1 + a_{22} y_2 + a_{23} y_3, \\
b_3 &= a_{31} y_1 + a_{32} y_2 + a_{33} y_3,
\end{aligned} \tag{A.2.5}$$

is written in the form

$$b_r = a_{rs} y_s, \tag{A.2.6}$$

where the index r is called the free index since it is not repeated in any single term. The appearance of a free index, by convention (2), means that Eq. (A.2.6) represents a system of three simultaneous equations. The same index cannot be used more than twice in the same term.

Equation (A.2.6) may be called a "tensor" equation. At the outset of this course of study, the quantities to be studied will be restricted to that part of tensor analysis commonly known as Cartesian tensors. This means that the frame of reference to be used is restricted to rectangular Cartesian coordinates. If a completely general curvilinear coordinate frame of reference is used, then the concepts of covariant and contravariant tensors become necessary. The power and usefulness of tensor analysis lie in the extreme ease with which equations formulated in rectangular Cartesian coordinates can be transformed to any kind of coordinate system.

A.3   THE e-SYMBOL AND THE KRONECKER DELTA [1], [2]

A symbol, which will prove very useful in dealing with determinants, is the completely skew-symmetric system of the third order, which is denoted by e_{rst}. The components of this system by definition have only the following three definite numerical values:

(a)   0 when any two of the indices are equal,
(b) $+1$ when rst is an even permutation of 1, 2, 3,
(c) -1 when rst is an odd permutation of 1, 2, 3.

The permutation of the three numbers 1, 2, 3 is accomplished by a finite number of interchanges of any pair of the numbers, and hence the number of interchanges required to bring about a given sequence of the numbers from the initial arrangement of 1, 2, 3 will be either odd or even. Thus the arrangement 312 requires two interchanges, whereas 132 requires only one interchange. Hence $e_{312} = 1$, and $e_{132} = -1$.

The Kronecker delta can be generated from the e-symbol with the aid of the indicial and summation notation

$$\delta_{rm} = \tfrac{1}{2} e_{rsp} e_{msp}. \tag{A.3.1}$$

This symbol δ_{rm}, called the Kronecker delta (a bit of trivia on which to attach a great man's name), has the following numerical values:

(a)   1 if the two indices are equal,
(b)   0 if the two indices are unequal.

A.4   COORDINATE TRANSFORMATIONS AND DETERMINANTS [1] [2]

Let Oy_1, Oy_2, Oy_3 and $O\tilde{y}_1$, $O\tilde{y}_2$, $O\tilde{y}_3$ be two rectangular Cartesian coordinate systems with a common origin O. The angles between Oy_1 and $O\tilde{y}_2$, for example, will be denoted by $\widehat{y_1\tilde{y}_2}$ or $\widehat{\tilde{y}_2 y_1}$. Similarly, the angle between Oy_n and $O\tilde{y}_m$ will be denoted by $\widehat{y_n\tilde{y}_m}$ or $\widehat{\tilde{y}_m y_n}$. Figure A.1 shows the angles which axis $O\tilde{y}_2$ makes with axes Oy_1, Oy_2, and Oy_3. A point p with coordinates y_1, y_2, y_3 in the first coordinate frame will have coordinates in the other frame given by

$$\begin{aligned}
\tilde{y}_1 &= y_1 \cos \widehat{\tilde{y}_1 y_1} + y_2 \cos \widehat{\tilde{y}_1 y_2} + y_3 \cos \widehat{\tilde{y}_1 y_3} = y_n l_{\tilde{1}n}, \\
\tilde{y}_2 &= y_1 \cos \widehat{\tilde{y}_2 y_1} + y_2 \cos \widehat{\tilde{y}_2 y_2} + y_3 \cos \widehat{\tilde{y}_2 y_3} = y_n l_{\tilde{2}n}, \\
\tilde{y}_3 &= y_1 \cos \widehat{\tilde{y}_3 y_1} + y_2 \cos \widehat{\tilde{y}_3 y_2} + y_3 \cos \widehat{\tilde{y}_3 y_3} = y_n l_{\tilde{3}n},
\end{aligned} \tag{A.4.1}$$

where $l_{\tilde{r}n}$ is called the direction cosine and is the cosine of the angle between axes $\tilde{y}_r$ and y_n. This can be compactly written as

$$\tilde{y}_m = y_n l_{\tilde{m}n}. \tag{A.4.2}$$

In this special case of Fig. A.1 it is obvious that the coordinate transformation given by Eq. (A.4.2) is reversible, that is, Eq. (A.4.2) can be solved to yield

$$y_n = \tilde{y}_m l_{n\tilde{m}}. \tag{A.4.3}$$

More formally, Eq. (A.4.2) can be solved only if the determinant of the coefficients of y_1, y_2, y_3 is not zero. The determinant will be denoted by θ, and is represented by

$$\theta = |l_{\tilde{m}n}| = \begin{vmatrix} l_{\tilde{1}1} & l_{\tilde{1}2} & l_{\tilde{1}3} \\ l_{\tilde{2}1} & l_{\tilde{2}2} & l_{\tilde{2}3} \\ l_{\tilde{3}1} & l_{\tilde{3}2} & l_{\tilde{3}3} \end{vmatrix}. \tag{A.4.4}$$

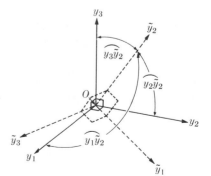

FIG. A.1   Two sets of rectangular Cartesian coordinate systems.

It can be verified through expansion of the determinant by rows that the value of the determinant is given by

$$\theta = |l_{\tilde{m}n}| = e_{mnp}l_{\tilde{1}m}l_{\tilde{2}n}l_{\tilde{3}p}. \tag{A.4.5}$$

If the cofactor of the term $l_{\tilde{m}n}$ in the determinant (Eq. A.4.4) is denoted by $\Theta_{\tilde{m}n}$, the expansion in terms of the rth column is given by

$$\delta_{mr}\theta = \Theta_{\tilde{n}m}l_{\tilde{n}r}, \tag{A.4.6}$$

and in terms of the sth row, by

$$\delta_{ms}\theta = \Theta_{\tilde{s}n}l_{\tilde{m}n}. \tag{A.4.7}$$

A.5 CURVILINEAR COORDINATES

The ideas in the preceding section can be easily extended to general curvilinear coordinates, x_m. These may be cylindrical, spherical, elliptical, or even more exotic frames. But all can be considered as a case of the general functional transformation

$$x_m = x_m(y_1, y_2, y_3), \tag{A.5.1}$$

where

y_m are the rectangular Cartesian coordinates,

x_m are general curvilinear coordinates,

$x_m(y_1, y_2, y_3)$ are single-valued functions of y_1, y_2, y_3.

For the transformation to be reversible, the determinant

$$\left|\frac{\partial x_m}{\partial y_s}\right| = e_{mnp}\frac{\partial x_1}{\partial y_m}\frac{\partial x_2}{\partial y_n}\frac{\partial x_3}{\partial y_p} \tag{A.5.2}$$

must be different from zero. Then Eq. (A.5.2) can be solved to yield

$$y_m = y_m(x_1, x_2, x_3). \tag{A.5.3}$$

The surfaces $x_m = $ const are called the coordinate surfaces, and the intersections of these coordinate surfaces are called the coordinate curves. Through each point in a Euclidean space there pass three coordinate curves. The intersections of the two surfaces $x_2 = $ const and $x_3 = $ const are curves along which only x_1 varies. Thus the x_m coordinate curves are those points in space for which only x_m is changing. In general, these curves are not straight lines, which is the justification for the name "curvilinear coordinates."

Similar relations must hold if another curvilinear system, $\tilde{x}_r$, is introduced. The functional transformations can be written as

$$\tilde{x}_r = \tilde{x}_r(x_1, x_2, x_3), \tag{A.5.4}$$

and

$$x_r = x_r(\tilde{x}_1, \tilde{x}_2, \tilde{x}_3). \tag{A.5.5}$$

If Eq. (A.5.4) is substituted into Eq. (A.5.5), we obtain

$$x_r = x_r[\tilde{x}_1(x_1, x_2, x_3), \tilde{x}_2(x_1, x_2, x_3), \tilde{x}_3(x_1, x_2, x_3)].$$ (A.5.6)

The partial derivative of this last equation yields

$$\frac{\partial x_r}{\partial x_s} = \frac{\partial x_r}{\partial \tilde{x}_m} \frac{\partial \tilde{x}_m}{\partial x_s}.$$ (A.5.7)

Since the curvilinear coordinates x_1, x_2, x_3 are independent, then if $r \neq s$, the left-hand side of Eq. (A.5.7) is zero. If $r = s$, then the left-hand side has the value "unity." This means that the left-hand side of Eq. (A.5.7) can be replaced by the Kronecker delta, that is,

$$\delta_{rs} = \frac{\partial x_r}{\partial \tilde{x}_m} \frac{\partial \tilde{x}_m}{\partial x_s}.$$ (A.5.8)

In a similar manner we find that

$$\delta_{rs} = \frac{\partial \tilde{x}_r}{\partial x_m} \frac{\partial x_m}{\partial \tilde{x}_s}.$$ (A.5.9)

If these results are applied to the pair of rectangular Cartesian frames, shown in Fig. A.1, we obtain

$$\delta_{rs} = l_{\tilde{m}r} l_{\tilde{m}s}.$$ (A.5.10)

One of these equations is the result,

$$(l_{\tilde{1}1})^2 + (l_{\tilde{2}1})^2 + (l_{\tilde{3}1})^2 = 1,$$ (A.5.11)

which is the familiar statement that the sum of the squares of the direction cosines of a line (in this example the line is the axis Oy_1) with respect to a rectangular frame $(O\tilde{y}_1, O\tilde{y}_2, O\tilde{y}_3)$ is equal to unity.

Another result contained in Eq. (A.5.10) is

$$l_{\tilde{1}1} l_{\tilde{1}2} + l_{\tilde{2}1} l_{\tilde{2}2} + l_{\tilde{3}1} l_{\tilde{3}2} = 0,$$ (A.5.12)

which can be recognized as the orthogonality condition for axes $O\tilde{y}_1$ and $O\tilde{y}_2$.

There are two other relations of the form given by Eq. (A.5.11) and two others of the form given by Eq. (A.5.12).

A.6 INVARIANTS, VECTORS, AND TENSORS [1], [2], [3]

The field of solid mechanics is intimately concerned with sets of quantities in relation to a system of coordinate frames. It is of prime importance to understand the manner in which these sets of quantities change if the original coordinate frame is transformed to another coordinate frame.

The simplest set of quantities is a single number which is usually referred to as a *scalar*. Such a quantity has the same value regardless of the coordinate frame which is used. Thus, if $f(x_1, x_2, x_3)$ denotes its value in one coordinate frame and $g(\tilde{x}_1, \tilde{x}_2, \tilde{x}_3)$ its value in another, then

$$f(x_1, x_2, x_3) = g(\tilde{x}_1, \tilde{x}_2, \tilde{x}_3)$$ (A.6.1)

defines what is variously referred to as a scalar, or an invariant, or a tensor of zero order. The labels "scalar," "invariant," or "tensor of zero order" have the same meaning. Familiar invariants are energy, the speed of light, and the length of a line element.

A familiar tensor of order one is the set of differentials dy_1, dy_2, dy_3. In another rectangular Cartesian frame $\tilde{y}_1$, $\tilde{y}_2$, $\tilde{y}_3$, the differentials are related to those in the first frame by

$$d\tilde{y}_m = \frac{\partial \tilde{y}_m}{\partial y_r} dy_r. \tag{A.6.2}$$

Note that the partial derivative $\partial \tilde{y}_m/\partial y_r$ is the direction cosine $l_{\tilde{m}r}$ (see Eq. A.4.1). Functions, such as the differentials, which transform in this manner are called tensors of order one, or vectors. If u_1, u_2, u_3 are the components of a first-order tensor, then under a transformation of coordinates, the tensor behaves according to the relation

$$\tilde{u}_m(\tilde{y}_1, \tilde{y}_2, \tilde{y}_3) = \frac{\partial \tilde{y}_m}{\partial y_r} u_r(y_1, y_2, y_3). \tag{A.6.3}$$

If Eq. (A.6.3) is multiplied by $\partial y_n/\partial \tilde{y}_m$, and summed on m from 1 to 3, we derive

$$\frac{\partial y_n}{\partial \tilde{y}_m} \tilde{u}_m = \frac{\partial y_n}{\partial \tilde{y}_m} \frac{\partial \tilde{y}_m}{\partial y_r} u_r = \delta_{nr} u_r = u_n,$$

through the use of Eq. (A.5.8). This result is the inverse transformation

$$u_n(y_1, y_2, y_3) = \frac{\partial y_n}{\partial \tilde{y}_m} \tilde{u}_m(\tilde{y}_1, \tilde{y}_2, \tilde{y}_3). \tag{A.6.4}$$

It should be observed that Eq. (A.6.4) has the same structure as Eq. (A.6.3).

A simple way to introduce tensors of the second order is to multiply two first-order tensors. Thus the second-order tensor w_{mn} can be formed by multiplying the tensors u_m and v_n:

$$w_{mn} = u_m v_n. \tag{A.6.5}$$

In general, a second-order tensor has nine independent components. If the coordinates are changed, then, since u_m and v_n obey the tensor transformation law, Eq. (A.6.3) or (A.6.4), we find that

$$\tilde{u}_m(\tilde{y}_1, \tilde{y}_2, \tilde{y}_3)\tilde{v}_n(\tilde{y}_1, \tilde{y}_2, \tilde{y}_3) = \frac{\partial \tilde{y}_m}{\partial y_r} \frac{\partial \tilde{y}_n}{\partial y_s} u_r(y_1, y_2, y_3)v_s(y_1, y_2, y_3). \tag{A.6.6}$$

Upon the introduction of Eq. (A.6.5), we obtain the transformation law for second-order tensors,

$$\tilde{w}_{mn}(\tilde{y}_1, \tilde{y}_2, \tilde{y}_3) = \frac{\partial \tilde{y}_m}{\partial y_r} \frac{\partial \tilde{y}_n}{\partial y_s} w_{rs}(y_1, y_2, y_3). \tag{A.6.7}$$

Systems of second-order quantities which transform according to Eq. (A.6.7) are called tensors of the second order. Familiar examples which will be studied in detail are the Cartesian strain and stress tensors.

The pattern for the transformation law of higher-order tensors can be seen by examining Eqs. (A.6.3) and (A.6.7). Thus a third-order tensor obeys the transformation law

$$\widetilde{w}_{mnp}(\tilde{y}_1, \tilde{y}_2, \tilde{y}_3) = \frac{\partial \tilde{y}_m}{\partial y_r} \frac{\partial \tilde{y}_n}{\partial y_s} \frac{\partial \tilde{y}_p}{\partial y_t} w_{rst}(y_1, y_2, y_3). \tag{A.6.8}$$

It should be noted that the components of a tensor in the new variables $\tilde{y}_r$, are linear combinations of the components in the old variables y_m. Consequently, if all the components of a tensor vanish in any coordinate frame, the components are zero in all coordinate frames. Or, if all the components are zero at some particular point in space, then all the components are zero at this point for all coordinate systems. These are extremely important properties of tensors.

One other tensor operation needs to be discussed, an operation called *contraction*. A second-order tensor has been defined as one which obeys the transformation law, Eq. (A.6.7). If, in the second-order tensor w_{mn}, the components w_{11}, w_{22}, and w_{33} are summed or added, the resulting quantity is represented by

$$w_{mm} = w_{11} + w_{22} + w_{33} \tag{A.6.9}$$

in accordance with the indicial and summation convention. Note carefully that w_{mm} does not contain any free indices, that is, w_{mm} is obtained by setting $m = n$ in w_{mn}; hence m becomes a dummy index. This process whereby two indices are set equal to each other in order to obtain a new quantity is called "contraction." In general, contraction reduces the order of the original tensor. Thus, a third-order tensor if contracted becomes a first-order tensor. Further, if the contraction process is carried to the point where there are no longer any free indices, then the resulting tensor is one of zero order, i.e., the resulting quantity is an invariant. For example, w_{mm} is an invariant, as can be demonstrated by setting $m = n$ in Eq. (A.6.7). Then

$$\widetilde{w}_{mm} = \frac{\partial \tilde{y}_m}{\partial y_r} \frac{\partial \tilde{y}_m}{\partial y_s} w_{rs} = w_{rs} l_{\widetilde{m}r} l_{\widetilde{m}s}, \tag{A.6.10}$$

and with the introduction of Eq. (A.5.10) this becomes

$$\widetilde{w}_{mm} = \delta_{rs} w_{rs} = w_{rr}. \tag{A.6.11}$$

Thus it has been demonstrated that $\widetilde{w}_{mm}$ in the coordinate system $\tilde{y}_1, \tilde{y}_2, \tilde{y}_3$ has the same value as w_{rr} in the coordinate system y_1, y_2, y_3; hence $\widetilde{w}_{mm}$ is an invariant.

A.7 VECTOR ANALYSIS [3], [4], [5]

In the previous section we defined a first-order tensor or vector in terms of its behavior under a transformation of coordinates. The branch of mathematics which deals with first-order tensors is called *vector analysis*. Many of the physical quantities encountered in mechanics, such as the displacement of a point, the velocity of a particle or a mechanical force, are vectors and require both magnitude and direction for complete specification. Thus a vector is any physical quantity which can be represented by a directed line segment. The following seven sections will define the essential operations of vector analysis. Vectors will be denoted by bold face letters, for example, **a**.

A.8 VECTOR ADDITION

Two vectors **a** and **b** are defined as equal if they are parallel, have the same sense, and are of equal length (see Fig. A.2). Equality is shown by writing

$$\mathbf{a} = \mathbf{b}. \tag{A.8.1}$$

Note that two vectors which are equal do not have to occupy the same points in space. The length or absolute magnitude is denoted by $|\mathbf{a}|$, and is always taken as positive.

Let **a** and **b** represent two vectors (see Fig. A.3). Then vector addition is defined as the geometrical sum of **a** and **b** and is written as

$$\mathbf{c} = \mathbf{a} + \mathbf{b}. \tag{A.8.2}$$

From Fig. A.3 we see that (1) the addition of vectors produces another vector, and (2) the order of addition is unimportant, i.e., vector addition is commutative:

$$\mathbf{a} + \mathbf{b} = \mathbf{b} + \mathbf{a}. \tag{A.8.3}$$

The process of addition can be extended to three or more vectors. Vector addition is also associative:

$$(\mathbf{a} + \mathbf{b}) + \mathbf{c} = \mathbf{a} + (\mathbf{b} + \mathbf{c}). \tag{A.8.4}$$

To complete the definition of vector addition, it is necessary to define a null vector which is represented by 0. If

$$\mathbf{a} + \mathbf{b} = 0, \tag{A.8.5}$$

then

$$\mathbf{b} = -\mathbf{a}, \tag{A.8.6}$$

which means that **b** is a vector parallel to **a** and of the same length but of opposite sense. The process of vector subtraction, for example $\mathbf{a} - \mathbf{b}$, means the addition of $(-\mathbf{b})$ to **a** (see Fig. A.4).

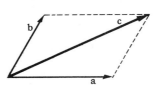

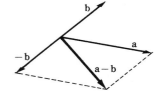

FIG. A.2 Equality of vectors.   FIG. A.3 Addition of two vectors.   FIG. A.4 Subtraction of two vectors.

A.9 SCALAR MULTIPLICATION

The multiplication of a vector **a** by a real number m is defined to result in a vector **c** parallel to **a**, and m times as long, with a sense the same as **a** or opposite, according to whether m is positive or negative. Scalar multiplication is symbolized by

$$\mathbf{c} = m\mathbf{a}, \tag{A.9.1}$$

and obeys the commutative, associative, and distributive laws:

$$m\mathbf{a} = \mathbf{a}m, \quad (mn)\mathbf{a} = m(n\mathbf{a}), \quad (m + n)\mathbf{a} = m\mathbf{a} + n\mathbf{a}. \tag{A.9.2}$$

A.10 UNIT VECTORS

A unit vector is one which has unit length or magnitude. The most important set of unit vectors comprises those which are oriented along the axes of a rectangular Cartesian coordinate system. They will be denoted by i_1, i_2, and i_3, and are referred to as the base vectors associated with the rectangular Cartesian coordinate system.

A.11 CARTESIAN REPRESENTATION OF A VECTOR

The components of a vector **a** are any vectors whose sum is equal to **a**. The most convenient components are those parallel to a rectangular Cartesian coordinate system. Let **a** be a vector with its tail at point O, the origin of y_1, y_2, y_3 (see Fig. A.5). Then **a** is completely determined by specifying its components a_1, a_2, a_3 on the axes y_1, y_2, y_3. This is symbolized by writing

$$\mathbf{a} = a_n \mathbf{i}_n = a_1 \mathbf{i}_1 + a_2 \mathbf{i}_2 + a_3 \mathbf{i}_3. \tag{A.11.1}$$

In this figure vector **a** is represented by the directed line segment $\overrightarrow{OP}$ and the three rectangular components of **a** are represented by $\overrightarrow{OP_1}$, $\overrightarrow{OP_2}$, and $\overrightarrow{OP_3}$, respectively. Let $\widehat{ay_1}$, $\widehat{ay_2}$, $\widehat{ay_3}$ represent the angles between OP and the axes y_1, y_2, y_3. Then the three components are obtained by multiplying the length of **a** by the corresponding direction cosines:

$$OP_1 = a_1 = |\mathbf{a}| \cos \widehat{ay_1}, \tag{A.11.2}$$

$$OP_2 = a_2 = |\mathbf{a}| \cos \widehat{ay_2}, \tag{A.11.3}$$

$$OP_3 = a_3 = |\mathbf{a}| \cos \widehat{ay_3}. \tag{A.11.4}$$

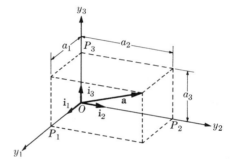

Fig. A.5 Rectangular Cartesian components of a vector.

The application of the Pythagorean Theorem gives the length of **a** in terms of its components; thus

$$(OP)^2 = (OP_1)^2 + (OP_2)^2 + (OP_3)^2, \quad \text{or} \quad |\mathbf{a}|^2 = (a_1)^2 + (a_2)^2 + (a_3)^2. \tag{A.11.5}$$

By substituting Eqs. (A.11.2), (A.11.3), and (A.11.4) into Eq. (A.11.5) we obtain

$$(\cos \widehat{ay_1})^2 + (\cos \widehat{ay_2})^2 + (\cos \widehat{ay_3})^2 = 1, \tag{A.11.6}$$

a result previously derived (cf. Eq. A.5.11).

The notion of the rectangular components of a vector can be generalized to represent the rectangular components of any point Q. Let y_1, y_2, y_3 be the rectangular Cartesian coordinates of point Q, and let **r** denote the line segment called the position vector, which connects the origin of the axis system, O, to Q. Then

$$\mathbf{r} = y_n \mathbf{i}_n = y_1 \mathbf{i}_1 + y_2 \mathbf{i}_2 + y_3 \mathbf{i}_3. \tag{A.11.7}$$

In this manner a position vector can be attached to all points in space.

A.12 THE SCALAR PRODUCT

The processes of vector addition and multiplication by a scalar have been defined to yield new quantities. Another vector operation, which involves just two vectors, is called the *scalar product*. The scalar product of two vectors **a** and **b** is defined as the scalar quantity formed by the product of the lengths of these vectors and the cosine of the angle between them:

$$\mathbf{a} \cdot \mathbf{b} = |\mathbf{a}||\mathbf{b}| \cos \widehat{ab}, \qquad (A.12.1a)$$

where $\widehat{ab}$ represents the angle between **a** and **b**. The scalar product may be interpreted geometrically as

$$\mathbf{a} \cdot \mathbf{b} = |\mathbf{a}| \text{ (component of } \mathbf{b} \text{ on } \mathbf{a}),$$

or

$$\mathbf{a} \cdot \mathbf{b} = |\mathbf{b}| \text{ (component of } \mathbf{a} \text{ on } \mathbf{b}), \qquad (A.12.1b)$$

as illustrated in Fig. A.6. The scalar product is, therefore, a number which is positive, zero, or negative depending on whether the angle between the two vectors is acute, right, or obtuse, providing neither vector is a null vector. Hence, if neither vector is a null vector, then the equation

$$\mathbf{a} \cdot \mathbf{b} = 0 \qquad (A.12.2)$$

means that vector **a** is perpendicular to vector **b**.

It can be seen from Eqs. (A.12.1b) that scalar multiplication obeys the commutative law,

$$\mathbf{a} \cdot \mathbf{b} = \mathbf{b} \cdot \mathbf{a}. \qquad (A.12.3)$$

In addition, the geometrical interpretation shown in Fig. A.7 indicates that

$$|\mathbf{a}| \text{ (component of } (\mathbf{b} + \mathbf{c}) \text{ on } \mathbf{a}) = |\mathbf{a}| \text{ (component of } \mathbf{b} \text{ on } \mathbf{a})$$
$$+ |\mathbf{a}| \text{ (component of } \mathbf{c} \text{ on } \mathbf{a}), \qquad (A.12.4a)$$

or

$$\mathbf{a} \cdot (\mathbf{b} + \mathbf{c}) = \mathbf{a} \cdot \mathbf{b} + \mathbf{a} \cdot \mathbf{c}, \qquad (A.12.4b)$$

which is the distributive law of scalar multiplication.

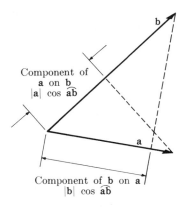

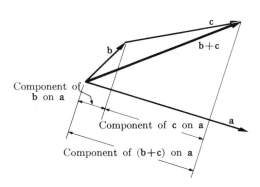

FIG. A.6 Geometrical representation of scalar product.

FIG. A.7 Geometrical representation of distributive law.

Let two vectors **a** and **b** be written in component form:

$$\mathbf{a} = a_n \mathbf{i}_n, \tag{A.12.5}$$

$$\mathbf{b} = b_n \mathbf{i}_n. \tag{A.12.6}$$

Then by virtue of the distributive law of multiplication the scalar product of two vectors is equal to the sum of the products of their corresponding rectangular components:

$$\mathbf{a} \cdot \mathbf{b} = (a_n \mathbf{i}_n) \cdot (b_m \mathbf{i}_m) = a_n b_m \delta_{nm} = a_n b_n. \tag{A.12.7}$$

In obtaining the result shown by Eq. (A.12.7), we have utilized the special property of scalar products of the base vectors, $\mathbf{i}_1, \mathbf{i}_2, \mathbf{i}_3$:

$$\mathbf{i}_n \cdot \mathbf{i}_m = \delta_{nm}. \tag{A.12.8}$$

Also, if $\mathbf{i}_n$ and $\mathbf{j}_r$ are unit vectors along the y_n- and $\tilde{y}_r$-axes, respectively, we have from Eq. (A.12.1)

$$\mathbf{i}_n \cdot \mathbf{j}_r = \mathbf{j}_r \cdot \mathbf{i}_n = l_{\bar{r}n}, \tag{A.12.9}$$

where $l_{\bar{r}n} = \cos(\widehat{\tilde{y}_r y_n})$ is the cosine of the angle between the y_n and $\tilde{y}_r$ axes.

The scalar product is useful in determining the angle between two vectors. Thus

$$\cos \widehat{\mathbf{ab}} = \frac{a_n b_n}{|\mathbf{a}||\mathbf{b}|}. \tag{A.12.10}$$

It is also useful in determining the length of a vector because the scalar product of a vector by itself yields

$$|\mathbf{a}|^2 = \mathbf{a} \cdot \mathbf{a} = a_n a_m \delta_{nm} = a_n a_n. \tag{A.12.11}$$

A.13 THE VECTOR PRODUCT

Scalar multiplication has been shown to yield a scalar. The *vector product*, which also involves only two vectors, of **a** and **b** is symbolized as **a** × **b** (read as **a** cross **b**), and is defined to yield a new vector:

$$\mathbf{a} \times \mathbf{b} = |\mathbf{a}||\mathbf{b}|\mathbf{n} \sin \widehat{\mathbf{ab}}, \tag{A.13.1}$$

where **n** is a unit vector perpendicular to both **a** and **b**, with a sense such that **a**, **b**, and **n** form a right-handed system. The meaning of this definition can be seen from Fig. A.8.

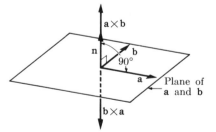

Fig. A.8 Geometrical representation of vector product.

If neither **a** nor **b** is a null vector, then the equation

$$\mathbf{a} \times \mathbf{b} = 0 \tag{A.13.2}$$

means that vector **a** is parallel to vector **b**.

The requirement that the new vector **a** × **b** form a right-handed system with **a** and **b** in that particular order means that vector multiplication is not commutative, but instead:

$$\mathbf{a} \times \mathbf{b} = -\mathbf{b} \times \mathbf{a}. \tag{A.13.3}$$

The process of vector multiplication applied to the unit vectors i_1, i_2, i_3 yields

$$i_1 \times i_2 = i_3, \tag{A.13.4}$$

$$i_2 \times i_3 = i_1, \tag{A.13.5}$$

$$i_3 \times i_1 = i_2, \tag{A.13.6}$$

$$i_1 \times i_1 = i_2 \times i_2 = i_3 \times i_3 = 0, \tag{A.13.7}$$

or in a more compact form

$$i_m \times i_n = e_{mnp} i_p. \tag{A.13.8}$$

The vector product can also be cast into component form as follows:

$$\mathbf{a} \times \mathbf{b} = (a_m i_m) \times (b_n i_n),$$
$$\mathbf{a} \times \mathbf{b} = (a_2 b_3 - a_3 b_2) i_1 + (a_3 b_1 - a_1 b_3) i_2 + (a_1 b_2 - a_2 b_1) i_3. \tag{A.13.9}$$

A useful device to help in the calculation of the components of $\mathbf{a} \times \mathbf{b}$ is to write

$$\mathbf{a} \times \mathbf{b} = \begin{vmatrix} i_1 & i_2 & i_3 \\ a_1 & a_2 & a_3 \\ b_1 & b_2 & b_3 \end{vmatrix}. \tag{A.13.10}$$

It is seen from a comparison of Eqs. (A.13.9) and (A.13.10) that the components of $\mathbf{a} \times \mathbf{b}$ are the cofactors of i_1, i_2, i_3 in the expansion of the determinant. In summation notation the vector product $\mathbf{a} \times \mathbf{b}$ is written as

$$\mathbf{a} \times \mathbf{b} = e_{mnr} a_m b_n i_r. \tag{A.13.11}$$

It is now a simple task to prove that vector multiplication is distributive:

$$\mathbf{a} \times (\mathbf{b} + \mathbf{c}) = \mathbf{a} \times \mathbf{b} + \mathbf{a} \times \mathbf{c}, \tag{A.13.12}$$

or, in component form,

$$\mathbf{a} \times (\mathbf{b} + \mathbf{c}) = a_m i_m \times (b_n + c_n) i_n. \tag{A.13.13}$$

By making use of Eq. (A.13.9) we find that

$$\mathbf{a} \times (\mathbf{b} + \mathbf{c}) = a_m (b_n + c_n) i_m \times i_n$$
$$= a_m b_n e_{mnp} i_p + a_m c_n e_{mnp} i_p. \tag{A.13.14}$$

The right-hand side according to Eq. (A.13.11) is seen to be

$$\mathbf{a} \times \mathbf{b} + \mathbf{a} \times \mathbf{c}.$$

Thus the distributive law is proved.

Some additional relations concerning vector multiplication can be included here. If m and n are two scalar quantities, then

$$(m\mathbf{a}) \times (n\mathbf{b}) = (mn)\mathbf{a} \times \mathbf{b}. \tag{A.13.15}$$

By repeated application of the distributive law, the vector product of two vector sums can be expanded as in ordinary algebra, provided the order of the factors is not altered. Thus, for example,

$$(\mathbf{a} + \mathbf{b}) \times (\mathbf{c} + \mathbf{d}) = \mathbf{a} \times \mathbf{c} + \mathbf{a} \times \mathbf{d} + \mathbf{b} \times \mathbf{c} + \mathbf{b} \times \mathbf{d}. \qquad (A.13.16)$$

If vector **c** is not a null vector, then the equation

$$\mathbf{a} \times \mathbf{c} = \mathbf{b} \times \mathbf{c}, \qquad (A.13.17)$$

which can be rearranged to read

$$(\mathbf{a} - \mathbf{b}) \times \mathbf{c} = 0, \qquad (A.13.18)$$

means that either **a** — **b** is a null vector, or **a** — **b** and **c** are parallel. It is seen that the process of cancellation cannot be applied to vector **c** in Eq. (A.13.7).

A.14   THE TRIPLE SCALAR PRODUCT

The scalar product of the vector **a** × **b** and a third vector **c** is symbolized as (**a** × **b**) · **c** or sometimes as [**abc**]. It is easily demonstrated (see Fig. A.9) that the triple scalar product is numerically equal to the volume of a parallelepiped having the vectors **a**, **b**, **c** as concurrent edges. Its sign is positive or negative depending on whether the three vectors form a right-handed or left-handed set. If none of the three vectors is a null vector, then the equation

$$(\mathbf{a} \times \mathbf{b}) \cdot \mathbf{c} = 0 \qquad (A.14.1)$$

means that the three vectors are coplanar.

In terms of the rectangular components, the triple scalar product can be obtained from the expansion of the determinant formed by the components:

$$(\mathbf{a} \times \mathbf{b}) \cdot \mathbf{c} = \begin{vmatrix} a_1 & a_2 & a_3 \\ b_1 & b_2 & b_3 \\ c_1 & c_2 & c_3 \end{vmatrix}. \qquad (A.14.2)$$

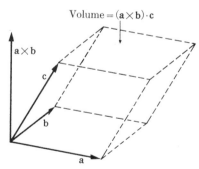

FIG. A.9 Geometrical representation of triple scalar product.

The product of two triple scalar products can be shown (albeit not easily) to be given by

$$[\mathbf{uvw}][\mathbf{abc}] = \begin{vmatrix} \mathbf{u} \cdot \mathbf{a} & \mathbf{u} \cdot \mathbf{b} & \mathbf{u} \cdot \mathbf{c} \\ \mathbf{v} \cdot \mathbf{a} & \mathbf{v} \cdot \mathbf{b} & \mathbf{v} \cdot \mathbf{c} \\ \mathbf{w} \cdot \mathbf{a} & \mathbf{w} \cdot \mathbf{b} & \mathbf{w} \cdot \mathbf{c} \end{vmatrix}. \qquad (A.14.3)$$

(See Ref. A.3, Section 24.)

A.15 EXTREME-VALUE PROBLEMS IN DIFFERENTIAL CALCULUS [4]

In developing some of the basic equations in solid mechanics, we become involved in the subsidiary problem of having to maximize or minimize a function of many variables. For example, we might begin with a set of general expressions for all the components of stress acting on an element of a body. Our task might be to find the maximum value of a particular component σ_{nn}, and the direction cosines of the plane on which $\sigma_{mn(\max)}$ acts.

We apply elementary partial differential calculus to the function of many variables in a manner analogous to the procedure used to find the local maxima and minima of a function of one variable. If $f(x_1, x_2, \ldots, x_n)$ is a function of n independent variables, we say that an *extreme value* of f is defined by

$$\frac{\partial f}{\partial x_i} = 0, \qquad i = 1, 2, \ldots, n. \tag{A.15.1}$$

That is, an extreme value of f occurs at any point where the rates of change of f with respect to each of the independent variables simultaneously vanish, as expressed by the set of n simultaneous equations (A.15.1).

As an example, consider the function of two independent variables:

$$f(x, y) = 4 - x^2 - y^2. \tag{A.15.2}$$

Its extreme values are found in a straightforward manner by solving the conditions (A.15.1) for the values $(x, y)_{\text{ext}}$, and introducing these back into f. Thus

$$\partial f/\partial x = -2x = 0, \qquad \partial f/\partial y = -2y = 0,$$
$$(x, y)_{\text{ext}} = (0, 0), \qquad f_{\text{ext}}(0, 0) = 4. \tag{A.15.3}$$

Since f is a function of only two independent variables, we can plot f on an axis perpendicular to the xy-plane, as illustrated by Fig. A.10. We can easily visualize the general function of two variables as a surface of "hills," "valleys," and "saddle points,"

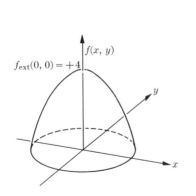

FIG. A.10 A surface representing a function $f(x, y)$.

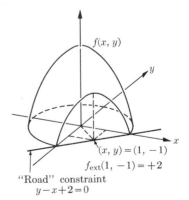

FIG. A.11 Intersection of $f(x, y)$ and a plane perpendicular to the xy-plane.

where the crest of a hill, the bottom of a valley, and the seat of a saddle point correspond to max-max, min-min, and max-min or min-max extreme values of the function. The function

$$f(x, y) = 4 - x^2 - y^2$$

is an inverted parabolic cone, and it may be thought of as a single "hill" with a crest 4 units above the xy-plane and directly above $(0, 0)$. The "hill-and-valley" concept will be useful in the ensuing discussion of more complicated extreme-value problems.

A.16   SOME EXAMPLES OF EXTREME-VALUE PROBLEMS WITH SIMPLE CONSTRAINTS

Suppose we add a condition to the simple extreme-value problem. We desire the locations $(x_1, \ldots, x_n)$ of the extreme values of $f(x_1, x_2, \ldots, x_n)$, *subject to the constraint that these locations also satisfy the equation*

$$g(x_1, x_2, \ldots, x_n) = 0. \tag{A.16.1}$$

We can think of the constraint expressed by Eq. (A.16.1) as some sort of hypersurface in the n-dimensional space of the problem. Satisfying the constraint is then equivalent to making a plot of the values of $f(x_1, \ldots, x_n)$ at locations (x_i) of the intersection of the surfaces defined by f and g. The extreme values of $f(x_i)$ on this intersection may be altogether different from the values we found when we were allowed to roam freely over all points on f.

As an example, let it be required to find the extreme values of

$$f(x, y) = 4 - x^2 - y^2$$

subject to the constraint   (A.16.2)

$$g(x, y) = y - x + 2 = 0.$$

Returning to the "hill-and-valley" analogy, we again visualize $f(x, y)$ as the parabolic "hill" shown in Fig. A.10. The constraint may be solved to yield

$$y = x - 2, \tag{A.16.3}$$

which is the equation of a straight line in the xy-plane, passing through the points $(0, -2)$ and $(2, 0)$. If we visualize the constraint as a "road" which follows the xy-plane and the part of $f(x, y)$ above the xy-plane, we can immediately make the general conclusion that we shall find a new extreme value for $f(x, y)$ in this case.

Figure A.11 shows the "hill" with the "road" constraint superimposed. For clarity, we have created a vertical "cliff" on the hill between the road and the observer. It is evident from the figure that the constraint is a surface, a plane perpendicular to the xy-plane. The constraint plane does not pass through the extreme values of the surface,

$$f_{\text{ext}}(0, 0) = 4,$$

found in the previous section.

To obtain the constrained extreme value of $f(x, y)$, we must satisfy the equation $g(x, y) = 0$ before applying the extremalizing conditions $\partial f / \partial x = 0$ and $\partial f / \partial y = 0$.

Equation (A.16.3) may be introduced directly into the first of Eqs. (A.16.2) to yield

$$f(x, y)|_g = 4 - x^2 - (x - 2)^2, \qquad (A.16.4)$$

where the symbol $|_g$ after $f(x, y)$ indicates that the expression for f which follows has been constrained to $g(x, y) = 0$. We observe that $f|_g$ is a function of the single independent variable x; thus the extremalizing conditions reduce to

$$\frac{df}{dx} = 0 = -2x - 2(x - 2),$$

which yields

$$x_{\text{ext}} = +1, \qquad y_{\text{ext}} = x - 2 = -1, \qquad (A.16.5)$$

as the location of the constrained extreme value of $f(x, y)$. The value itself is obtained by introducing Eqs. (A.16.5) back into $f(x, y)$:

$$f_{\text{ext}}(1, -1) = 4 - 1 - 1 = 2. \qquad (A.16.6)$$

A.17   THE LAGRANGE MULTIPLIER METHOD [4]

Let us consider again the parabolic "hill" which we examined in the previous section. This time we are asked to find the extreme values of

$$f(x, y) = 4 - x^2 - y^2$$

subject to the constraint

$$g(x, y) = xy + 1 = 0. \qquad (A.17.1)$$

It is possible to solve Eqs. (A.17.1) for the constrained extreme value of f by the procedure developed in Section A.16. However, let us instead add λ times $g(x, y)$ to $f(x, y)$, where λ has the properties of an independent variable:

$$F(x, y) = f(x, y) + \lambda g(x, y), \qquad (A.17.2)$$

or

$$F(x, y) = 4 - x^2 - y^2 + \lambda(xy + 1). \qquad (A.17.3)$$

We now apply the extremalizing formulas of elementary partial differential calculus to Eq. (A.17.3), considering x, y, and λ as independent variables:

$$\frac{\partial F}{\partial x} = 0, \qquad \frac{\partial F}{\partial y} = 0, \qquad \frac{\partial F}{\partial \lambda} = 0. \qquad (A.17.4)$$

The results of the calculations are:

$$\text{(a)} \quad \frac{\partial F}{\partial x} = -2x + \lambda y = 0,$$

$$\text{(b)} \quad \frac{\partial F}{\partial y} = -2y + \lambda x = 0, \qquad (A.17.5)$$

$$\text{(c)} \quad \frac{\partial F}{\partial \lambda} = xy + 1 = 0.$$

We can now solve Eqs. (A.17.5) for x, y, and λ:

$$\text{from (c),}\quad x = -\frac{1}{y};$$

$$\text{from (a),}\quad \frac{2}{y} + \lambda y = 0 \Rightarrow \lambda = -\frac{2}{y^2};$$

$$\text{from (b),}\quad -2y + \frac{2}{y^2}\left(\frac{1}{y}\right) = 0 \Rightarrow y^4 = 1.$$

The real solutions thus become

$$x = -1, +1, \quad y = +1, -1, \quad \lambda = -2, +2. \tag{A.17.6}$$

Introduction of the values $(x, y) = (-1, +1)$ or $(+1, -1)$ into $f(x, y)$ yields the solution

$$f_{\text{ext}}(x, y)\big|_g = +2 \quad \text{at} \quad \begin{cases} (+1, -1), \\ (-1, +1). \end{cases} \tag{A.17.7}$$

Figure A.12 illustrates the intersection of the constraint surface (a hyperbolic cylinder) with the portion of $f(x, y)$ which lies above the xy-plane. Calculation of $f_{\text{ext}}\big|_g$ by the method of Section A.16 will yield the answer in Eq. (A.17.7).

In the above method of solution, we began by creating an extra (and apparently useless) variable, λ, called a Lagrange multiplier. The normal procedure of first eliminating as many independent variables as possible (Section A.16) is more efficient when the problem to be solved is simple. However, the advantages of the systematic Lagrange multiplier method will become apparent in problems having many independent variables and multiple nonlinear constraints.

We shall first consider a quite general extreme-value problem. Let

$$f = f(x_1, x_2, \ldots, x_n)$$

be a given function of the n independent variables $\{x_i\}$, and let

$$\begin{aligned} g_1 &= g_1(x_1, x_2, \ldots, x_n) = 0, \\ g_2 &= \qquad \cdots \qquad\quad = 0, \\ &\;\vdots \qquad\qquad\qquad\;\; \vdots \\ g_m &= g_m(x_1, x_2, \ldots, x_n) = 0 \end{aligned} \tag{A.17.8}$$

be given equations of constraint. We are required to find the extreme values of f subject simultaneously to all m of the constraints, $g_j = 0$. We observe that the prob-

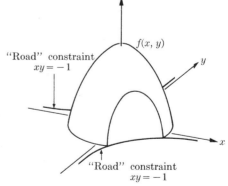

Fig. A.12   Intersection of $f(x, y)$ and a hyperbolic cylinder.

lem is trivial if $m = n$, since the constraint equations will then confine the examination to one point in the n-space. If $m > n$, the problem is overconstrained, i.e., there may be no points of f which will satisfy all the constraints. Thus we can restrict our consideration to problems in which $m < n$ without losing any generality.

In the previous section, we saw that extreme-value problems of the type outlined above can be solved, in theory, by straightforward elimination by substitution of each of the constraint equations before the extremalizing conditions are applied. In practice, the number and complexity of the constraint equations may be such that it is not possible to make the substitutions in a systematic manner. We then turn to the Lagrange multiplier method, which provides the systematic approach for the general extreme-value problem.

The steps in the Lagrange method (outlined here without proof) are as follows (see Ref. A.4, Sections 3.10 and 3.11):

(a) Multiply each constraint by an undetermined parameter, called a Lagrange multiplier.
(b) Form the sum of the products of constraints and multipliers by summing over all the constraints.
(c) Add the above sum to the function f, which is to be extremalized, and call this sum F.
(d) Apply the extremalizing conditions by setting equal to zero the partial derivatives of F with respect to each independent variable *and each Lagrange multiplier.*
(e) Solve the resulting equations for the locations and values of the extreme values of f.

Example. Let it be required to find the extreme values of

$$f(x_1, x_2, x_3, x_4) = (x_1)^2 + 2(x_2)^2 + 3(x_3)^2 + 4(x_4)^2 \qquad \text{(A.17.9)}$$

subject to the constraints

$$\begin{aligned} g_1 &= x_1 + x_2 - 3 = 0, \\ g_2 &= x_2 + x_3 - 5 = 0, \\ g_3 &= x_3 + x_4 - 7 = 0. \end{aligned} \qquad \text{(A.17.10)}$$

Solution: We must define three Lagrange multipliers λ_1, λ_2, and λ_3 to form

$$\lambda_1 g_1 + \lambda_2 g_2 + \lambda_3 g_3.$$

The function $F(x_1, x_2, x_3, x_4)$ then becomes

$$F = f + \sum_{i=1}^{3} \lambda_i g_i,$$

or

$$\begin{aligned} F = (x_1)^2 + 2(x_2)^2 + 3(x_3)^2 + 4(x_4)^2 + \lambda_1(x_1 + x_2 - 3) \\ + \lambda_2(x_2 + x_3 - 5) + \lambda_3(x_3 + x_4 - 7). \end{aligned} \qquad \text{(A.17.11)}$$

Applying the extremalizing conditions, we find four equations by setting $\partial F/\partial x_i = 0$:

$$\frac{\partial F}{\partial x_1} = 2x_1 + \lambda_1 = 0, \qquad\qquad \frac{\partial F}{\partial x_2} = 4x_2 + \lambda_1 + \lambda_2 = 0,$$

$$\text{(A.17.12)}$$

$$\frac{\partial F}{\partial x_3} = 6x_3 + \lambda_2 + \lambda_3 = 0, \qquad\qquad \frac{\partial F}{\partial x_4} = 8x_4 + \lambda_3 = 0,$$

and three additional equations by setting $\partial F/\partial \lambda_i = 0$:

$$\frac{\partial F}{\partial \lambda_1} = x_1 + x_2 - 3 = 0,$$

$$\frac{\partial F}{\partial \lambda_2} = x_2 + x_3 - 5 = 0, \tag{A.17.13}$$

$$\frac{\partial F}{\partial \lambda_3} = x_3 + x_4 - 7 = 0.$$

Comparison of Eqs. (A.17.13) with Eqs. (A.17.10) shows that by taking $\partial F/\partial \lambda_i = 0$, we have obtained the constraint equations again.

The seven Eqs. (A.17.12) and (A.17.13) may be solved systematically by first eliminating the λ_i from Eqs. (A.17.12). We find that

$$\lambda_1 = -2x_1, \qquad \lambda_2 = 2x_1 - 4x_2,$$
$$\lambda_3 = -6x_3 - 2x_1 + 4x_2, \qquad \lambda_3 = -8x_4. \tag{A.17.14}$$

The third and fourth of Eqs. (A.17.14) can now be combined with Eqs. (A.17.13) to get a set of four simultaneous equations in the x_i:

$$\begin{aligned}
x_1 - 2x_2 + 3x_3 - 4x_4 &= 0, \\
x_1 + x_2 \qquad\qquad &= 3, \\
x_2 + x_3 \qquad &= 5, \\
x_3 + x_4 &= 7.
\end{aligned} \tag{A.17.15}$$

Solution of Eqs. (A.17.15) by the methods of ordinary algebra yields the location of a single extreme value of f:

$$x_1 = 2, \qquad x_2 = 1, \qquad x_3 = 4, \qquad x_4 = 3, \tag{A.17.16}$$

and introduction of Eqs. (A.17.16) into Eq. (A.17.9) for f yields

$$f_{\text{ext}}(2, 1, 4, 3) = +90. \tag{A.17.17}$$

A.18   ELEMENTS OF THE CALCULUS OF VARIATIONS [6]

We are all familiar with the problem of finding a maximum or a minimum of a simple function $y(x)$. Graphically, such a function is shown in Fig. A.13. For a relative maximum or minimum to occur, the slope dy/dx must be zero.

Let us now consider the problem of determining a function $y(x)$ which makes the integral

$$I = \int_a^b F(x, y, y') \, dx \tag{A.18.1}$$

an extremum, i.e., stationary, and which satisfies the prescribed end conditions,

$$y(a) = y_a \qquad \text{and} \qquad y(b) = y_b. \tag{A.18.2}$$

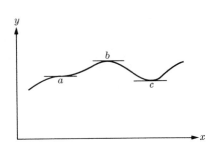

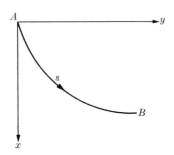

FIG. A.13   Maximum and minimum of a   FIG. A.14   Coordinate system for the
function $y(x)$.   brachistochrone problem.

The integrand $F(x, y, y')$ is a function of x, y, and the first derivative dy/dx, which for
convenience will be denoted by y'.

As an example of F, let us consider the famous problem of the brachistochrone which
Johann Bernoulli formulated in 1696. Imagine a particle constrained to slide without
friction along a certain curve joining a point A to a lower point B. Along which curve
will the time required for the descent be the least if this particle is allowed to fall under
the influence of gravity?

Let s be the length of the curve measured from A (see Fig. A.14), v the velocity,
and t the time. We wish to determine the curve which will make the total time T
of descent from A to B a minimum. The arc length is calculable from

$$ds = v\, dt, \tag{A.18.3}$$

$$ds = \sqrt{dx^2 + dy^2} = \sqrt{1 + y'^2}\, dx. \tag{A.18.4}$$

A freely falling body gains velocity according to

$$v = 2gx, \tag{A.18.5}$$

where g is the acceleration due to gravity. Then from the preceding three equations,

$$dt = \frac{ds}{v} = \frac{\sqrt{1 + y'^2}}{\sqrt{2gx}}\, dx. \tag{A.18.6}$$

The total time is the integral of Eq. (A.18.6),

$$I = \frac{1}{\sqrt{2g}} \int_{x=0}^{x=x_B} \frac{\sqrt{1 + y'^2}}{\sqrt{x}}\, dx. \tag{A.18.7}$$

By comparing Eq. (A.18.7) with Eq. (A.18.1), we find that the functional of interest
in the brachistochrone problem is

$$F = \frac{\sqrt{1 + y'^2}}{\sqrt{x}}. \tag{A.18.8}$$

We want to develop the ideas which will enable us to solve this particular problem as
well as many others.

Let us suppose that $y(x)$ is the actual minimizing function. This is schematically illustrated in Fig. A.15. The method of attack will be to investigate what happens to the integral I when neighboring functions are used in F. These functions are constructed by adding to $y(x)$ a function $\epsilon\eta(x)$, where ϵ is a constant which can take on different quantitative values, and $\eta(x)$ is an arbitrary function of x which vanishes at $x = a$ and $x = b$, that is,

$$\eta(a) = 0, \quad \text{and} \quad \eta(b) = 0. \quad (A.18.9)$$

Thus $y(x) + \epsilon\eta(x)$ will satisfy the end conditions. A neighboring function is shown as a dashed curve in Fig. A.15. The integral I which is obtained by replacing y by $y + \epsilon\eta$ is then a function of ϵ once η is chosen, since y is the actual minimizing function. We take note of this dependence on ϵ by writing

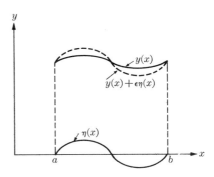

$$I(\epsilon) = \int_a^b F(x, y + \epsilon\eta, y' + \epsilon\eta') \, dx. \quad (A.18.10)$$

Fig. A.15 Function $y(x)$ and a neighboring function.

This integral takes on its minimum value when ϵ is zero, but this is possible only if

$$\frac{dI(\epsilon)}{d\epsilon} = 0 \quad \text{when} \quad \epsilon = 0. \quad (A.18.11)$$

Our manipulations will become a bit easier if we set

$$F_e = F(x, y + \epsilon\eta, y' + \epsilon\eta') \quad (A.18.12)$$

and

$$Y(x) = y(x) + \epsilon\eta(x). \quad (A.18.13)$$

Then the differential of Y with respect to x is

$$Y' = y'(x) + \epsilon\eta'(x). \quad (A.18.14)$$

Next we differentiate F_e with respect to ϵ to obtain

$$\frac{dF_e}{d\epsilon} = \frac{\partial F_e}{\partial Y}\frac{\partial Y}{\partial \epsilon} + \frac{\partial F}{\partial Y'}\frac{\partial Y'}{\partial \epsilon}. \quad (A.18.15)$$

Note that F_e is treated as a composite function of Y and Y'. From Eqs. (A.18.13) and (A.18.14) we find that

$$\frac{\partial Y}{\partial \epsilon} = \eta(x), \quad (A.18.16)$$

$$\frac{\partial Y'}{\partial \epsilon} = \eta'(x), \quad (A.18.17)$$

so that Eq. (A.18.17) takes on the form

$$\frac{dF_e}{d\epsilon} = \frac{\partial F_e}{\partial Y}\,\eta(x) + \frac{\partial F_e}{\partial Y'}\,\eta'(x). \tag{A.18.18}$$

It follows that

$$\frac{dI}{d\epsilon} = \int_a^b \left\{ \frac{\partial F_e}{\partial Y}\,\eta(x) + \frac{\partial F_e}{\partial Y'}\,\eta'(x) \right\} dx = 0. \tag{A.18.19}$$

The second term can be transformed through integration by parts:

$$\int_a^b \frac{\partial F_e}{\partial Y'}\,\eta'(x)\,dx = \frac{\partial F_e}{\partial Y'}\,\eta(x) \Big]_a^b - \int_a^b \frac{d}{dx}\left(\frac{\partial F_e}{\partial Y'} \right) \eta(x)\,dx. \tag{A.18.20}$$

Since $\eta(a) = 0$ and $\eta(b) = 0$, the integrated term vanished. Thus

$$\int_a^b \left\{ \frac{\partial F_e}{\partial Y}\,\eta(x) - \frac{d}{dx}\left(\frac{\partial F_e}{\partial Y'} \right) \eta(x) \right\} dx = 0. \tag{A.18.21}$$

This derivative must vanish when ϵ approaches 0. But as ϵ approaches 0, F_e approaches F, Y approaches y, and Y' approaches y'. Hence, Eq. (A.18.21) becomes

$$\int_a^b \left\{ \frac{\partial F}{\partial y} - \frac{d}{dx}\left(\frac{\partial F}{\partial y'} \right) \right\} \eta(x)\,dx = 0. \tag{A.18.22}$$

There is a lemma called the Fundamental Lemma of the Calculus of Variations which says that if $\eta(x)$ is well behaved, then for the integral to be zero,

$$\frac{\partial F}{\partial y} - \frac{d}{dx}\left(\frac{\partial F}{\partial y'} \right) = 0. \tag{A.18.23}$$

This is called *Euler's* equation, and is a necessary condition for $y(x)$ to make I a minimum or maximum. We remind ourselves that the partial derivatives have been taken by assuming that x, y, and y' are three independent variables. But in computing d/dx we must recall that y and y' are functions of x. The last term can be expanded as follows:

$$\frac{d}{dx}\left(\frac{\partial F}{\partial y'} \right) = \frac{\partial^2 F}{\partial x\,\partial y'} + \frac{\partial^2 F}{\partial y\,\partial y'}\frac{dy}{dx} + \frac{\partial^2 F}{\partial y'^2}\frac{d^2 y}{dx^2}. \tag{A.18.24}$$

Euler's equation becomes

$$\frac{\partial F}{\partial y} - \frac{\partial^2 F}{\partial x\,\partial y'} - \frac{\partial^2 F}{\partial y\,\partial y'}\frac{dy}{dx} - \frac{\partial^2 F}{\partial y'^2}\frac{d^2 y}{dx^2} = 0. \tag{A.18.25}$$

This is a second-order differential equation, and hence its solution contains two arbitrary constants. These are to be determined by the requirement that the curve pass through the end-points A and B.

In the *brachistochrone* problem,

$$F = \frac{\sqrt{1 + y'^2}}{\sqrt{x}}.$$

(A.18.26)

Hence

$$\frac{\partial F}{\partial y} = 0,$$

(A.18.27)

$$\frac{\partial F}{\partial y'} = \frac{y'}{\sqrt{x(1 + y'^2)}}.$$

(A.18.28)

Euler's equation becomes

$$-\frac{d}{dx}\left(\frac{y'}{\sqrt{x(1 + y'^2)}}\right) = 0.$$

(A.18.29)

The solution is a curve called the brachistochrone.

The variational procedure can be formalized by introducing the variational notation. In our treatment of the simplest case, we were seeking a function $y(x)$ which makes a certain integral I an extremum. The function

$$Y = y(x) + \epsilon\eta(x)$$

was formed. This can be rewritten as

$$\epsilon\eta(x) = Y - y.$$

(A.18.30)

The quantity $\epsilon\eta(x)$ is called the variation of y and will be denoted by δy, that is,

$$\delta y \equiv \epsilon\eta(x).$$

(A.18.31)

Similarly, we can write

$$\epsilon\eta' = Y' - y'.$$

(A.18.32)

This is the variation of y' and will be denoted by $\delta y'$, that is,

$$\delta y' \equiv \epsilon\eta'.$$

(A.18.33)

It follows from these definitions that

$$\frac{d}{dx}(\delta y) = \delta\left(\frac{dy}{dx}\right).$$

(A.18.34)

That is, the operators δ and d/dx are commutative.

The integrand we considered previously was $F = F(x, y, y')$. This integrand for a *fixed* value of x depends only upon the function $y(x)$ and its derivative $y'(x)$. Note that y is the dependent variable, and x is the independent variable. Now when $y(x)$ and $y'(x)$ are varied, F will also vary. We denote the variation of F by δF. Since F

for a fixed value of x depends on $y(x)$ and $y'(x)$, the variation of F can be written as

$$\delta F = \frac{\partial F}{\partial y} \delta y + \frac{\partial F}{\partial y'} \delta y'. \tag{A.18.35}$$

It can be shown that the laws governing the variation are the same as those governing differentiation. For example,

$$\delta(F_1 F_2) = F_1 \, \delta F_2 + F_2 \, \delta F_1, \tag{A.18.36}$$

$$\delta\left(\frac{F_1}{F_2}\right) = \frac{F_2 \, \delta F_1 - F_1 \, \delta F_2}{F_2^2}, \tag{A.18.37}$$

where F_1 and F_2 are two different functions of x, y, and y'.

For the integral I, the variation has the following meaning:

$$\delta I = \delta \int_a^b F(x, y, y') \, dx = \int_a^b \delta F \, dx. \tag{A.18.38}$$

A necessary condition that the integral I be stationary is that its first variation vanish, that is,

$$\delta I = \int_a^b \delta F(x, y, y') \, dx = 0. \tag{A.18.39}$$

We can easily show this by writing

$$\int_a^b \delta F \, dx = \int_a^b \left[\frac{\partial F}{\partial y} \delta y + \frac{\partial F}{\partial y'} \delta y' \right] dx = 0. \tag{A.18.40}$$

Integrating the second term by parts, we have

$$\int_a^b \frac{\partial F}{\partial y'} \delta\left(\frac{dy}{dx}\right) dx = \int_a^b \frac{\partial F}{\partial y'} \frac{d}{dx} (\delta y) \, dx = \frac{\partial F}{\partial y'} \delta y \Big]_a^b - \int_a^b \frac{d}{dx}\left(\frac{\partial F}{\partial y'}\right) \delta y \, dx. \tag{A.18.41}$$

Thus

$$\int_a^b \delta F \, dx = \int_a^b \left[\frac{\partial F}{\partial y} - \frac{d}{dx} \frac{\partial F}{\partial y'} \right] \delta y \, dx + \frac{\partial F}{\partial y'} \delta y \Big]_a^b = 0. \tag{A.18.42}$$

This is precisely what we had previously (see Eqs. A.18.20 and A.18.22).

Let us consider next a more general case in which the integral is of the form

$$I = \iint_R F(x, y, u, v, u_x, u_y, v_x, v_y) \, dx \, dy, \tag{A.18.43}$$

where the subscripts indicate differentiation, that is, $u_x = \partial u / \partial x$. Here x, y are the independent variables, the region R could be a body in a state of plane stress, and u and v would be the displacements. Then the derivatives u_x, u_y, etc., are related to the components of strain. The condition for an extremum is again

$$\delta I = 0. \tag{A.18.44}$$

This becomes

$$\delta I = \iint \delta F \, dx \, dy = \iint_R \left\{ \left(\frac{\partial F}{\partial u} \delta u + \frac{\partial F}{\partial u_x} \delta u_x + \frac{\partial F}{\partial u_y} \delta u_y \right) \right.$$
$$\left. + \left(\frac{\partial F}{\partial v} \delta v + \frac{\partial F}{\partial v_x} \delta v_x + \frac{\partial F}{\partial v_y} \delta v_y \right) \right\} dx \, dy = 0. \qquad \text{(A.18.45)}$$

The terms involving δu_x, δu_y, δv_x, and δv_y must be integrated by parts. We obtain

$$\delta I = \iint_R \left\{ \left[\frac{\partial F}{\partial u} - \frac{\partial}{\partial x} \left(\frac{\partial F}{\partial u_x} \right) - \frac{\partial}{\partial y} \left(\frac{\partial F}{\partial u_y} \right) \right] \delta u \right.$$
$$+ \left[\frac{\partial F}{\partial v} - \frac{\partial}{\partial x} \left(\frac{\partial F}{\partial v_x} \right) - \frac{\partial}{\partial y} \left(\frac{\partial F}{\partial v_y} \right) \right] \delta v \right\} dx \, dy$$
$$+ \int \frac{\partial F}{\partial u_x} \delta u \, dy \Big]_c^d + \int \frac{\partial F}{\partial u_y} \delta u \, dx \Big]_a^b + \int \frac{\partial F}{\partial v_x} \delta v \, dy \Big]_c^d$$
$$+ \int \frac{\partial F}{\partial v_y} \delta v \, dx \Big]_a^b = 0. \qquad \text{(A.18.46)}$$

The necessary conditions (i.e. Euler equations) become, since δu and δv are independent,

$$\frac{\partial F}{\partial u} - \frac{\partial}{\partial x} \left(\frac{\partial F}{\partial u_x} \right) - \frac{\partial}{\partial y} \left(\frac{\partial F}{\partial u_y} \right) = 0, \qquad \text{(A.18.47)}$$

$$\frac{\partial F}{\partial v} - \frac{\partial}{\partial x} \left(\frac{\partial F}{\partial v_x} \right) - \frac{\partial}{\partial y} \left(\frac{\partial F}{\partial v_y} \right) = 0. \qquad \text{(A.18.48)}$$

A.19   THE DIVERGENCE THEOREM [3], [4], [7]

Consider a volume V enclosed by a surface boundary A. If a vector $\mathbf{P}(y_1, y_2, y_3)$ and its derivatives with respect to y_1, y_2, and y_3 are continuous at all points in V, then

$$\iiint \nabla \cdot \mathbf{P} \, dV = \iint \mathbf{P} \cdot \mathbf{n} \, dA, \qquad \text{(A.19.1)}$$

where in the rectangular Cartesian coordinates the operator ∇ is defined by the relation

$$\nabla = \mathbf{i}_m \frac{\partial}{\partial y_m}, \qquad \text{(A.19.2)}$$

and $\mathbf{n}$ is a unit vector along the direction normal to the surface. This statement is known as the Divergence Theorem. In the literature it is sometimes referred to as Green's Theorem or Gauss' Theorem. It can also be written in scalar form as

$$\iiint \frac{\partial p_m}{\partial y_m} \, dV = \iint p_m n_m \, dA. \qquad \text{(A.19.3)}$$

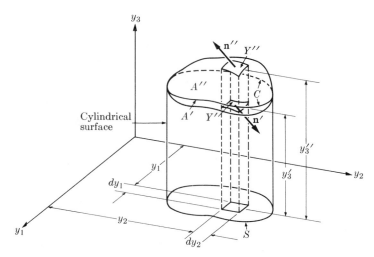

FIG. A.16   A surface represented by A' and A'' and its projection on the y_1y_2-plane.

Here the vectors $\mathbf{P}$ and $\mathbf{n}$ are written in terms of their components,

$$\mathbf{P} = p_m \mathbf{i}_m, \tag{A.19.4}$$

$$\mathbf{n} = n_m \mathbf{i}_m. \tag{A.19.5}$$

To prove the theorem let us first prove that

$$\iiint \frac{\partial p_3}{\partial y_3} \, dV = \iint p_3 n_3 \, dA \tag{A.19.6}$$

for a simple case where the boundary surface A is pierced by a line parallel to the y_3-axis in, at most, two points. As illustrated in Fig. A.16, the entire surface A can be divided into two parts, A' and A'', by a closed curve C which is also on a cylindrical surface with generators parallel to the y_3-axis. The projection of this cylinder on the y_1y_2-plane encloses the area S. We introduce an element of volume which forms a column parallel to the y_3-axis, and has a projection $dy_1 \, dy_2$ on the y_1y_2-plane. The element intersects A' and A'' at $Y'(y_1, y_2, y_3')$ and $Y''(y_1, y_2, y_3'')$, respectively. Thus

$$\iiint \frac{\partial p_3}{\partial y_3} \, dV = \iint_S \left[\int_{y_3'}^{y_3''} \frac{\partial p_3}{\partial y_3} \, dy_3 \right] dy_1 \, dy_2$$

$$= \iint_S [p_3(y_1, y_2, y_3'') - p_3(y_1, y_2, y_3')] \, dy_1 \, dy_2. \tag{A.19.7}$$

Let dA' and dA'' be, respectively, the areas on the two surfaces cut by this vertical column, and $\mathbf{n}'$ and $\mathbf{n}''$ be the corresponding outward unit normals. Then from Eq.

(6.4.13) we have

$$dy_1 \, dy_2 = -dA'n_3' = dA''n_3'', \qquad (A.19.8)$$

where n_3' and n_3'' are the direction cosines which connect the y_3-axis with the surface normals $\mathbf{n}'$ and $\mathbf{n}''$, respectively. The negative sign is used in the second term because the normal $\mathbf{n}'$ is in a direction opposite to the positive y_3-axis. Thus Eq. (A.19.7) becomes

$$\iiint \frac{\partial p_3}{\partial y_3} \, dV = \iint\limits_{A'} p_3(y_1, y_2, y_3')n_3' \, dA + \iint\limits_{A''} p_3(y_1, y_2, y_3'')n_3'' \, dA, \qquad (A.19.9)$$

which can be written in the form of Eq. (A.19.6). In the case where A is pierced by a vertical line in more than two points, the total volume V can be subdivided into a number of regions $V_1, V_2, \ldots$ by a series of surface areas $T_1, T_2, \ldots$ such that the boundary of each of the regions will be intersected by a vertical line in at most two points. Equation (A.19.6) then can be applied to each region. When the individual equations are added, the surface integrals over $T_1, T_2, \ldots$ will cancel; hence Eq. (A.19.6) is also applicable to the total volume.

In a similar manner we can prove that

$$\iiint \frac{\partial p_1}{\partial y_1} \, dV = \iint p_1 n_1 \, dA \qquad (A.19.10)$$

and

$$\iiint \frac{\partial p_2}{\partial y_2} \, dV = \iint p_2 n_2 \, dA. \qquad (A.19.11)$$

Equations (A.19.6), (A.19.10), and (A.19.11) can then be combined to yield

$$\iiint \frac{\partial p_m}{\partial y_m} \, dV = \iint p_m n_m \, dA, \qquad (A.19.12)$$

which is the Divergence Theorem.

A.20 MATRIX ALGEBRA [6], [8], [9]

In Section A.2 the summation and indicial notation for linear homogeneous functions was introduced. There is an alternative way, called the *matrix notation*, to represent such functions. This matrix notation possesses certain pictorial qualities which are helpful in visualizing the structure of systems of simultaneous equations. In addition, the numerical solution of equations is aided by the use of matrices.

For example, let us consider a system of simultaneous linear algebraic equations

$$\begin{aligned}
K_{11}q_1 + K_{12}q_2 + \cdots + K_{1n}q_n &= Q_1, \\
K_{21}q_1 + \cdots \qquad \cdots + K_{2n}q_n &= Q_2, \\
\vdots \qquad\qquad\qquad \vdots \\
K_{n1}q_1 + \cdots \qquad \cdots + K_{nn}q_n &= Q_n,
\end{aligned} \qquad (A.20.1)$$

which in summation form is

$$\sum_{i=1}^{n} K_{ij}q_j = Q_i, \qquad i = 1, 2, \ldots, n. \tag{A.20.2}$$

In matrix notation, this is displayed in the form

$$\begin{bmatrix} K_{11} & K_{12} & \ldots & K_{1n} \\ K_{21} & \ldots & \ldots & K_{2n} \\ \vdots & & & \vdots \\ K_{n1} & \ldots & \ldots & K_{nn} \end{bmatrix} \begin{bmatrix} q_1 \\ q_2 \\ \vdots \\ q_n \end{bmatrix} = \begin{bmatrix} Q_1 \\ Q_2 \\ \vdots \\ Q_n \end{bmatrix}, \tag{A.20.3}$$

or more compactly as

$$[K_{ij}]\{q_j\} = \{Q_i\}, \tag{A.20.4}$$

or as

$$[K]\{q\} = \{Q\}. \tag{A.20.5}$$

In general, a matrix $[A]$ or $[A_{ij}]$ represents an array of numbers arranged in m rows and n columns, such as

$$[A]_{m \times n} = \begin{bmatrix} A_{11} & A_{12} & \ldots & A_{1n} \\ A_{21} & A_{22} & \ldots & A_{2n} \\ \vdots & & & \vdots \\ A_{m1} & A_{m2} & \ldots & A_{mn} \end{bmatrix}. \tag{A.20.6}$$

Thus A_{ij} is the typical element of the matrix $[A]$. Note that the first subscript locates the row, and the second subscript the column of the element. A matrix having m rows and n columns is said to have an order m by n. A matrix with equal numbers of rows and columns is called a *square matrix*, and is denoted by

$$\begin{bmatrix} A_{11} & A_{12} & \ldots & A_{1n} \\ A_{21} & \ldots & \ldots & A_{2n} \\ \vdots & & & \vdots \\ A_{n1} & \ldots & \ldots & A_{nn} \end{bmatrix} = [A]. \tag{A.20.7}$$

A matrix which has m elements arranged in a single column is called a *column matrix*, and is usually denoted by

$$\begin{bmatrix} A_1 \\ A_2 \\ \vdots \\ A_m \end{bmatrix} = \{A\}. \tag{A.20.8}$$

A matrix which has n elements arranged in a single row is called a *row matrix* and is usually denoted by

$$[A_1 A_2 \ldots A_n] = \lfloor A \rfloor. \tag{A.20.9}$$

Important elementary operations of matrices are given as follows:

A.21 EQUALITY

Two matrices $[A]$ and $[B]$ are equal, for example,

$$[A]_{m \times n} = [B]_{m \times n}, \tag{A.21.1}$$

only if

$$A_{ij} = B_{ij}, \tag{A.21.2}$$

that is, each element of A is equal to the similarly located element of B.

A.22 ADDITION AND SUBTRACTION

The operations of addition and subtraction are defined only for matrices of the same order. The definitions are for addition:

$$[A]_{m \times n} + [B]_{m \times n} = [C]_{m \times n}, \tag{A.22.1}$$

$$A_{ij} + B_{ij} = C_{ij}, \tag{A.22.2}$$

and for subtraction:

$$[A]_{m \times n} - [B]_{m \times n} = [D]_{m \times n}, \tag{A.22.3}$$

$$A_{ij} - B_{ij} = D_{ij}. \tag{A.22.4}$$

Thus it is observed that corresponding elements are either added or subtracted in the operations of matrix addition and subtraction.

A.23 SCALAR MULTIPLICATION

The operation denoted by

$$l[A]_{m \times n} = [B]_{m \times n}, \tag{A.23.1}$$

where l is a constant is defined by

$$l A_{ij} = B_{ij}, \tag{A.23.2}$$

that is, every element in the matrix $[A]$ is multiplied by the scalar quantity l.

A.24 MATRIX MULTIPLICATION

At first glance the operation of matrix multiplication will appear complicated; a little bit of practice is necessary to gain proficiency.

The multiplication of two matrices $[A]$ and $[B]$ to yield a matrix $[C]$ must be performed in a specified order, that is,

$$[A]_{m \times n}[B]_{n \times p} = [C]_{m \times p}. \tag{A.24.1}$$

The number of columns in $[A]$ must equal the number of rows in $[B]$. To further define the process of multiplication, let us suppose that $[A]$ and $[B]$ are matrices of the orders denoted by

$$[A]_{2\times3} = \begin{bmatrix} A_{11} & A_{12} & A_{13} \\ A_{21} & A_{22} & A_{23} \end{bmatrix},$$

$$[B]_{3\times3} = \begin{bmatrix} B_{11} & B_{12} & B_{13} \\ B_{21} & B_{22} & B_{23} \\ B_{31} & B_{32} & B_{33} \end{bmatrix}.$$

(A.24.2)

Then, for example, two of the elements of $[C]$ are given by

$$C_{11} = A_{11}B_{11} + A_{12}B_{21} + A_{13}B_{31},$$
$$C_{21} = A_{21}B_{11} + A_{22}B_{21} + A_{23}B_{31}.$$

(A.24.3)

The element C_{ik} is defined by the row-by-column rule, which can be stated as

$$C_{ik} = \sum_{j=1}^{n} A_{ij}B_{jk}.$$

(A.24.4)

Thus, if the order of the matrix $[A]$ is m by n, and the order of the matrix $[B]$, n by p, then the order of the product $[C]$ is m by p.

Matrix multiplication is not commutative, that is,

$$[A][B] \neq [B][A].$$

(A.24.5)

Thus, to be specific, the operation $[A][B]$ should be stated as "the premultiplication of $[B]$ by $[A]$," or "the postmultiplication of $[A]$ by $[B]$."

A.25 UNIT MATRIX

A unit matrix $[I]$ is of the form

$$[I] = \begin{bmatrix} 1 & 0 & 0 & \cdots & 0 \\ 0 & 1 & 0 & \cdots & 0 \\ 0 & 0 & 1 & \cdots & 0 \\ \vdots & & & & \vdots \\ 0 & 0 & 0 & \cdots & 1 \end{bmatrix}.$$

(A.25.1)

It is a square matrix with all the diagonal terms equal to unity while all the off-diagonal terms are zero. Thus, premultiplication or postmultiplication by a unit matrix does not alter the original matrix.

A.26 INVERSION OF MATRICES

Unlike the other elementary algebraic operations, the process of division in matrix algebra is not similar to the corresponding process with numbers. Division is accomplished by inverting matrices. The inverse of a matrix $[A]$ is denoted by $[A]^{-1}$

and is defined by both conditions:

$$[A][A]^{-1} = [I] \tag{A.26.1}$$

and

$$[A]^{-1}[A] = [I], \tag{A.26.2}$$

and thus in order to have an inverse, $[A]$ must by necessity be a square matrix.

When the equation

$$[A]\{B\} = \{C\} \tag{A.26.3}$$

is premultiplied by the inverse of $[A]$, we obtain

$$[I]\{B\} = [A]^{-1}\{C\}, \tag{A.26.4}$$

or simply

$$\{B\} = [A]^{-1}\{C\}. \tag{A.26.5}$$

Let $\{C\}$ be a column matrix with the ith element equal to unity while all other elements are equal to zero. Then $\{B\}$ is identical to the ith column of the $[A]^{-1}$ matrix. This shows that the elements of an inverse matrix of order n by n can be obtained by solving n sets of simultaneous equations in the form of Eq. (A.26.5), with the value of unity placed at the appropriate location in the $\{C\}$ matrix.

REFERENCES FOR APPENDIX

(1) A. J. McCONNELL, *Applications of Tensor Analysis*, Dover Publications, New York, 1957.
(2) I. S. SOKOLNIKOFF, *Tensor Analysis*, John Wiley and Sons, New York, 1951.
(3) L. BRAND, *Vector and Tensor Analysis*, John Wiley and Sons, New York, 1947.
(4) I. S. SOKOLNIKOFF and R. M. REDHEFFER, *Mathematics of Physics and Modern Engineering*, McGraw-Hill, New York, 1958.
(5) G. B. THOMAS, *Calculus and Analytic Geometry*, Addison-Wesley, Reading, Mass., 2nd Ed., 1960.
(6) F. B. HILDEBRAND, *Methods of Applied Mathematics*, Prentice-Hall, New York, 1952.
(7) F. B. HILDEBRAND, *Advanced Calculus for Engineers*, Prentice-Hall, New York, 1948.
(8) SAM PERLIS, *Theory of Matrices*, Addison-Wesley, Reading, Mass., 1952.
(9) R. A. FRAZIER, W. J. DUNCAN, A. R. COLLAR, *Elementary Matrices*, MacMillan, New York, 1947.

Bibliography

CHAPTER 1

1.1   F. Cajori, *Sir Isaac Newton's Mathematical Principles*, Univ. of Calif. Press, Berkeley, Calif., 1960. This is a revision of the original English translation of Newton's *Principia* [1.2]* made by Andrew Motte in 1729. Of particular interest are a series of prefaces which describe the philosophy and motivation for the work, and an Appendix which provides historical notes of the controversies surrounding some of Newton's discoveries and ideas.

1.2   I. Newton, *Philosophiae Naturalis Principia Mathematica*, Jussn Societatis ac Typis Josephi Streater Anno MDCLXXXVII. This is the original treatise in Latin.

1.3   E. Mach, *The Science of Mechanics*, 6th Ed., Open Court Publishing, La Salle, Ill., 1960. This is an English translation of Mach's book which bears the subtitle "A Critical and Historical Account of Its Development." The introduction to this, the Sixth American Edition, provides some interesting and challenging views on the study of mechanics.

1.4   J. L. Synge and B. A. Griffith, *Principles of Mechanics*, 3rd Ed., McGraw-Hill, New York, 1959. Section 1.1 has provided the ideas for Chapter 1 of this book.

CHAPTER 2

2.1   F. Cajori, *Sir Isaac Newton's Mathematical Principles* [1.1]. It is interesting and informative to read Newton's original statements and explanations of the laws of motion. He expresses these in terms of three laws and six corollaries on pages 13 to 28.

2.2   E. Mach, *The Science of Mechanics* [1.3]. Chapter I which is entitled, "The Development of the Principles of Statics," is an excellent presentation not only of the historical background but of the content of statics. The writing is clear and detailed with a minimum of equations. Of especial interest are pages 44–59 of Chapter I which bears the subtitle, "The Principle of the Composition of Forces." Also, the reader will gain additional insight by reading pages 226–247, "The Achievements of Newton."

---

* These bracketed numbers refer to references in this bibliography.

2.3   J. L. Synge and B. A. Griffith, *Principles of Mechanics* [1.4].   Chapters 1, 2, 3, and 10 may be helpful in that the contents of Chapter 2 of this book are presented in more expanded form.

2.4   S. H. Crandall and N. C. Dahl, *An Introduction to the Mechanics of Solids*, McGraw-Hill, New York, 1959.   This is one of the best of the recent books on solid mechanics.   Chapter 1 of Crandall and Dahl parallels Chapter 2 of the *Statics of Deformable Bodies*.   There is a wealth of imaginative problems.

2.5   Sir Horace Lamb, *Statics*, 3rd Ed., Cambridge Univ. Press, New York, 1928. The first edition of this book was published in 1912.   It emphasizes geometrical methods as applied to statics.   Vectors are used, but it is not until Chapter VI that the scalar product is introduced.

2.6   L. Brand, *Vectorial Mechanics*, John Wiley and Sons, New York, 1930.   This is one of the earliest books in English which makes extensive use of vector analysis for statics.   Chapters II, III, IV, and V are pertinent to the material covered in our book.

CHAPTER 3

3.1   C. H. Norris and J. B. Wilbur, *Elementary Structural Analysis*, 2nd Ed., McGraw-Hill, New York, 1960.   The material contained in pages 60–153 is a very lucid and systematic presentation of the statics of beams and trusses.   There are many illustrative examples, and accompanying these are discussions of the fine points of the solutions.

3.2   A. S. Niles and J. S. Newell, *Airplane Structures*, 4th Ed., John Wiley and Sons, New York, 1954.   This book, which first appeared in 1929, was one of the earliest texts to be directed toward the structural problems of flying vehicles.   Chapter II which deals with beams and Chapter VIII which deals with trusses present slightly different approaches to these two topics.   For example, the statics of beams is treated by a series of propositions.

3.3   S. H. Crandall and N. C. Dahl, *An Introduction to the Mechanics of Solids*. Read comments in [2.4].   Chapters 2 and 3 of this book parallel Chapters 3 and 4 of our book.

3.4   L. Brand, *Vectorial Mechanics*.   Read comments in [2.6].

CHAPTER 4

4.1   C. H. Norris and J. B. Wilbur, *Elementary Structural Analysis* [3.1].   Pages 400 through 408 provide an excellent presentation of a method of approach to statically indeterminate problems.   The examples which come after page 408 require more knowledge than the reader would have at this time.

4.2   S. H. Crandall and N. C. Dahl, *An Introduction to the Mechanics of Solids*. Read comments in [3.3].

4.3   F. R. Shanley, *Strength of Materials*, McGraw-Hill, New York, 1957.   The author approaches structural analysis from the standpoint of force transmission.   His ideas on force transmission are presented in Sections 1.12, 13.1, 13.2, 14.2, 15.1, and 17.1.   This book stresses what might be termed a "physical-intuitive" approach to the mechanics of deformable bodies.

4.4   W. M. Fife and J. B. Wilbur, *Theory of Statically Indeterminate Structures*, McGraw-Hill, New York, 1937. This is one of the earliest books in this country to treat indeterminate structures in a systematic and coherent manner.

4.5   C. L. M. H. Navier, *Resumé des leçons données à l'école des ponts and chaussées sur l'application de la méchanique a l'établissement des construction et des machines*, Paris, 1833. This book was first published in 1826 and extensively revised by Navier in 1833 for the second edition. Under the editorship of St. Venant it was further revised in 1863 (Navier died in 1836). Navier was the first to recognize that the elasticity of the structure will supply the additional equations for the solution of problems for which the elementary equations of statics do not supply enough equations, i.e., he was the first to develop a general method of solving statically indeterminate problems.

CHAPTER 5

5.1   A. E. H. Love, *A Treatise on the Mathematical Theory of Elasticity*, 4th Ed., 1927, 1st Amer. Printing, 1944, Dover Publications, New York. The first edition of this book was published in 1892 by the Cambridge University Press. Everyone interested in elasticity should have a copy in his bookcase, even though it is difficult to read because the notation and phraseology do not conform to current usage. Chapter I is concerned with the analysis of strain.

5.2   F. D. Murnaghan, "Finite Deformations of An Elastic Solid," *American Journal of Mathematics* **59**, 235–260 (1937). This is one of the earliest tensor treatments of strain.

5.3   S. Timoshenko and J. N. Goodier, *Theory of Elasticity*, 2nd Ed., McGraw-Hill, New York, 1951. The first edition of this book appeared in 1934, and for more than two decades has been the standard reference on elasticity in the United States. It progresses from the simple to the more complicated problems, so that not until page 221 is the analysis of strain in three dimensions discussed. The marvelous physical intuition which blessed the early elasticians is evident in this book. A deficiency of the book from a modern pedagogic point of view is that vectors and tensors are not used. Nevertheless, the *Theory of Elasticity* remains a valuable source-book and belongs on every structural engineer's bookshelf. Strain is discussed in Sections 5, 8, 72, 73, 74, 75, 77, and 78.

5.4   A. E. Green and W. Zerna, *Theoretical Elasticity*, Oxford Univ. Press, New York, 1954. This book has had a tremendous influence in shaping the modern approach to elasticity. It has shown the way to the tensor formulation of the equations of elasticity and has demonstrated the power of these methods to handle complicated geometrical problems. In fact, geometric nonlinearities in elasticity, as well as non-orthogonal curvilinear coordinates, can be easily included in the equations (however, the solution still presents difficulties). *Theoretical Elasticity* is written almost entirely in terms of general tensors (not the Cartesian ones used in *Statics of Deformable Bodies*. Consequently, a great deal of study by the reader is required if any benefit is to be gained. This is not a book for beginners; even the expert unfamiliar with tensors will find it difficult. However, both beginner and expert should be cognizant of this work. Sections 2.1, 2.2, and 2.3 present a general treatment parallel to Chapter 5 of *Statics of Deformable Bodies*.

5.5   V. V. Novozhilov, *Foundations of the Nonlinear Theory of Elasticity*, Graylock Press, New York, 1953. Chapter I of this book, comprising some 60 pages, is a very lucid account of the geometry of strain. The treatment, which uses vectors sparingly and tensors not at all, is in terms of rectangular Cartesian coordinates. Novozhilov's discussion of the transition from large strains to small to linear is the best to be found in the literature. The author is one of the well-known Russian elasticians. The book, originally published in Russian in 1948, has been beautifully translated into English.

5.6   I. S. Sokolnikoff, *Mathematical Theory of Elasticity*, 2nd Ed., McGraw-Hill, New York, 1956. This is one of the earliest books (the first edition appeared in 1946) written in English to approach the theory of elasticity by first developing the three-dimensional equations and then specializing to simpler problems. Chapters I, II, and III parallel Chapters 4, 5, and 6 of the *Statics of Deformable Bodies*.

5.7   B. de Saint-Venant, "Mémoire sur l'équilibre des corps solides, dans les limites de leur élasticité, et les conditions de leur résistance, quand les déplacements éprouvés par leurs pointes ne sont pas trés petits," *Comptes Rendus de l'Académie des Science*, Vol. XXIV, pp. 260–263 (1847). (Memoir on the "Equilibrium of Solid Bodies within the Elastic Limit and on the Conditions of Resistance When the Displacements Experienced by Points in the Body Are Not Very Small.") St. Venant was the first to notice that the relative elongations can be small even though the angles of rotation are numerically large. He recognized the geometrical meanings of the components of the strain tensor.

CHAPTER 6

6.1   A. E. H. Love, *A Treatise on the Mathematical Theory of Elasticity*. Read comments for [5.1]. Chapter 2 presents the analysis of stress.

6.2   S. Timoshenko and J. N. Goodier, *Theory of Elasticity*. Read comments for [5.3]. Stress in this book is discussed in Sections 2, 3, 4, 11, 13, 67, 68, 69, 70, 71, 76, and 79.

6.3   A. E. Green and W. Zerna, *Theoretical Elasticity*. Read comments for [5.4]. Stress is discussed in Sections 2.4, 2.5, and 5.2.

6.4   V. V. Novozhilov, *Foundations of the Nonlinear Theory of Elasticity*. Read comments for [5.5]. Stress is discussed in Chapter 2.

6.5   I. S. Sokolnikoff, *Mathematical Theory of Elasticity*. Read comments for [5.6]. Stress is discussed in Chapter 2.

6.6   A. L. Cauchy, "Recherches sur l'équilibre et le mouvement intérieur des corps solides ou fluides, élastiques ou non élastiques," *Bulletin des Sciences à la Societé Philomathique*, 1823. Cauchy was the first to show that stress at a point can be expressed in terms of six components. He showed the significance of the principal planes of stress. Also, he was the first to formulate the stress-strain relations in the form which is used in modern times.

CHAPTER 7

7.1   A. E. H. Love, *A Treatise on the Mathematical Theory of Elasticity*. Read comments for [5.1]. Sections 6.4 and 6.6 discuss the generalized Hooke's law. Additionally, the first 37 pages of Love's book give an excellent historical summary of elasticity.

7.2   S. Timoshenko and J. N. Goodier, *Theory of Elasticity*. Read comments for [5.3]. Section 6 presents the stress-strain relations for isotropic materials. Note that Timoshenko's results differ from ours by a factor of $\frac{1}{2}$ in three of the relations.

7.3.   A. E. Green and W. Zerna, *Theoretical Elasticity*. Read comments for [5.4]. A general discussion of stress-strain relations for homogeneous isotropic bodies is presented in Section 2.8. Anisotropic materials for linear strain (Green and Zerna use "infinitesimal" for "linear") are discussed in Section 5.3.

7.4   V. V. Novozhilov, *Foundations of the Nonlinear Theory of Elasticity*. Read comments for [5.5]. The stress-strain relations are discussed in Chapter 3.

7.5   I. S. Sokolnikoff, *Mathematical Theory of Elasticity*. Read comments for [5.6]. Sections 20, 21, 22, and 23 are on the subject of the stress-strain relations.

7.6   H. B. Huntington, "The Elastic Constants of Crystals," *Solid State Physics—Advances in Research and Application*, Vol. 7, pp. 213–351. Academic Press, New York, 1958. This is a comprehensive article which approaches the stress-strain relations from the viewpoint of physics of the solid state.

7.7   R. F. S. Hearmon, *An Introduction to Applied Anisotropic Elasticity*, Oxford Univ. Press, New York, 1961. The first 44 pages of this little book discuss the stress-strain relations for anisotropic elasticity.

7.8   I. Todhunter and K. Pearson, *A History of the Theory of Elasticity and of the Strength of Materials*, Volumes I, II, Dover Publications, New York, 1960. The first volume of this set was published by Cambridge University Press in 1886. These volumes will be of interest to those who are interested in a critical review of the development of elasticity. There are more than 2200 pages in this history.

7.9   S. Timoshenko, *History of the Theory of Elasticity*, McGraw-Hill, New York, 1953. This is a very readable book on the history of elasticity.

7.10   C. Truesdell and R. A. Toupin, "The Classical Field Theories," *Encyclopedia of Physics*, Vol. III/1, pp. 226–793, Springer, New York, 1960. This is an exhaustive treatise on the exact and general theory of the continuous field. Included as a special case is the subject which we have called the statics of deformable bodies. This article is not for the beginner in the field. As a matter of fact, even the expert is first advised to read a portion of its appendix on mathematics before tackling the article.

CHAPTER 8

8.1   W. Prager and P. G. Hodge, Jr., *Theory of Perfectly Plastic Solids*, John Wiley and Sons, New York, 1951. Read Chapter 1 on "Basic Concepts." Note again that the conventional definition of shear strain is adopted in Prager and Hodge, hence three of the stress-strain relations differ from ours by a factor of $\frac{1}{2}$.

8.2   O. Hoffman and G. Sachs, *Introduction to the Theory of Plasticity for Engineers*, McGraw-Hill, New York, 1953. Read Chapter 5, "Plastic Stress-strain Relations," and Chapter 6, "Experimental Data."

8.3   A. Nadai, *Theory of Flow and Fracture of Solids*, Vol. I, 2nd Ed., McGraw-Hill, New York, 1950. This book contains detailed discussions of both the mathematical treatment and experimental evidences of plastic deformation of solids. Read Part II of the text.

8.4   F. A. McClintock and A. S. Argon, *An Introduction to the Mechanical Behavior of Materials*, Addison-Wesley, Reading, Mass., in press. Chapters 5, 6, and 7 contain detailed discussions of plasticity in single crystals and polycrystals and of elastic stress-strain relations.

8.5   R. Hill, *Mathematical Theory of Elasticity*, Oxford Univ. Press, New York, 1950. The first chapter contains a brief summary of historical background, and the second presents the foundations of the theory of plasticity.

8.6   H. W. Hayden, W. G. Moffatt and J. Wulff, *Mechanical Behavior of Materials* (Vol. III of a series on "Structure and Properties of Materials"), John Wiley and Sons, New York, 1965. For readers who are interested in the properties of solids from the microscopic point of view, this text book is a good reference.

8.7   A. G. Guy, *Elements of Physical Metallurgy*, 2nd Ed., Chapter 9, "Plasticity of Metals," Addison-Wesley, Reading, Mass., 1959. This chapter presents a discussion of the plastic properties of solids from the microscopic point of view.

CHAPTER 9

9.1   R. L. Bisplinghoff, H. Ashley, and R. L. Halfman, *Aeroelasticity*, Addison-Wesley, Reading, Mass., 1955. Chapter 2 contains a detailed treatment on the determination by energy methods of the stiffness and flexibility influence coefficients of airplane wing structures.

9.2   C. H. Norris, and J. B. Wilbur [3.1]. Chapter 12 presents a different development of the Principle of Virtual Work and includes numerous illustrative examples on the deflections of trusses, beams, and frames. Sections 13.7 and 13.8 contain a discussion of the solution of statically indeterminate structures by the Theorem of Least Work.

9.3   S. Timoshenko and J. N. Goodier [5.3]. Chapter 6 of this text presents strain-energy methods for continuous elastic bodies. The principles referred to are the Principle of Virtual Work, Castigliano's Second Theorem, and the Theorem of Least Work. Plane stress problems are used in the derivation of these principles and in illustrative examples.

9.4   I. S. Sokolnikoff [5.6]. Pages 377–401 present an introduction to the calculus of variation and derivations of the Theorems of Minimum Potential and Complementary Energies, and other related energy principles.

9.5   H. L. Langhaar, *Energy Methods in Applied Mechanics*, John Wiley and Sons, New York, 1962. This book contains energy methods for statics and dynamics of deformable bodies. Chapter 3 presents elements of the calculus of variations, and Chapter 4 presents energy methods for deflections of structures and the solution of statically indeterminate structures.

9.6   N. J. Hoff, *The Analysis of Structures*, John Wiley and Sons, New York, 1956. The complete text contains the treatment of energy principles for static analysis of structures. It also presents many methods derived from the principles of virtual work, potential energy, and complementary energy.

Nomenclature,

Symbolism, and Assorted Marks

A	area
$\mathbf{a}$	acceleration vector
$\mathbf{C}$	couple
C_{mn}	flexibility influence coefficient
dS	differential element of length in deformed body
ds	differential element of length in undeformed body
E	Young's modulus
E_{mnpr}	elasticity tensor
e_{rst}	permutation symbol
F	axial force in a slender beam
$\mathbf{F}$	force vector, body force vector
F_n	component of $\mathbf{F}$ with respect to $\mathbf{i}_n$
G	shear modulus
G_{mn}	metric tensor in deformed body
g_{mn}	metric tensor in undeformed body
I	section moment of inertia of a slender beam
I_1, I_2, I_3	stress invariants
$\mathbf{i}_1, \mathbf{i}_2, \mathbf{i}_3$	unit vectors along y_1, y_2, y_3, respectively
J_1, J_2, J_3	strain invariants
$\mathbf{j}_1, \mathbf{j}_2, \mathbf{j}_3$	unit vectors along $\tilde{y}_1, \tilde{y}_2, \tilde{y}_3$, respectively
K	bulk modulus
K_{mn}	stiffness influence coefficient
k	spring constant
L	length of bar
$l_{m\tilde{r}}$	direction cosine of the angle between y_m- and $\tilde{y}_r$-axes
M	bending moment in a slender beam
$\mathbf{M}$	moment vector
M_1, M_2, M_3	components of moment vector $\mathbf{M}$ with respect to $\mathbf{i}_1, \mathbf{i}_2, \mathbf{i}_3$
m	mass
$\mathbf{n}$	unit normal vector

$\mathbf{P}$	force vector
p	distributed load in a slender beam
P_n	components of force vector $\mathbf{P}$ with respect to $\mathbf{i}_n$
Q	generalized force
q	generalized displacement
$\mathbf{R}$	position vector to point P in deformed body
$\mathbf{r}$	position vector to point p in undeformed body
R_n	components of position vector $\mathbf{R}$, with respect to $\mathbf{i}_n$
S	shear force in a slender beam
S_{mnpr}	compliance tensor
T	torsional moment on a slender beam; thrust; tension force
U	strain energy
U^*	strain energy density
$\hat{U}^*$	distortion energy density (deviator strain energy)
$\hat{U}^*$	dilatation energy density (spherical strain energy)
U_c	complementary strain energy
$\mathbf{u}$	displacement vector connecting point p in undeformed body to P in deformed body
u_n	components of displacement vector $\mathbf{u}$ with respect to $\mathbf{i}_n$
V	volume
w	deflection of a slender beam
$\mathbf{X}$	redundant force
y_n	rectangular Cartesian axes. Also components of position vector $\mathbf{r}$, with respect to $\mathbf{i}_n$
α	coefficient of thermal expansion
γ	shear strain
γ_{mn}	strain tensor
$\hat{\gamma}_{mn}$	components of deviator strain tensor
$\hat{\gamma}_{mn}$	components of spherical strain tensor
ΔT	change in temperature
δ	deflection
δ_{mn}	Kronecker delta
ϵ_{mn}	linear strain tensor
$\bar{\epsilon}$	equivalent strain
λ	one of the two Lamé constants
$\lambda_1, \lambda_2, \lambda_3$	direction cosines
μ	shear modulus, also one of the two Lamé constants $(= G)$
ν	Poisson's ratio
π_c	total complementary energy
π_p	total potential energy
σ	normal stress
$\bar{\sigma}$	equivalent stress
$\boldsymbol{\sigma}$	stress vector (force per unit area)
$\boldsymbol{\sigma}^*$	prescribed surface stress

$\boldsymbol{\sigma}_m$	stress vector acting on element of area for which y_m is a constant
σ_o	yield stress in simple tension
σ_{mn}	stress tensor in rectangular, Cartesian coordinates (force per unit area)
$\hat{\sigma}_{mn}$	components of deviator stress tensor
$\hat{\hat{\sigma}}_{mn}$	components of spherical stress tensor
τ	shear stress
ϕ	Airy Stress Function
$(\ )^e$	indicates elastic strain components
$(\ )^p$	indicates plastic strain components
$(\ )_n(\ )_n$	repeated subscripts are to be summed
$(\ )_n$	subscript indicating values $1, 2, 3, \ldots$
$(\sim)$	pronounced "til'de." Usually indicates another coordinate system
$(\frown)$	deviatoric components
$(\widehat{\frown})$	spherical components
$(\ )^{(\mathrm{I})}, (\ )_{\mathrm{II}}$	Roman numeral subscripts or superscripts indicate quantity is a principal one
$(\ ) \cdot (\ )$	indicates scalar product
$(\ ) \times (\ )$	indicates cross product

Index

Index

A CATALOG OF SELECTED
DOVER BOOKS
IN SCIENCE AND MATHEMATICS

QUALITATIVE THEORY OF DIFFERENTIAL EQUATIONS, V.V. Nemytskii and V.V. Stepanov. Classic graduate-level text by two prominent Soviet mathematicians covers classical differential equations as well as topological dynamics and erqodic theory. Bibliographies. 523pp. 5⅜ × 8½.   65954-2 Pa. $10.95

MATRICES AND LINEAR ALGEBRA, Hans Schneider and George Phillip Barker. Basic textbook covers theory of matrices and its applications to systems of linear equations and related topics such as determinants, eigenvalues and differential equations. Numerous exercises. 432pp. 5⅜ × 8½.   66014-1 Pa. $8.95

QUANTUM THEORY, David Bohm. This advanced undergraduate-level text presents the quantum theory in terms of qualitative and imaginative concepts, followed by specific applications worked out in mathematical detail. Preface. Index. 655pp. 5⅜ × 8½.   65969-0 Pa. $10.95

ATOMIC PHYSICS (8th edition), Max Born. Nobel laureate's lucid treatment of kinetic theory of gases, elementary particles, nuclear atom, wave-corpuscles, atomic structure and spectral lines, much more. Over 40 appendices, bibliography. 495pp. 5⅜ × 8½.   65984-4 Pa. $11.95

ELECTRONIC STRUCTURE AND THE PROPERTIES OF SOLIDS: The Physics of the Chemical Bond, Walter A. Harrison. Innovative text offers basic understanding of the electronic structure of covalent and ionic solids, simple metals, transition metals and their compounds. Problems. 1980 edition. 582pp. 6⅛ × 9¼.   66021-4 Pa. $14.95

BOUNDARY VALUE PROBLEMS OF HEAT CONDUCTION, M. Necati Özisik. Systematic, comprehensive treatment of modern mathematical methods of solving problems in heat conduction and diffusion. Numerous examples and problems. Selected references. Appendices. 505pp. 5⅜ × 8½.   65990-9 Pa. $11.95

A SHORT HISTORY OF CHEMISTRY (3rd edition), J.R. Partington. Classic exposition explores origins of chemistry, alchemy, early medical chemistry, nature of atmosphere, theory of valency, laws and structure of atomic theory, much more. 428pp. 5⅜ × 8½. (Available in U.S. only)   65977-1 Pa. $10.95

A HISTORY OF ASTRONOMY, A. Pannekoek. Well-balanced, carefully reasoned study covers such topics as Ptolemaic theory, work of Copernicus, Kepler, Newton, Eddington's work on stars, much more. Illustrated. References. 521pp. 5⅜ × 8½.   65994-1 Pa. $11.95

PRINCIPLES OF METEOROLOGICAL ANALYSIS, Walter J. Saucier. Highly respected, abundantly illustrated classic reviews atmospheric variables, hydrostatics, static stability, various analyses (scalar, cross-section, isobaric, isentropic, more). For intermediate meteorology students. 454pp. 6⅛ × 9¼. 65979-8 Pa. $12.95

RELATIVITY, THERMODYNAMICS AND COSMOLOGY, Richard C. Tolman. Landmark study extends thermodynamics to special, general relativity; also applications of relativistic mechanics, thermodynamics to cosmological models. 501pp. 5⅜ × 8½. 65383-8 Pa. $11.95

APPLIED ANALYSIS, Cornelius Lanczos. Classic work on analysis and design of finite processes for approximating solution of analytical problems. Algebraic equations, matrices, harmonic analysis, quadrature methods, much more. 559pp. 5⅜ × 8½. 65656-X Pa. $11.95

SPECIAL RELATIVITY FOR PHYSICISTS, G. Stephenson and C.W. Kilmister. Concise elegant account for nonspecialists. Lorentz transformation, optical and dynamical applications, more. Bibliography. 108pp. 5⅜ × 8½. 65519-9 Pa. $3.95

INTRODUCTION TO ANALYSIS, Maxwell Rosenlicht. Unusually clear, accessible coverage of set theory, real number system, metric spaces, continuous functions, Riemann integration, multiple integrals, more. Wide range of problems. Undergraduate level. Bibliography. 254pp. 5⅜ × 8½. 65038-3 Pa. $7.00

INTRODUCTION TO QUANTUM MECHANICS With Applications to Chemistry, Linus Pauling & E. Bright Wilson, Jr. Classic undergraduate text by Nobel Prize winner applies quantum mechanics to chemical and physical problems. Numerous tables and figures enhance the text. Chapter bibliographies. Appendices. Index. 468pp. 5⅜ × 8½. 64871-0 Pa. $9.95

ASYMPTOTIC EXPANSIONS OF INTEGRALS, Norman Bleistein & Richard A. Handelsman. Best introduction to important field with applications in a variety of scientific disciplines. New preface. Problems. Diagrams. Tables. Bibliography. Index. 448pp. 5⅜ × 8½. 65082-0 Pa. $10.95

MATHEMATICS APPLIED TO CONTINUUM MECHANICS, Lee A. Segel. Analyzes models of fluid flow and solid deformation. For upper-level math, science and engineering students. 608pp. 5⅜ × 8½. 65369-2 Pa. $12.95

ELEMENTS OF REAL ANALYSIS, David A. Sprecher. Classic text covers fundamental concepts, real number system, point sets, functions of a real variable, Fourier series, much more. Over 500 exercises. 352pp. 5⅜ × 8½. 65385-4 Pa. $8.95

PHYSICAL PRINCIPLES OF THE QUANTUM THEORY, Werner Heisenberg. Nobel Laureate discusses quantum theory, uncertainty, wave mechanics, work of Dirac, Schroedinger, Compton, Wilson, Einstein, etc. 184pp. 5⅜ × 8½.
60113-7 Pa. $4.95

INTRODUCTORY REAL ANALYSIS, A.N. Kolmogorov, S.V. Fomin. Translated by Richard A. Silverman. Self-contained, evenly paced introduction to real and functional analysis. Some 350 problems. 403pp. 5⅜ × 8½. 61226-0 Pa. $7.95

PROBLEMS AND SOLUTIONS IN QUANTUM CHEMISTRY AND PHYSICS, Charles S. Johnson, Jr. and Lee G. Pedersen. Unusually varied problems, detailed solutions in coverage of quantum mechanics, wave mechanics, angular momentum, molecular spectroscopy, scattering theory, more. 280 problems plus 139 supplementary exercises. 430pp. 6½ × 9¼. 65236-X Pa. $10.95

ASYMPTOTIC METHODS IN ANALYSIS, N.G. de Bruijn. An inexpensive, comprehensive guide to asymptotic methods—the pioneering work that teaches by explaining worked examples in detail. Index. 224pp. 5⅜ × 8½.   64221-6 Pa. $5.95

OPTICAL RESONANCE AND TWO-LEVEL ATOMS, L. Allen and J.H. Eberly. Clear, comprehensive introduction to basic principles behind all quantum optical resonance phenomena. 53 illustrations. Preface. Index. 256pp. 5⅜ × 8½.

65533-4 Pa. $6.95

COMPLEX VARIABLES, Francis J. Flanigan. Unusual approach, delaying complex algebra till harmonic functions have been analyzed from real variable viewpoint. Includes problems with answers. 364pp. 5⅜ × 8½.   61388-7 Pa. $7.95

ATOMIC SPECTRA AND ATOMIC STRUCTURE, Gerhard Herzberg. One of best introductions; especially for specialist in other fields. Treatment is physical rather than mathematical. 80 illustrations. 257pp. 5⅜ × 8½.   60115-3 Pa. $4.95

APPLIED COMPLEX VARIABLES, John W. Dettman. Step-by-step coverage of fundamentals of analytic function theory—plus lucid exposition of 5 important applications: Potential Theory; Ordinary Differential Equations; Fourier Transforms; Laplace Transforms; Asymptotic Expansions. 66 figures. Exercises at chapter ends. 512pp. 5⅜ × 8½.   64670-X Pa. $10.95

ULTRASONIC ABSORPTION: An Introduction to the Theory of Sound Absorption and Dispersion in Gases, Liquids and Solids, A.B. Bhatia. Standard reference in the field provides a clear, systematically organized introductory review of fundamental concepts for advanced graduate students, research workers. Numerous diagrams. Bibliography. 440pp. 5⅜ × 8½.   64917-2 Pa. $8.95

UNBOUNDED LINEAR OPERATORS: Theory and Applications, Seymour Goldberg. Classic presents systematic treatment of the theory of unbounded linear operators in normed linear spaces with applications to differential equations. Bibliography. 199pp. 5⅜ × 8½.   64830-3 Pa. $7.00

LIGHT SCATTERING BY SMALL PARTICLES, H.C. van de Hulst. Comprehensive treatment including full range of useful approximation methods for researchers in chemistry, meteorology and astronomy. 44 illustrations. 470pp. 5⅜ × 8½.   64228-3 Pa. $9.95

CONFORMAL MAPPING ON RIEMANN SURFACES, Harvey Cohn. Lucid, insightful book presents ideal coverage of subject. 334 exercises make book perfect for self-study. 55 figures. 352pp. 5⅜ × 8¼.   64025-6 Pa. $8.95

OPTICKS, Sir Isaac Newton. Newton's own experiments with spectroscopy, colors, lenses, reflection, refraction, etc., in language the layman can follow. Foreword by Albert Einstein. 532pp. 5⅜ × 8½.   60205-2 Pa. $8.95

GENERALIZED INTEGRAL TRANSFORMATIONS, A.H. Zemanian. Graduate-level study of recent generalizations of the Laplace, Mellin, Hankel, K. Weierstrass, convolution and other simple transformations. Bibliography. 320pp. 5⅜ × 8½.   65375-7 Pa. $7.95

THE ELECTROMAGNETIC FIELD, Albert Shadowitz. Comprehensive undergraduate text covers basics of electric and magnetic fields, builds up to electromagnetic theory. Also related topics, including relativity. Over 900 problems. 768pp. 5⅝ × 8¼. 65660-8 Pa. $15.95

FOURIER SERIES, Georgi P. Tolstov. Translated by Richard A. Silverman. A valuable addition to the literature on the subject, moving clearly from subject to subject and theorem to theorem. 107 problems, answers. 336pp. 5⅜ × 8½. 63317-9 Pa. $7.95

THEORY OF ELECTROMAGNETIC WAVE PROPAGATION, Charles Herach Papas. Graduate-level study discusses the Maxwell field equations, radiation from wire antennas, the Doppler effect and more. xiii + 244pp. 5⅜ × 8½. 65678-0 Pa. $6.95

DISTRIBUTION THEORY AND TRANSFORM ANALYSIS: An Introduction to Generalized Functions, with Applications, A.H. Zemanian. Provides basics of distribution theory, describes generalized Fourier and Laplace transformations. Numerous problems. 384pp. 5⅜ × 8½. 65479-6 Pa. $8.95

THE PHYSICS OF WAVES, William C. Elmore and Mark A. Heald. Unique overview of classical wave theory. Acoustics, optics, electromagnetic radiation, more. Ideal as classroom text or for self-study. Problems. 477pp. 5⅜ × 8½. 64926-1 Pa. $10.95

CALCULUS OF VARIATIONS WITH APPLICATIONS, George M. Ewing. Applications-oriented introduction to variational theory develops insight and promotes understanding of specialized books, research papers. Suitable for advanced undergraduate/graduate students as primary, supplementary text. 352pp. 5⅜ × 8½. 64856-7 Pa. $8.50

A TREATISE ON ELECTRICITY AND MAGNETISM, James Clerk Maxwell. Important foundation work of modern physics. Brings to final form Maxwell's theory of electromagnetism and rigorously derives his general equations of field theory. 1,084pp. 5⅜ × 8½. 60636-8, 60637-6 Pa., Two-vol. set $19.00

AN INTRODUCTION TO THE CALCULUS OF VARIATIONS, Charles Fox. Graduate-level text covers variations of an integral, isoperimetrical problems, least action, special relativity, approximations, more. References. 279pp. 5⅜ × 8½. 65499-0 Pa. $6.95

HYDRODYNAMIC AND HYDROMAGNETIC STABILITY, S. Chandrasekhar. Lucid examination of the Rayleigh-Benard problem; clear coverage of the theory of instabilities causing convection. 704pp. 5⅜ × 8¼. 64071-X Pa. $12.95

CALCULUS OF VARIATIONS, Robert Weinstock. Basic introduction covering isoperimetric problems, theory of elasticity, quantum mechanics, electrostatics, etc. Exercises throughout. 326pp. 5⅜ × 8½. 63069-2 Pa. $7.95

DYNAMICS OF FLUIDS IN POROUS MEDIA, Jacob Bear. For advanced students of ground water hydrology, soil mechanics and physics, drainage and irrigation engineering and more. 335 illustrations. Exercises, with answers. 784pp. 6⅛ × 9¼. 65675-6 Pa. $19.95

NUMERICAL METHODS FOR SCIENTISTS AND ENGINEERS, Richard Hamming. Classic text stresses frequency approach in coverage of algorithms, polynomial approximation, Fourier approximation, exponential approximation, other topics. Revised and enlarged 2nd edition. 721pp. 5⅜ × 8½.
65241-6 Pa. $14.95

THEORETICAL SOLID STATE PHYSICS, Vol. I: Perfect Lattices in Equilibrium; Vol. II: Non-Equilibrium and Disorder, William Jones and Norman H. March. Monumental reference work covers fundamental theory of equilibrium properties of perfect crystalline solids, non-equilibrium properties, defects and disordered systems. Appendices. Problems. Preface. Diagrams. Index. Bibliography. Total of 1,301pp. 5⅜ × 8½. Two volumes.　　　Vol. I 65015-4 Pa. $12.95
Vol. II 65016-2 Pa. $12.95

OPTIMIZATION THEORY WITH APPLICATIONS, Donald A. Pierre. Broadspectrum approach to important topic. Classical theory of minima and maxima, calculus of variations, simplex technique and linear programming, more. Many problems, examples. 640pp. 5⅜ × 8½.　　　65205-X Pa. $12.95

THE MODERN THEORY OF SOLIDS, Frederick Seitz. First inexpensive edition of classic work on theory of ionic crystals, free-electron theory of metals and semiconductors, molecular binding, much more. 736pp. 5⅜ × 8½.
65482-6 Pa. $14.95

ESSAYS ON THE THEORY OF NUMBERS, Richard Dedekind. Two classic essays by great German mathematician: on the theory of irrational numbers; and on transfinite numbers and properties of natural numbers. 115pp. 5⅜ × 8½.
21010-3 Pa. $4.95

THE FUNCTIONS OF MATHEMATICAL PHYSICS, Harry Hochstadt. Comprehensive treatment of orthogonal polynomials, hypergeometric functions, Hill's equation, much more. Bibliography. Index. 322pp. 5⅜ × 8½.　　65214-9 Pa. $8.95

NUMBER THEORY AND ITS HISTORY, Oystein Ore. Unusually clear, accessible introduction covers counting, properties of numbers, prime numbers, much more. Bibliography. 380pp. 5⅜ × 8½.　　　65620-9 Pa. $8.95

THE VARIATIONAL PRINCIPLES OF MECHANICS, Cornelius Lanczos. Graduate level coverage of calculus of variations, equations of motion, relativistic mechanics, more. First inexpensive paperbound edition of classic treatise. Index. Bibliography. 418pp. 5⅜ × 8½.　　　65067-7 Pa. $10.95

MATHEMATICAL TABLES AND FORMULAS, Robert D. Carmichael and Edwin R. Smith. Logarithms, sines, tangents, trig functions, powers, roots, reciprocals, exponential and hyperbolic functions, formulas and theorems. 269pp. 5⅜ × 8½.　　　60111-0 Pa. $5.95

THEORETICAL PHYSICS, Georg Joos, with Ira M. Freeman. Classic overview covers essential math, mechanics, electromagnetic theory, thermodynamics, quantum mechanics, nuclear physics, other topics. First paperback edition. xxiii + 885pp. 5⅜ × 8½.　　　65227-0 Pa. $17.95

HANDBOOK OF MATHEMATICAL FUNCTIONS WITH FORMULAS, GRAPHS, AND MATHEMATICAL TABLES, edited by Milton Abramowitz and Irene A. Stegun. Vast compendium: 29 sets of tables, some to as high as 20 places. 1,046pp. 8 × 10½. 61272-4 Pa. $21.95

MATHEMATICAL METHODS IN PHYSICS AND ENGINEERING, John W. Dettman. Algebraically based approach to vectors, mapping, diffraction, other topics in applied math. Also generalized functions, analytic function theory, more. Exercises. 448pp. 5⅜ × 8¼. 65649-7 Pa. $8.95

A SURVEY OF NUMERICAL MATHEMATICS, David M. Young and Robert Todd Gregory. Broad self-contained coverage of computer-oriented numerical algorithms for solving various types of mathematical problems in linear algebra, ordinary and partial, differential equations, much more. Exercises. Total of 1,248pp. 5⅜ × 8½. Two volumes. Vol. I 65691-8 Pa. $13.95
Vol. II 65692-6 Pa. $13.95

TENSOR ANALYSIS FOR PHYSICISTS, J.A. Schouten. Concise exposition of the mathematical basis of tensor analysis, integrated with well-chosen physical examples of the theory. Exercises. Index. Bibliography. 289pp. 5⅜ × 8½.
65582-2 Pa. $7.95

INTRODUCTION TO NUMERICAL ANALYSIS (2nd Edition), F.B. Hildebrand. Classic, fundamental treatment covers computation, approximation, interpolation, numerical differentiation and integration, other topics. 150 new problems. 669pp. 5⅜ × 8½. 65363-3 Pa. $13.95

INVESTIGATIONS ON THE THEORY OF THE BROWNIAN MOVEMENT, Albert Einstein. Five papers (1905–8) investigating dynamics of Brownian motion and evolving elementary theory. Notes by R. Fürth. 122pp. 5⅜ × 8½.
60304-0 Pa. $3.95

NUMERICAL METHODS FOR SCIENTISTS AND ENGINEERS, Richard Hamming. Classic text stresses frequency approach in coverage of algorithms, polynomial approximation, Fourier approximation, exponential approximation, other topics. Revised and enlarged 2nd edition. 721pp. 5⅜ × 8½. 65241-6 Pa. $14.95

AN INTRODUCTION TO STATISTICAL THERMODYNAMICS, Terrell L. Hill. Excellent basic text offers wide-ranging coverage of quantum statistical mechanics, systems of interacting molecules, quantum statistics, more. 523pp. 5⅜ × 8½. 65242-4 Pa. $10.95

ELEMENTARY DIFFERENTIAL EQUATIONS, William Ted Martin and Eric Reissner. Exceptionally, clear comprehensive introduction at undergraduate level. Nature and origin of differential equations, differential equations of first, second and higher orders. Picard's Theorem, much more. Problems with solutions. 331pp. 5⅜ × 8½. 65024-3 Pa. $8.95

STATISTICAL PHYSICS, Gregory H. Wannier. Classic text combines thermodynamics, statistical mechanics and kinetic theory in one unified presentation of thermal physics. Problems with solutions. Bibliography. 532pp. 5⅜ × 8½.
65401-X Pa. $10.95

ORDINARY DIFFERENTIAL EQUATIONS, Morris Tenenbaum and Harry Pollard. Exhaustive survey of ordinary differential equations for undergraduates in mathematics, engineering, science. Thorough analysis of theorems. Diagrams. Bibliography. Index. 818pp. 5⅜ × 8½. 64940-7 Pa. $15.95

STATISTICAL MECHANICS: Principles and Applications, Terrell L. Hill. Standard text covers fundamentals of statistical mechanics, applications to fluctuation theory, imperfect gases, distribution functions, more. 448pp. 5⅜ × 8½. 65390-0 Pa. $9.95

ORDINARY DIFFERENTIAL EQUATIONS AND STABILITY THEORY: An Introduction, David A. Sánchez. Brief, modern treatment. Linear equation, stability theory for autonomous and nonautonomous systems, etc. 164pp. 5⅜ × 8¼. 63828-6 Pa. $4.95

THIRTY YEARS THAT SHOOK PHYSICS: The Story of Quantum Theory, George Gamow. Lucid, accessible introduction to influential theory of energy and matter. Careful explanations of Dirac's anti-particles, Bohr's model of the atom, much more. 12 plates. Numerous drawings. 240pp. 5⅜ × 8½. 24895-X Pa. $5.95

ORDINARY DIFFERENTIAL EQUATIONS, I.G. Petrovski. Covers basic concepts, some differential equations and such aspects of the general theory as Euler lines, Arzel's theorem, Peano's existence theorem, Osgood's uniqueness theorem, more. 45 figures. Problems. Bibliography. Index. xi + 232pp. 5⅜ × 8½. 64683-1 Pa. $6.00

GREAT EXPERIMENTS IN PHYSICS: Firsthand Accounts from Galileo to Einstein, edited by Morris H. Shamos. 25 crucial discoveries: Newton's laws of motion, Chadwick's study of the neutron, Hertz on electromagnetic waves, more. Original accounts clearly annotated. 370pp. 5⅜ × 8½. 25346-5 Pa. $8.95

INTRODUCTION TO PARTIAL DIFFERENTIAL EQUATIONS WITH APPLICATIONS, E.C. Zachmanoglou and Dale W. Thoe. Essentials of partial differential equations applied to common problems in engineering and the physical sciences. Problems and answers. 416pp. 5⅜ × 8½. 65251-3 Pa. $9.95

BURNHAM'S CELESTIAL HANDBOOK, Robert Burnham, Jr. Thorough guide to the stars beyond our solar system. Exhaustive treatment. Alphabetical by constellation: Andromeda to Cetus in Vol. 1; Chamaeleon to Orion in Vol. 2; and Pavo to Vulpecula in Vol. 3. Hundreds of illustrations. Index in Vol. 3. 2,000pp. 6⅛ × 9¼. 23567-X, 23568-8, 23673-0 Pa., Three-vol. set $38.85

ASYMPTOTIC EXPANSIONS FOR ORDINARY DIFFERENTIAL EQUATIONS, Wolfgang Wasow. Outstanding text covers asymptotic power series, Jordan's canonical form, turning point problems, singular perturbations, much more. Problems. 384pp. 5⅜ × 8½. 65456-7 Pa. $8.95

AMATEUR ASTRONOMER'S HANDBOOK, J.B. Sidgwick. Timeless, comprehensive coverage of telescopes, mirrors, lenses, mountings, telescope drives, micrometers, spectroscopes, more. 189 illustrations. 576pp. 5⅜ × 8¼. 24034-7 Pa. $8.95

SPECIAL FUNCTIONS, N.N. Lebedev. Translated by Richard Silverman. Famous Russian work treating more important special functions, with applications to specific problems of physics and engineering. 38 figures. 308pp. 5⅜ × 8½.
60624-4 Pa. $6.95

OBSERVATIONAL ASTRONOMY FOR AMATEURS, J.B. Sidgwick. Mine of useful data for observation of sun, moon, planets, asteroids, aurorae, meteors, comets, variables, binaries, etc. 39 illustrations 384pp. 5⅜ × 8¼. (Available in U.S. only)
24033-9 Pa. $5.95

INTEGRAL EQUATIONS, F.G. Tricomi. Authoritative, well-written treatment of extremely useful mathematical tool with wide applications. Volterra Equations, Fredholm Equations, much more. Advanced undergraduate to graduate level. Exercises. Bibliography. 238pp. 5⅜ × 8½.
64828-1 Pa. $6.95

CELESTIAL OBJECTS FOR COMMON TELESCOPES, T.W. Webb. Inestimable aid for locating and identifying nearly 4,000 celestial objects. 77 illustrations. 645pp. 5⅜ × 8½.
20917-2, 20918-0 Pa., Two-vol. set $12.00

MODERN NONLINEAR EQUATIONS, Thomas L. Saaty. Emphasizes practical solution of problems; covers seven types of equations. ". . . a welcome contribution to the existing literature. . . ."—*Math Reviews.* 490pp. 5⅜ × 8½. 64232-1 Pa. $9.95

FUNDAMENTALS OF ASTRODYNAMICS, Roger Bate et al. Modern approach developed by U.S. Air Force Academy. Designed as a first course. Problems, exercises. Numerous illustrations. 455pp. 5⅜ × 8½. 60061-0 Pa. $8.95

INTRODUCTION TO LINEAR ALGEBRA AND DIFFERENTIAL EQUATIONS, John W. Dettman. Excellent text covers complex numbers, determinants, orthonormal bases, Laplace transforms, much more. Exercises with solutions. Undergraduate level. 416pp. 5⅜ × 8½. 65191-6 Pa. $8.95

INCOMPRESSIBLE AERODYNAMICS, edited by Bryan Thwaites. Covers theoretical and experimental treatment of the uniform flow of air and viscous fluids past two-dimensional aerofoils and three-dimensional wings; many other topics. 654pp. 5⅜ × 8½. 65465-6 Pa. $14.95

INTRODUCTION TO DIFFERENCE EQUATIONS, Samuel Goldberg. Exceptionally clear exposition of important discipline with applications to sociology, psychology, economics. Many illustrative examples; over 250 problems. 260pp. 5⅜ × 8½. 65084-7 Pa. $6.95

LAMINAR BOUNDARY LAYERS, edited by L. Rosenhead. Engineering classic covers steady boundary layers in two- and three-dimensional flow, unsteady boundary layers, stability, observational techniques, much more. 708pp. 5⅜ × 8½.
65646-2 Pa. $15.95

LECTURES ON CLASSICAL DIFFERENTIAL GEOMETRY, Second Edition, Dirk J. Struik. Excellent brief introduction covers curves, theory of surfaces, fundamental equations, geometry on a surface, conformal mapping, other topics. Problems. 240pp. 5⅜ × 8½. 65609-8 Pa. $6.95

ROTARY-WING AERODYNAMICS, W.Z. Stepniewski. Clear, concise text covers aerodynamic phenomena of the rotor and offers guidelines for helicopter performance evaluation. Originally prepared for NASA. 537 figures. 640pp. 6⅛ × 9¼.
64647-5 Pa. $14.95

DIFFERENTIAL GEOMETRY, Heinrich W. Guggenheimer. Local differential geometry as an application of advanced calculus and linear algebra. Curvature, transformation groups, surfaces, more. Exercises. 62 figures. 378pp. 5⅜ × 8½.
63433-7 Pa. $7.95

INTRODUCTION TO SPACE DYNAMICS, William Tyrrell Thomson. Comprehensive, classic introduction to space-flight engineering for advanced undergraduate and graduate students. Includes vector algebra, kinematics, transformation of coordinates. Bibliography. Index. 352pp. 5⅜ × 8½.   65113-4 Pa. $8.00

A SURVEY OF MINIMAL SURFACES, Robert Osserman. Up-to-date, in-depth discussion of the field for advanced students. Corrected and enlarged edition covers new developments. Includes numerous problems. 192pp. 5⅜ × 8½.
64998-9 Pa. $8.00

ANALYTICAL MECHANICS OF GEARS, Earle Buckingham. Indispensable reference for modern gear manufacture covers conjugate gear-tooth action, gear-tooth profiles of various gears, many other topics. 263 figures. 102 tables. 546pp. 5⅜ × 8½.   65712-4 Pa. $11.95

SET THEORY AND LOGIC, Robert R. Stoll. Lucid introduction to unified theory of mathematical concepts. Set theory and logic seen as tools for conceptual understanding of real number system. 496pp. 5⅜ × 8¼.   63829-4 Pa. $8.95

A HISTORY OF MECHANICS, René Dugas. Monumental study of mechanical principles from antiquity to quantum mechanics. Contributions of ancient Greeks, Galileo, Leonardo, Kepler, Lagrange, many others. 671pp. 5⅜ × 8½.
65632-2 Pa. $14.95

FAMOUS PROBLEMS OF GEOMETRY AND HOW TO SOLVE THEM, Benjamin Bold. Squaring the circle, trisecting the angle, duplicating the cube: learn their history, why they are impossible to solve, then solve them yourself. 128pp. 5⅜ × 8½.   24297-8 Pa. $3.95

MECHANICAL VIBRATIONS, J.P. Den Hartog. Classic textbook offers lucid explanations and illustrative models, applying theories of vibrations to a variety of practical industrial engineering problems. Numerous figures. 233 problems, solutions. Appendix. Index. Preface. 436pp. 5⅜ × 8½.   64785-4 Pa. $8.95

CURVATURE AND HOMOLOGY, Samuel I. Goldberg. Thorough treatment of specialized branch of differential geometry. Covers Riemannian manifolds, topology of differentiable manifolds, compact Lie groups, other topics. Exercises. 315pp. 5⅜ × 8½.   64314-X Pa. $6.95

HISTORY OF STRENGTH OF MATERIALS, Stephen P. Timoshenko. Excellent historical survey of the strength of materials with many references to the theories of elasticity and structure. 245 figures. 452pp. 5⅜ × 8½.   61187-6 Pa. $9.95

CATALOG OF DOVER BOOKS

GEOMETRY OF COMPLEX NUMBERS, Hans Schwerdtfeger. Illuminating, widely praised book on analytic geometry of circles, the Moebius transformation, and two-dimensional non-Euclidean geometries. 200pp. 5⅝ × 8¼.
63830-8 Pa. $6.95

MECHANICS, J.P. Den Hartog. A classic introductory text or refresher. Hundreds of applications and design problems illuminate fundamentals of trusses, loaded beams and cables, etc. 334 answered problems. 462pp. 5⅜ × 8½. 60754-2 Pa. $8.95

TOPOLOGY, John G. Hocking and Gail S. Young. Superb one-year course in classical topology. Topological spaces and functions, point-set topology, much more. Examples and problems. Bibliography. Index. 384pp. 5⅝ × 8¼.
65676-4 Pa. $7.95

STRENGTH OF MATERIALS, J.P. Den Hartog. Full, clear treatment of basic material (tension, torsion, bending, etc.) plus advanced material on engineering methods, applications. 350 answered problems. 323pp. 5⅜ × 8½. 60755-0 Pa. $7.50

ELEMENTARY CONCEPTS OF TOPOLOGY, Paul Alexandroff. Elegant, intuitive approach to topology from set-theoretic topology to Betti groups; how concepts of topology are useful in math and physics. 25 figures. 57pp. 5⅜ × 8½.
60747-X Pa. $2.95

ADVANCED STRENGTH OF MATERIALS, J.P. Den Hartog. Superbly written advanced text covers torsion, rotating disks, membrane stresses in shells, much more. Many problems and answers. 388pp. 5⅜ × 8½. 65407-9 Pa. $8.95

COMPUTABILITY AND UNSOLVABILITY, Martin Davis. Classic graduate-level introduction to theory of computability, usually referred to as theory of recurrent functions. New preface and appendix. 288pp. 5⅜ × 8½. 61471-9 Pa. $6.95

GENERAL CHEMISTRY, Linus Pauling. Revised 3rd edition of classic first-year text by Nobel laureate. Atomic and molecular structure, quantum mechanics, statistical mechanics, thermodynamics correlated with descriptive chemistry. Problems. 992pp. 5⅜ × 8½. 65622-5 Pa. $18.95

AN INTRODUCTION TO MATRICES, SETS AND GROUPS FOR SCIENCE STUDENTS, G. Stephenson. Concise, readable text introduces sets, groups, and most importantly, matrices to undergraduate students of physics, chemistry, and engineering. Problems. 164pp. 5⅜ × 8½. 65077-4 Pa. $5.95

THE HISTORICAL BACKGROUND OF CHEMISTRY, Henry M. Leicester. Evolution of ideas, not individual biography. Concentrates on formulation of a coherent set of chemical laws. 260pp. 5⅜ × 8½. 61053-5 Pa. $6.00

THE PHILOSOPHY OF MATHEMATICS: An Introductory Essay, Stephan Körner. Surveys the views of Plato, Aristotle, Leibniz & Kant concerning propositions and theories of applied and pure mathematics. Introduction. Two appendices. Index. 198pp. 5⅜ × 8½. 25048-2 Pa. $5.95

THE DEVELOPMENT OF MODERN CHEMISTRY, Aaron J. Ihde. Authoritative history of chemistry from ancient Greek theory to 20th-century innovation. Covers major chemists and their discoveries. 209 illustrations. 14 tables. Bibliographies. Indices. Appendices. 851pp. 5⅜ × 8½. 64235-6 Pa. $15.95

THE FOUR-COLOR PROBLEM: Assaults and Conquest, Thomas L. Saaty and Paul G. Kainen. Engrossing, comprehensive account of the century-old combinatorial topological problem, its history and solution. Bibliographies. Index. 110 figures. 228pp. 5⅜ × 8½. 65092-8 Pa. $6.00

CATALYSIS IN CHEMISTRY AND ENZYMOLOGY, William P. Jencks. Exceptionally clear coverage of mechanisms for catalysis, forces in aqueous solution, carbonyl- and acyl-group reactions, practical kinetics, more. 864pp. 5⅜ × 8½. 65460-5 Pa. $18.95

PROBABILITY: An Introduction, Samuel Goldberg. Excellent basic text covers set theory, probability theory for finite sample spaces, binomial theorem, much more. 360 problems. Bibliographies. 322pp. 5⅜ × 8½. 65252-1 Pa. $7.95

LIGHTNING, Martin A. Uman. Revised, updated edition of classic work on the physics of lightning. Phenomena, terminology, measurement, photography, spectroscopy, thunder, more. Reviews recent research. Bibliography. Indices. 320pp. 5⅜ × 8¼. 64575-4 Pa. $7.95

PROBABILITY THEORY: A Concise Course, Y.A. Rozanov. Highly readable, self-contained introduction covers combination of events, dependent events, Bernoulli trials, etc. Translation by Richard Silverman. 148pp. 5⅜ × 8¼. 63544-9 Pa. $4.50

THE CEASELESS WIND: An Introduction to the Theory of Atmospheric Motion, John A. Dutton. Acclaimed text integrates disciplines of mathematics and physics for full understanding of dynamics of atmospheric motion. Over 400 problems. Index. 97 illustrations. 640pp. 6 × 9. 65096-0 Pa. $16.95

STATISTICS MANUAL, Edwin L. Crow, et al. Comprehensive, practical collection of classical and modern methods prepared by U.S. Naval Ordnance Test Station. Stress on use. Basics of statistics assumed. 288pp. 5⅜ × 8½. 60599-X Pa. $6.00

WIND WAVES: Their Generation and Propagation on the Ocean Surface, Blair Kinsman. Classic of oceanography offers detailed discussion of stochastic processes and power spectral analysis that revolutionized ocean wave theory. Rigorous, lucid. 676pp. 5⅜ × 8½. 64652-1 Pa. $14.95

STATISTICAL METHOD FROM THE VIEWPOINT OF QUALITY CONTROL, Walter A. Shewhart. Important text explains regulation of variables, uses of statistical control to achieve quality control in industry, agriculture, other areas. 192pp. 5⅜ × 8½. 65232-7 Pa. $6.00

THE INTERPRETATION OF GEOLOGICAL PHASE DIAGRAMS, Ernest G. Ehlers. Clear, concise text emphasizes diagrams of systems under fluid or containing pressure; also coverage of complex binary systems, hydrothermal melting, more. 288pp. 6½ × 9¼. 65389-7 Pa. $8.95

STATISTICAL ADJUSTMENT OF DATA, W. Edwards Deming. Introduction to basic concepts of statistics, curve fitting, least squares solution, conditions without parameter, conditions containing parameters. 26 exercises worked out. 271pp. 5⅜ × 8½. 64685-8 Pa. $7.95

DE RE METALLICA, Georgius Agricola. The famous Hoover translation of greatest treatise on technological chemistry, engineering, geology, mining of early modern times (1556). All 289 original woodcuts. 638pp. 6¾ × 11.
60006-8 Clothbd. $15.95

SOME THEORY OF SAMPLING, William Edwards Deming. Analysis of the problems, theory and design of sampling techniques for social scientists, industrial managers and others who find statistics increasingly important in their work. 61 tables. 90 figures. xvii + 602pp. 5⅜ × 8½. 64684-X Pa. $14.95

THE VARIOUS AND INGENIOUS MACHINES OF AGOSTINO RAMELLI: A Classic Sixteenth-Century Illustrated Treatise on Technology, Agostino Ramelli. One of the most widely known and copied works on machinery in the 16th century. 194 detailed plates of water pumps, grain mills, cranes, more. 608pp. 9 × 12.
25497-6 Clothbd. $34.95

LINEAR PROGRAMMING AND ECONOMIC ANALYSIS, Robert Dorfman, Paul A. Samuelson and Robert M. Solow. First comprehensive treatment of linear programming in standard economic analysis. Game theory, modern welfare economics, Leontief input-output, more. 525pp. 5⅜ × 8½. 65491-5 Pa. $12.95

ELEMENTARY DECISION THEORY, Herman Chernoff and Lincoln E. Moses. Clear introduction to statistics and statistical theory covers data processing, probability and random variables, testing hypotheses, much more. Exercises. 364pp. 5⅜ × 8½. 65218-1 Pa. $8.95

THE COMPLEAT STRATEGYST: Being a Primer on the Theory of Games of Strategy, J.D. Williams. Highly entertaining classic describes, with many illustrated examples, how to select best strategies in conflict situations. Prefaces. Appendices. 268pp. 5⅜ × 8½. 25101-2 Pa. $5.95

MATHEMATICAL METHODS OF OPERATIONS RESEARCH, Thomas L. Saaty. Classic graduate-level text covers historical background, classical methods of forming models, optimization, game theory, probability, queueing theory, much more. Exercises. Bibliography. 448pp. 5⅜ × 8¼. 65703-5 Pa. $12.95

CONSTRUCTIONS AND COMBINATORIAL PROBLEMS IN DESIGN OF EXPERIMENTS, Damaraju Raghavarao. In-depth reference work examines orthogonal Latin squares, incomplete block designs, tactical configuration, partial geometry, much more. Abundant explanations, examples. 416pp. 5⅜ × 8¼.
65685-3 Pa. $10.95

THE ABSOLUTE DIFFERENTIAL CALCULUS (CALCULUS OF TENSORS), Tullio Levi-Civita. Great 20th-century mathematician's classic work on material necessary for mathematical grasp of theory of relativity. 452pp. 5⅜ × 8½.
63401-9 Pa. $9.95

VECTOR AND TENSOR ANALYSIS WITH APPLICATIONS, A.I. Borisenko and I.E. Tarapov. Concise introduction. Worked-out problems, solutions, exercises. 257pp. 5⅜ × 8¼. 63833-2 Pa. $6.95

TENSOR CALCULUS, J.L. Synge and A. Schild. Widely used introductory text covers spaces and tensors, basic operations in Riemannian space, non-Riemannian spaces, etc. 324pp. 5⅜ × 8¼. 63612-7 Pa. $7.00

A CONCISE HISTORY OF MATHEMATICS, Dirk J. Struik. The best brief history of mathematics. Stresses origins and covers every major figure from ancient Near East to 19th century. 41 illustrations. 195pp. 5⅜ × 8½. 60255-9 Pa. $7.95

A SHORT ACCOUNT OF THE HISTORY OF MATHEMATICS, W.W. Rouse Ball. One of clearest, most authoritative surveys from the Egyptians and Phoenicians through 19th-century figures such as Grassman, Galois, Riemann. Fourth edition. 522pp. 5⅜ × 8½. 20630-0 Pa. $9.95

HISTORY OF MATHEMATICS, David E. Smith. Non-technical survey from ancient Greece and Orient to late 19th century; evolution of arithmetic, geometry, trigonometry, calculating devices, algebra, the calculus. 362 illustrations. 1,355pp. 5⅜ × 8½. 20429-4, 20430-8 Pa., Two-vol. set $21.90

THE GEOMETRY OF RENÉ DESCARTES, René Descartes. The great work founded analytical geometry. Original French text, Descartes' own diagrams, together with definitive Smith-Latham translation. 244pp. 5⅜ × 8½. 60068-8 Pa. $6.00

THE ORIGINS OF THE INFINITESIMAL CALCULUS, Margaret E. Baron. Only fully detailed and documented account of crucial discipline: origins; development by Galileo, Kepler, Cavalieri; contributions of Newton, Leibniz, more. 304pp. 5⅜ × 8½. (Available in U.S. and Canada only) 65371-4 Pa. $7.95

THE HISTORY OF THE CALCULUS AND ITS CONCEPTUAL DEVELOPMENT, Carl B. Boyer. Origins in antiquity, medieval contributions, work of Newton, Leibniz, rigorous formulation. Treatment is verbal. 346pp. 5⅜ × 8½. 60509-4 Pa. $6.95

THE THIRTEEN BOOKS OF EUCLID'S ELEMENTS, translated with introduction and commentary by Sir Thomas L. Heath. Definitive edition. Textual and linguistic notes, mathematical analysis. 2500 years of critical commentary. Not abridged. 1,414pp. 5⅜ × 8½. 60088-2, 60089-0, 60090-4 Pa., Three-vol. set $26.85

A HISTORY OF VECTOR ANALYSIS: The Evolution of the Idea of a Vectorial System, Michael J. Crowe. The first large-scale study of the history of vector analysis, now the standard on the subject. Unabridged republication of the edition published by University of Notre Dame Press, 1967, with second preface by Michael C. Crowe. Index. 278pp. 5⅜ × 8½. 64955-5 Pa. $7.00

THE HISTORICAL ROOTS OF ELEMENTARY MATHEMATICS, Lucas N.H. Bunt, Phillip S. Jones, and Jack D. Bedient. Fundamental underpinnings of modern arithmetic, algebra, geometry and number systems derived from ancient civilizations. 320pp. 5⅜ × 8½. 25563-8 Pa. $7.95

CALCULUS REFRESHER FOR TECHNICAL PEOPLE, A. Albert Klaf. Covers important aspects of integral and differential calculus via 756 questions. 566 problems, most answered. 431pp. 5⅜ × 8½. 20370-0 Pa. $7.95

CHALLENGING MATHEMATICAL PROBLEMS WITH ELEMENTARY SOLUTIONS, A.M. Yaglom and I.M. Yaglom. Over 170 challenging problems on probability theory, combinatorial analysis, points and lines, topology, convex polygons, many other topics. Solutions. Total of 445pp. 5⅜ × 8½. Two-vol. set.
Vol. I 65536-9 Pa. $5.95
Vol. II 65537-7 Pa. $5.95

FIFTY CHALLENGING PROBLEMS IN PROBABILITY WITH SOLUTIONS, Frederick Mosteller. Remarkable puzzlers, graded in difficulty, illustrate elementary and advanced aspects of probability. Detailed solutions. 88pp. 5⅜ × 8½.
65355-2 Pa. $3.95

EXPERIMENTS IN TOPOLOGY, Stephen Barr. Classic, lively explanation of one of the byways of mathematics. Klein bottles, Moebius strips, projective planes, map coloring, problem of the Koenigsberg bridges, much more, described with clarity and wit. 43 figures. 210pp. 5⅜ × 8½.
25933-1 Pa. $4.95

RELATIVITY IN ILLUSTRATIONS, Jacob T. Schwartz. Clear non-technical treatment makes relativity more accessible than ever before. Over 60 drawings illustrate concepts more clearly than text alone. Only high school geometry needed. Bibliography. 128pp. 6⅛ × 9¼.
25965-X Pa. $5.95

AN INTRODUCTION TO ORDINARY DIFFERENTIAL EQUATIONS, Earl A. Coddington. A thorough and systematic first course in elementary differential equations for undergraduates in mathematics and science, with many exercises and problems (with answers). Index. 304pp. 5⅜ × 8¼.
65942-9 Pa. $7.95

FOURIER SERIES AND ORTHOGONAL FUNCTIONS, Harry F. Davis. An incisive text combining theory and practical example to introduce Fourier series, orthogonal functions and applications of the Fourier method to boundary-value problems. 570 exercises. Answers and notes. 416pp. 5⅜ × 8½.
65973-9 Pa. $8.95

THE THOERY OF BRANCHING PROCESSES, Theodore E. Harris. First systematic, comprehensive treatment of branching (i.e. multiplicative) processes and their applications. Galton-Watson model, Markov branching processes, electron-photon cascade, many other topics. Rigorous proofs. Bibliography. 240pp. 5⅜ × 8½.
65952-6 Pa. $6.95

AN INTRODUCTION TO ALGEBRAIC STRUCTURES, Joseph Landin. Superb self-contained text covers "abstract algebra": sets and numbers, theory of groups, theory of rings, much more. Numerous well-chosen examples, exercises. 247pp. 5⅜ × 8½.
65940-2 Pa. $6.95

GAMES AND DECISIONS: Introduction and Critical Survey, R. Duncan Luce and Howard Raiffa. Superb non-technical introduction to game theory, primarily applied to social sciences. Utility theory, zero-sum games, n-person games, decision-making, much more. Bibliography. 509pp. 5⅜ × 8½.   65943-7 Pa. $10.95

Prices subject to change without notice.
Available at your book dealer or write for free Mathematics and Science Catalog to Dept. GI, Dover Publications, Inc., 31 East 2nd St., Mineola, N.Y. 11501. Dover publishes more than 175 books each year on science, elementary and advanced mathematics, biology, music, art, literary history, social sciences and other areas.